TEACHING SCIENTIFIC INQUIRY

Teaching Scientific Inquiry
Recommendations for Research and Implementation

Edited by

Richard A. Duschl
Rutgers University, USA

and

Richard E . Grandy
Rice University, USA

SENSE PUBLISHERS
ROTTERDAM / TAIPEI

A C.I.P. record for this book is available from the Library of Congress.

ISBN 978-90-8790-271-1 (paperback)
ISBN 978-90-8790-272-8 (hardback)

Published by: Sense Publishers,
P.O. Box 21858, 3001 AW Rotterdam, The Netherlands
http://www.sensepublishers.com

Printed on acid-free paper

All rights reserved © 2008 Sense Publishers

No part of this work may be reproduced, stored in a retrieval system, or transmitted in any form or by any means, electronic, mechanical, photocopying, microfilming, recording or otherwise, without written permission from the Publisher, with the exception of any material supplied specifically for the purpose of being entered and executed on a computer system, for exclusive use by the purchaser of the work.

TABLE OF CONTENTS

Preface ix
Introduction

Chapter 1. Reconsidering the Character and Role of Inquiry in School Science: Framing the Debates 1
Richard Duschl & Richard Grandy

Day 1 – Foundations of Science Education

Inquiry: The Child as Scientist
Chapter 2. In What Sense Can The Child Be Considered to Be a "Little Scientist"? 38
William Brewer

Chapter 3. Commentary: Three Questions about Development 50
Leona Schauble

Inquiry: How Science Works
Chapter 4. Model-Based Reasoning in Scientific Practice 57
Nancy Nersessian

Chapter 5. Commentary: Modeling Science Classrooms after Scientific Laboratories 80
Fouad Abd-El-Khalick

Inquiry: Knowledge as Social Processes
Chapter 6. Social Epistemology of Science 86
Miriam Solomon

Chapter 7. Commentary: Should the Sociology of Science Be Rated X? 95
Nancy Brickhouse

Inquiry: Conceptual Change and Constructivism
Chapter 8. Inquiry, Activity and Epistemic Practice 99
Gregory Kelly

TABLE OF CONTENTS

Chapter 9. Commentary on "Inquiry, Activity and Epistemic Practice"   118
John Rudolph

Day 1 Panel Discussion – Foundations of Science Education

Chapter 10. Philosophical Issues and Next Steps for Research   123

> *Stephen Stich*   123
> *Harvey Siegel*   129
> *Helen Longino*   134

Day 2 – Practice and Policy in Science Education

Inquiry: Epistemic Practices in Classrooms

Chapter 11. Identifying Inquiry and Conceptualizing Abilities   138
David Hammer, Rosemary Russ, Jamie Mikeska & Rachel Scherr

Chapter 12. Commentary: Exploring Children's Understanding of the Purpose and Value of Inquiry   157
William Sandoval

Inquiry: Engineering the Design of Learning Environments

Chapter 13. Engineering Pedagogical Reform: A Case Study of Technology Supported Inquiry   164
Daniel Edelson

Chapter 14. A Commentary on Engineering Pedagogical Reform   182
Janice Bordeaux

Inquiry: Learning to use Data, Models and Explanations

Chapter 15. Learning to Use Scientific Models: Multiple Dimensions of Conceptual Change   191
Clark Chinn & Ala Samarapungavan

Chapter 16. Commentary on Chinn's & Samarapungavan's Paper   226
Joseph Krajcik

Inquiry: Literacy Practices and Science Communication

Chapter 17. Reading as Inquiry   233
Stephen Norris & Linda Phillips

Chapter 18. Inquiry as Inscriptional Work: A Commentary on Norris & Phillips 263
Philip Bell

Day 2 Panel Discussion Practice & Policy in Science Education
Chapter 19. Practice and Next Steps for Educational Research 268

> *Drew Gitomer* 268
> *Cindy Hmelo-Silver* 272
> *Eugenia Etkina* 275
> *Mark Windschitl* 278

Reflections and Recommendations

Chapter 20. Codas

> What is Inquiry? To Whom Should It be Authentic? 284
> *Nancy Brickhouse*

> Continuing Conversation Regarding Inquiry 288
> *Greg Kelly*

> Our Challenges in Disrupting Popular Folk Theories of "Doing Science" 292
> *Mark Windschitl*

Chapter 21. Consensus: Expanding The Scientific Method and School Science 304
Richard Grandy & Richard Duschl

Bibliography 323

Authors Bios 351

Index 360

PREFACE

For science education, the last 50 years were a period of dynamic developments and debates. New curriculum programs, new laboratory-based and computer-based programs of study, new learning theories and new standards emerged several times over. Dynamic developments and debates have taken place in science and technology, too. New tools, new technologies and new theories have abounded in the sciences since the 1950s. When one examines such developments in the sciences, it is clear that we have learned how to learn. What we learn about nature informs and influences how we observe and measure nature, how we poke around and what questions and problems we choose to address. The boundaries between science and technology are blurring with the passage of each decade Scientific practices today have as much to say about solving human problems of hunger, disease, water quality and global habitability as it does about the pursuit of new basic knowledge and explanations.

The growth of scientific knowledge has been conceptual and methodological as well. The scientific advancements in the second half of the 20th century have fundamentally changed our beliefs about what it means to do science, to engage in scientific inquiry and to describe science as a way of knowing. The science studies disciplines too – philosophy of science, history of science, sociology of science, cognitive sciences, anthropology of science – have seen dynamic developments and debates about the nature of science and inquiry. Since the 1950s advances in the cognitive sciences have altered our beliefs about knowledge acquisition, meaning making, and conditions for school learning. The individualistic and domain general stances of Skinnerian behaviorism and Piagetian stage theory have been supplanted by cognitive, sociocultural and domain specific stances of constructivist and situated learning theory.

One might not be surprised to learn that a plethora of perspectives exist about what counts as images scientific inquiry, given such dynamics across the sciences, the science studies disciplines and learning science disciplines. But we were surprised and disturbed by the wide and often conflicting range of opinions and beliefs about inquiry found among researchers and in government science education policy documents.

This volume reports on a conference that sought to establish a consensus agenda for teaching inquiry in K-12 science. With sponsorship from the National Science Foundation, we brought together for two days 24 scholars who study science,

PREFACE

science learning/reasoning, science communication, and/or science teaching. Eight philosophers of science, cognitive and developmental psychologists and educational researchers participated as writers, commentators, or panelists. All participants received the plenary paper *Reconsidering the Character and Role of Inquiry in School Science: Framing the Debates* that appears as Chapter 1. Chapter 1 presents overviews of recent developments in 3 domains: of philosophy of science, psychology of learning and pedagogical practices that serve to frame perspectives about inquiry. Writers were assigned specific topics relevant to inquiry and to their personal areas of expertise. These topics appear as chapter headings. Commentators were selected from different disciplinary domains (e.g., cognitive scientists commenting on philosophers' papers and educational researchers commenting on cognitive scientists', etc.) in order to foster scholarly dialogue and consensus building. Draft manuscripts from writers and commentators were sent to participants 1 month prior to the conference. Ample time was provided for summary presentations of the papers and importantly for extended discussions of the papers. Each day ended with a panel presentation comprised of 4 persons who were asked to summarize themes, issues and omissions. Panelists' comments were transcribed and edited for inclusion in the volume. Day 1 topics and discussions were dedicated to "Philosophical Issues and Next Steps for Research" and Day 2 topics and discussions addressed "Policy, Practice and Next Steps for Educational Research". The organization of the volume reflects the Day 1 and Day 2 foci of the sessions. The final chapter in the volume reports on the consensus perspectives that emerged from the inquiry conference. Some readers may find it beneficial to begin reading there to get a sense of the themes and issues that emerged.

We thank the National Science Foundation and in particular Barry Sloan for supporting the conference. We also thank the conference participants, Julie Monet and Jessie Armendt for assistance with preparation of manuscript references and the volume index, Jana Curry for conference logistics and arrangement, Janice Bordeaux for suggesting the title, and Peter de Liefde for making this volume part of Sense Publishers portfolio of books.

Richard A. Duschl   Richard E. Grandy
Rutgers University   Rice University

RICHARD A. DUSCHL AND RICHARD E. GRANDY

RECONSIDERING THE CHARACTER AND ROLE OF INQUIRY IN SCHOOL SCIENCE: FRAMING THE DEBATES

INTRODUCTION

The purpose of this paper is to provide a structure for discussion of science education with the goal of synthesizing developments in

1. science studies, e.g., history, philosophy and sociology of science
2. the learning sciences, e.g., cognitive science, philosophy of mind, educational psychology, social psychology, computer sciences, linguistics, and
3. educational research focusing on the design of learning environments that promote inquiry and that facilitate dynamic assessments.

Taken together these three domains have reshaped our thinking about the role that inquiry has in science education programs. Over the past 50 years there have been dynamic changes in our conceptualizations of science, of learning, and of science learning environments. Such changes have important implications for how we interpret (1) the role of inquiry in K-12 science education programs and (2) the design of curriculum, instruction, and assessment models that strive to meet the NSES inquiry goals: Students should learn to do scientific inquiry; students should develop an understanding of scientific inquiry.

Although there is general agreement that important changes have taken place, it is less clear where there is consensus on the new directions and where there is not. Furthermore, it is not clear yet which of the areas of dissensus represent lacuna relevant to science education and which are irrelevant for science education, however central they may be to philosophy of science, psychology of science, history of science, or any other area of science studies. Our goal in this conference is to determine areas of consensus and areas where we lack consensus, and among the latter to identify which areas are crucial for the further development of science education and which are not.

Since the first NSF funded era of science education reform in the 1960s and 1970s, we see a shift from science as experimentation to science as explanation/model building and revision; from learning as a passive individualistic process to learning as an active individual and social process; from science teaching focusing on the management of learners' behaviors and "hands-on" materials to science teaching focusing on the management of learners' ideas, access to information, and interactions among learners. Some of the shifts have

Richard A. Duschl and Richard E. Grandy (eds.), Teaching Scientific Inquiry: Recommendations for Research and Implementation, 1–37.
© 2008 *Sense Publishers. All rights reserved.*

been motivated by new technological development but new theories about learning have contributed too.

One important change that has significant implications for the role of inquiry in school science concerns the realm of scientific observations. Over the last 100 years new technologies and new scientific theories have modified the nature of scientific observation from an enterprise dominated by *sense perception,* aided or unaided, to a *theory-driven* enterprise. We now know that what we see is influenced by what we know and how we "look"; scientific theories are inextricably involved in the design and interpretation of experimental methods.

New technologies and learning theories also have effected how we monitor, diagnose and nurture learning. Scientific databases like Geographical Information Systems (GIS) make it possible to engage in rich scientific inquiry without engaging in hands-on science involving the collection of data. Instead, the data are provided and the inquiry begins with the selection of information for analysis. This is one example of how science education has shifted from management of materials for collecting data to management of information for scrutinizing databases. Such a shift has implications regarding the manner in which interactions with phenomenon are designed and included in science lessons for all grade levels. Information in the guise of data, evidence, models and explanations represents, in an important sense, the new materials for school classrooms and laboratories. Taken together these developments in technologies and theories have implications for how we conceptualize the design and delivery of science curriculum materials for purposes of supporting students' learning as well as teachers' assessments for promoting learning.

The use of computer-supported instrumentation, information systems, data analysis techniques and scientific inquiry practices in general, has created a problem. The language of science in schools and in the media has not kept pace with the language of scientific practice – a practice that is decreasingly about experiments and increasingly about data and data modeling. In brief, one could argue that causal explanations grounded in control of variable experiments have largely been replaced by statistical/probabilistic explanations grounded in modeling experiments. The language of science in each experimental context is different. A reconsideration of the role of inquiry in school science must address this language gap and herein lays the importance of promoting scientific discourse practices. Examples of newly designed inquiry curriculum sequences that are striving to address the language gap include the LETUS program at Northwestern, the Learning by Design program at Georgia Tech, the SCOPE program at Berkeley and University of Washington, the WISE program at Berkeley, and the MUSE program at Wisconsin, among others.

These approaches adopt a model of science instruction that situates learning within design, problem or project contexts. The design, problem, or project based immersion units represent 4-6 week long lesson sequences that are situated within a compelling context to motivate students and to advance rigorous learning. Furthermore, in order to support learning, the immersion units typically contain

tasks that help make students thinking visible and thus provide teachers with valuable insights about how to give feedback to students in each of the three goal domains
- promoting the communication of scientific ideas,
- developing scientific reasoning,
- developing the ability to assess the epistemic status that can be attached to scientific claims.

The goal is to assist learners with both the construction and the evaluation of knowledge claims. Thus, by design, students are given extended opportunities to explore the relationships between evidence and explanation. To this end, inquiries are situated into longer thematic instructional sequences, where the theme is defined not by the conceptual structures of scientific content alone. Rather, the sequence of inquiries is designed to support acquisition and evaluation of evidence, as well as language and reasoning skills that promote progress toward a meaningful inquiry goal; e.g., the design, problem or project. The shift from a content/process focus of science education to an evidence/explanation focus has significant implications about the role of inquiry in school science.

The lesson sequence approach, referred to as full-inquiry or immersion units, stands in stark opposition to single lesson approaches that partition concepts and processes. Osborne and Freyberg (1985) report that students' understandings of the goals of lessons do not match teacher's goals for the same lessons. When students do not understand the goals of inquiry, negative consequences for student learning occur (Schauble, Glaser, Duschl, Schulz & John, 1995). Unfortunately, the single science lesson approach is the dominant practice found in schools. By situating science instruction and learning within a design-based, problem-based, or project-based context, to which members of the class have both individual and group responsibilities, a very different classroom learning environment develops. Specifically, the design of thematic instructional sequences allow us to approach closer to an understanding of the developmental landscapes located within domains of science learning; landscapes that do not presuppose a single developmental trajectory or path but do require a clear understanding of the conceptual, epistemic and social developmental goals within a unit of science instruction.

When we synthesize the learning sciences research (cf., Bransford, Brown & Cocking, 2000; Pellegrino, Churkowsky & Glaser, 2002), the science studies research (cf., Giere, 1988; Hull, 1988; Longino, 2002; Nersessian, 1999) and science education research (cf. Millar, Leach & Osborne, 2001; Minstrel & Van Zee, 2001), we learn that :
1. The incorporation and assessment of scientific inquiry in educational contexts should focus on three integrated domains:
- The conceptual structures and cognitive processes used when reasoning scientifically,
- The epistemic frameworks used when developing and evaluating scientific knowledge, and,

- The social processes and contexts that shape how knowledge is communicated, represented, argued and debated.
2. The conditions for science inquiry learning and assessment improve through the establishment of:
- Learning environments that promote student centered learning,
- Instructional sequences that promote integrating science learning across each of the 3 domains in (1),
- Activities and tasks that make students' thinking visible in each of the 3 domains, and
- Teacher designed assessment practices that monitor learning and provide feedback on thinking and learning in each of the three domains.

Full inquiry or immersion units promote the meaningful learning of difficult scientific concepts, the development of scientific thinking and reasoning, the development of epistemological criteria essential for evaluating the status of scientific claims and the development of social skills concerning the communication and representation of scientific ideas and information. Providing students with opportunities to engage with natural phenomenon and to link evidence to explanations is vital. In the sections to follow, we will examine, how research from science studies, the learning sciences, and educational research each argues for the reconceptualization and reorganization of inquiry instruction.

CHANGING IMAGES OF INQUIRY

Over the last 50 years, science education in the USA has been a strong focus of attention. Following on from World War II, steps were taken to sustain the scientific superiority that contributed to winning the war. New technologies and new frontiers of science defined post-war America. Policy makers agreed that in order to keep our technological and scientific edge science education in our pre-college schools and classrooms needed changes that would modernize both what was taught and how it was taught.

The task of overhauling high school science programs initially fell to the same scientists who contributed to our war effort (Rudolph, 2002; Duschl, 1990). The goal was to establish a curriculum that would develop in learners the capacity to think like a scientist and prepare for a career in science, mathematics and engineering. Thus, not surprisingly, the initial NSF-funded science curriculum (PSSC, BSCS, CHEMSTUDY) had a 'science for scientists' focus of instruction. Embedded within the 'science for scientists' approach was a commitment that students should be provided opportunities to engage with phenomena; that is to probe the natural world and conduct inquiries that would reveal the patterns of nature and the guiding conceptions of science. The goal was to downsize the role of the textbook in science teaching and elevate the role of investigative and laboratory experiences in science classrooms. That is, according to Joseph Schwab (1962), first director of BSCS, science education should be designed so that learning is an 'enquiry into enquiry' and not a rhetoric of conclusions, e.g., teaching what we know.

The commitment to inquiry and to lab investigation is a hallmark of USA science education. The development of curriculum materials that would engage students in the doing of science though required an investment in the infrastructure of schools for the building of science labs and for the training of teachers. What is important to note is that at the same time period (1955 to 1970) when scientists were leading the revamping of science education to embrace inquiry approaches, historians and philosophers of science were revamping ideas about the nature of scientific inquiry and cognitive psychologists were revamping ideas about learning. A reconsideration of the role of inquiry in school science, it can be argued, began approximately 50 years ago.

Unfortunately, the widespread reconsideration has also led to a proliferation of meanings associated with "inquiry". In a recent international set of symposium papers (Abd-El-Khalick et al., 2004), the following terms and phrases were used to characterize inquiry:
– scientific processes
– scientific method
– experimental approach
– problem solving
– conceiving problems
– formulating hypotheses
– designing experiments
– gathering and analyzing data
– drawing conclusions
– deriving conceptual understandings
– examining the limitations of scientific explanations
– methodological strategies
– knowledge as "temporary truths"
– practical work
– finding and exploring questions
– independent thinking
– creative inventing abilities
– hands-on activities

Whereas the 'science for scientists' approach to science education stressed teaching what we know and what methods to use, the new views of science and of psychology raise pressing issues of how we know what we know and why we believe certain statements rather than competing alternatives. The shift was a move from a curriculum position that asks, "what do we want students to know and what do they need to do to know it", to a curriculum position that asks, "what do we want students to be able to do and what do they need to know to do it". The NSES content goals for inquiry focus on student's abilities to pursue inquiry and to understand the nature of scientific inquiry. But once again we seem to find ourselves in the situation were science education has not kept pace with developments in science. That is, science education continues to be dominated by hypothetico-deductive views of science while philosophers of science have shown that scientific inquiry has other equally essential elements: theory development, conceptual change, and model-construction. This is not to imply that scientists no longer engage in experiments. Rather, the role of experiments is situated in theory and model building, testing and revising, and the character of experiments is situated in how we choose to conduct observations and measurements; i.e., data collection. The danger is privileging one aspect of doing science to the exclusion of others.

Developments in scientific theory coupled with concomitant advances in material sciences, engineering and technologies have given rise to radically new ways of observing nature and engaging with phenomenon. At the beginning of the 20th century scientists were debating the existence of atoms and genes, by the end of the century they were manipulating individual atoms and engaging in genetic engineering.

These developments have altered the nature of scientific inquiry and greatly complicated our images of what it means to engage in scientific inquiry. Where once scientific inquiry was principally the domain of unaided sense perception, today scientific inquiry is guided by highly theoretical beliefs that determine the very existence of observational events (e.g., neutrino capture experiments in the ice fields of Antarctica).

Historically, scientific inquiry has often been motivated by practical concerns, e.g., improvements in astronomy were largely driven and financed by the quest for a better calendar, and thermodynamics was primarily motivated by the desire for more efficient steam engines. But today scientific inquiry underpins the development of vastly more powerful new technologies and addresses more pressing social problems, e. g., finding clean renewable energy sources, feeding an exploding world population through genetically modified food technologies; stem cell research. In such pragmatic problem-based contexts, new scientific knowledge is as much a consequence of inquiry as the goal of inquiry.

Looking back there are several trends in science education that have altered our images of the role of the inquiry in science education:

- From a goal of providing science education for scientists, to providing science education for all.
- From an image of science education as what we know, to science education as teaching science as a way knowing.
- From an image of science education that emphasizes content and process goals to science education that stresses goals examining the relation between evidence and explanations.
- From an emphasis on individual science lessons that demonstrate concepts, to science lesson sequences that promote reasoning with and about concepts.
- From the study of science topics that examine current scientific thinking without regard for social context, to the study of science topics in social contexts.

and electrons which are each very small in relation to the size of the atom, he described the experiment as shooting electrons at a thin sheet of gold foil and *seeing* that most electrons pass through but some bounce straight back. Millikan (1965) in describing his classic oil drop experiment in which he measured the charge on the electron speaks also of *seeing* individual electrons. We would regard these as metaphorical, perhaps, but there is no question that from 1900 to 2000 science progressed from a stage where the existence of atoms was a debatable hypothesis to one where we can create images of individual atoms and we can manipulate them individually.

Tenet 2, the belief in inductive logic was important as part of the conception of scientific rationality. Based on the success in developing a deductive logic that was adequate for almost all mathematical purposes, the logical positivists saw it as a natural extension to provide an inductive logic for theory evaluation. The goal of Carnap, Hempel, Reichenbach and others was to provide an algorithm for theory evaluation. Given a formal representation of the theory and a formal representation of the data, the algorithm would provide *the rational degree of confirmation* the data confer on the theory. For example, the law that all copper melts at temperature t, would be represented in logical notation as $(\forall x)(Cx \supset Mxt)$ where C represents the property of being copper and M the relation that holds if x melts at temperature t. The formal sentence is transliterated into English as: "For all values of x, if x is copper, then x melts at temperature t." Fundamental problems beset the positivist project and they did not progress beyond very elementary examples in their discussions of confirmation and theory evaluation. (See Hempel, 1943, for the early version of the program, and Hempel, 1988, for a postmortem discussion of the successes and (mostly) failures of the program.)

The problem with Tenet 3 is not that it postulates a distinction between the context of discovery, the situation in which a theory is first discovered, and the context of justification. the presentation of the theory in its final axiomatized form, but that these were seen as exhausting the process of theory development. Perhaps the most important element Kuhn and others added to the problem mix is the recognition that most of the theory change that occurs in science is not final theory acceptance, but improvement and refinement of a theory. Ninety nine percent of what occurs in science is neither the context of discovery nor the context of justification, but the context of theory development, of conceptual modification. The dialogical processes of theory development and of dealing with anomalous data occupy a great deal of scientists' time and energy. The logical positivist's "context of justification" is a formal final point – the end of a journey; moreover, it is a destination few theories ever achieve, and so focus on it entirely misses the importance of the journey. Importantly, the journey involved in the growth of scientific knowledge reveals the ways in which scientists respond to new data, to new theories that interpret data, or to both. Some people describe this feature of the scientific process by saying that scientific claims are tentative; we prefer to say that science and scientists are *responsive*, thus avoiding the connotation that tentative claims are unsupported by evidence or scientific reasoning. (Tenets 4-7

LOGICAL POSITIVISM

We find it helpful to identify *seven main tenets of philosophy of science that* underly logical positivism:
1. There is an epistemologically significant distinction between observation language and theoretical language and that this distinction can be made in terms of syntax or grammar.
2. Some form of inductive logic would be found that would provide a formal criterion for theory evaluation,
3. There is an important dichotomy between contexts of discovery and contexts of justification
4. The individual scientist is the basic unit of analysis for understanding science s
5. Scientific development is cumulatively progressive.
6. Different scientific frameworks are commensurable.
7. Scientific theories can most usefully be thought of as sets of sentences in a formal language.

Our discussion of these tenets will necessarily be brief and in many cases simplify complicated issues. For more detailed discussions, see Hempel (1970), Suppe (1977), and Friedman (1999).

Tenet 1 posits a linguistic distinction between theoretical and observational terms in the languages of science. Over the years both in terms of internal developments in logical positivism and external criticisms philosophers of science recognized that the theory/observation language distinction can't be made on the basis of grammar alone. A statement about the mass of an object in the range of 20 grams to 2000 kilograms would be an observation statement, but a statement about the mass of a star or of an electron would be theoretical (Hempel, 1966). One of the most important external critics was Norwood Russell Hanson with his book *Patterns of Discovery*. This has led to the recognition that our ordinary perceptual language is theory laden, that we talk about the red shift of stars or the solubility of sugar, and this has led gradually to the recognition that a better description of what we are doing is creating models of data. This began in the 60s but has been accelerated with the development of computers so that much more powerful modeling techniques are now available and computational power is increased unimaginably. Modeling efforts such as the human genome project or efforts to understand global climate changes were inconceivable in 1970. New measurement techniques, indeed measurement of quantities that were not suspected, such as the geomagnetic patterns in the ocean floor, have led to the creation of new forms of data as well. This in turn has led to large databases which are available for classroom use.

Logical positivists were slow to recognize the shift in what counts as observational, which is not a matter of grammar but evolves historically as science changes. Moreover, scientists themselves describe the processes in terms that suit their goals. For example, when Rutherford discovered that atoms consist of nuclei

In this view of science, theories still played a central role, but they shared the stage with other elements of science, including a social dimension. Although Kuhn saw the scientific communities as essential elements in the cognitive functioning of science, his early work did not present a detailed analysis. The most recent movements in philosophy of science can be seen as filling in some of the gaps left by Kuhn's demolition of the basic tenets of logical positivism. This movement

1. emphasizes the role of models and data construction in the scientific process and demotes the role of theory;
2. sees the scientific community as an essential part of the scientific process;
3. sees the cognitive scientific processes as a distributed system that includes instruments.

Among the major figures in this movement are Nancy Cartwright (1983), Ron Giere (1988, 1999), Helen Longino (1990, 2002) Nancy Nersessian (1999), Patrick Suppe (1969), Fred Suppes (1989), and others.

One can summarize developments along a continuum where science has been conceived as an experiment-driven enterprise, a theory-driven enterprise, and a model-driven enterprise. The experiment-driven view of science emerged out of the early 20th century activities of the Vienna Circle and related movements in Berlin and Poland. Peano, Frege, Russell and Whitehead had apparently shown that all of mathematics could be reduced to a single syntactically formulated axiomatic system[1]. The commitment among a group of natural philosophers (e.g., Mach, Carnap, Hempel, Reichenbach) was that science like mathematics should be grounded in the new symbolic logic. The enterprise gave birth to the movements called logical positivism or logical empiricism and shaped the development of analytic philosophy and gave rise to the hypothetico-deductive conception of science. The image of scientific inquiry was that experiment led to new knowledge which accrued to established knowledge. How knowledge was discovered was not the philosophical agenda, only the justification of knowledge was important. This early 20th century perspective is referred to as the 'received view' of philosophy of science.

This conception of science is closely related to traditional explanations of "the scientific method." The steps in the method are:
− Make observations
− Formulate a hypothesis
− Deduce consequences from the hypothesis
− Make observations to test the consequences
− Accept or reject the hypothesis based on the observations.

It is important for our purposes not to simply reject logical positivism without understanding it. If we do so we risk both losing some of the insights and losing perspective on some of the oversimplifications that were involved. Similarly, we do not want to reject this conception of scientific method, but to radically supplement it.

- From a view of science that emphasizes observation and experimentation, to a view that stresses theory and model building and revision.
- From a view of scientific evidence principally derived from sense-perception (either direct or augmented) to a view that evidence is obtained from theory-driven observations.

The implications for teaching inquiry in school science are significant since these changes raise questions about
1. the amount of time allocated to interactions with basic scientific phenomena;
2. the depth and breadth of experiences learners bring with them to the science classroom; and
3. the kind of phenomena and experiences that stimulate science learning.

As stated above, the 1960s NSF sponsored revolution in science education focused on a science for scientists approach. Twenty years later after an enormous infusion of scientific knowledge into all walks of life, arguments for a science for all approach to science education began to emerge. We now see scientific knowledge as indispensable for participation in the workplace and in a modern democracy. However, the science education community has been slow to embrace new philosophical, psychological and pedagogical models that can inform the design of curriculum, instruction, and assessment frameworks that, in turn, inform the place of inquiry in science education.

CHANGING IMAGES OF SCIENCE IN SCIENCE STUDIES IN THE 20TH CENTURY

It is well beyond the scope of this paper to provide a comprehensive review of developments in the various areas of science studies; the interested reader can find useful summaries in Godfrey-Smith (2003), Duschl (1990, 1994) and Matthews (1994). In very broad brushstrokes, 20th century developments in science studies can be divided into three periods. In the first, logical positivism, with its emphasis on mathematical logic and the hypothetico-deductive method was dominant. Some of the major figures in the movement were Rudolf Carnap, Carl G. Hempel, Ernest Nagel, and Hans Reichenbach. In the 1950s and 60s, various writers questioned many of the fundamental assumptions of logical positivism and argued for the relevance of historical and psychological factors in understanding science. Thomas S. Kuhn is the best known of the figures in this movement, but there were numerous others, including Paul Feyerabend, (1993) Norwood Russell Hanson (1958), Mary Hesse (1966), and Stephen Toulmin (1959, 1961). Kuhn (1962/1996) introduced the conception of paradigm shifts in the original version of *Structure of Scientific Revolutions*, and then revised it in the postscript to the 1970 second edition, introducing the concept of a disciplinary matrix. One important aspect of Kuhn's work was the distinction between revolutionary and normal science. Revolutionary science involves significant conceptual changes, while normal science consists of "puzzle solving", of making nature fit into the boxes specified by the disciplinary matrix.

will be discussed later in the contexts in which they come into question instead of being presupposed.)

One of the most striking points that Kuhn made in *Structure* was that science textbooks are a monological presentation of *ex post facto* evidence for current theories and entirely suppress the dialogues concerning alternative conceptions and apparent evidence against the current theory. The more sophisticated and looser contemporary construal of the connection between theory and data recognizes that, when we have an apparent conflict between theory and data, we sometimes reject the theory. How and why to revise the theory is a complicated epistemic matter and since there is no algorithm for theory change, the changes almost always require dialogic justification within the context of the relevant social structure of expert scientists.

Sometimes, however, we reject the data, and again, contra logical positivism, since there is no privileged observation language and no theory choice algorithm, this decision is a matter of judgment. And sometimes we reject both theory and data. A historical example is Tycho Brahe who rejected both the Copernican and Ptolemaic theories of the solar system and created new instruments to make better observations. He hoped that his better observational data would support his own unique compromise which maintained the earth in the center of the universe, but which had all of the other planets (except the moon) circling the sun which circled the earth. Instead, his data provided evidence which led Kepler to his three laws of planetary motion, and moved one step further toward the Newtonian conclusion of the Copernican revolution (T S. Kuhn, 1957).

One important question for our context is at what stage of science education we introduce these complications about data revision, theory revision and their interaction. Like logical positivism, science education which teaches final theory has no place for these essential elements of the development of science. While knowledge of final (or at least current) theory is the main goal of science education for scientists, it should not be for science education for well-informed citizens.

The 3rd tenet, the dichotomy between context of discovery and context of justification, involves the contingent psychological and historical situation in which a theory was first conceived or discovered and contrasts it with the context in which a theory, presented in a formal language according to Tenet 2 was evaluated/justified for acceptability. Given this either/or perspective of end-point theory evaluation, it is significant to note there was no recognition of the important intermediary processes between these two stages of scientific development in the growth of scientific knowledge. In particular, we wish to stress the importance of the dialogic processes that shape and focus evidence as well as explanatory frameworks. Such processes are eminently, and simultaneously, cognitive, epistemic, and social in constitution and are critical to the design of science learning environments.

CONCEPTUAL CHANGE

In the first edition of *Structure*, Kuhn defined two kinds of scientific development in terms of *paradigms*. Normal science involves the articulation and refinement of a paradigm that is shared by the relevant scientific community; in revolutionary scientific change, one paradigm is rejected and another takes its place. One reason for the widespread influence of the book outside of the community of philosophers and historians was that the conception of a group or community guided by a paradigm seemed to have explanatory value in many settings. This use of the term has become firmly entrenched as a standard expression in English and appears in cartoons and business management courses although most of its contemporary users have no notion of its source.

However useful the term "paradigm" has proven in the general culture, it was the cause of considerable criticism in the reception of the book because critical readers perceived that he was using the term very variously and loosely. One critic (Masterman, 1974) taxonomized twenty-two distinguishable senses of the term in *Structure*. Kuhn disagreed with the precise count, but saw that clarification was required. Many of the criticisms were aired at two important conferences that focused heavily on his work. The first was held in London in 1965 and the second in Champaign IL in 1969. The proceedings of these were eventually published as *Criticism and the Growth of Knowledge* (ed. Lakatos and Musgrave, 1970) and *The Structure of Scientific Theories* (Suppe, 1977). As a result of these influences and further reflection, in the "Postscript" to the second edition of *Structure* in 1970 Kuhn expressed a desire to replace the term "paradigm" with two new terms, "exemplar" and "disciplinary matrix", which he believed expressed the two main distinct uses he had made of "paradigm". Much of the confusion regarding Kuhn's position is also caused by the fact that Kuhn never rewrote *Structure* clarifying which sense of 'paradigm' he intended in each context, he merely added the postscript. Thus, first time readers would first pass through the original representations of paradigms guiding conceptual change, and most likely not realizing the radical revisions presented in the postscript. Qualifications on the meaning of paradigm as that found in the postscript of Structure are also expressed in his contributions to the London and Illinois conferences.

Exemplars represent one new meaning of "paradigm" (and the final element of a disciplinary matrix) and Kuhn emphasizes that these are *concrete* examples of problem solutions. One of the most crucial points in his emphasis on exemplars is that they give guidance to future research *by example*, not by incorporating rules or explicit method. A scientific field or specialty is given its coherence partly by these shared examples, but it is given its diversity of approaches by the possibility of researchers interpreting those examples somewhat differently from one another. Researchers can all agree that they want to do for their field what Newton did for his, but they may disagree fairly radically about what that was, and therefore on what they intend to achieve. Some of the confusion in interpreting *Structure* was due to the fact that one sense of paradigms, exemplars, are an element of the other's sense, disciplinary matrices.

In a *disciplinary matrix* Kuhn intends to include at least seven elements: equations or other symbolic representations, instruments, standards of accuracy and experimental repeatability, metaphysical assumptions, values, and the domain of inquiry. The domain of inquiry, in turn, includes the problems which workers in the field regard as relevant but unsolved and the workers' shared exemplars, the problem solutions which are set forth as examples of good solutions of important problems. With the modifications provided by the disciplinary matrix, the growth of scientific knowledge was conceptualized as a problem solving activity. A philosophical concern arose though concerning the direction of growth and the factors that influenced growth. One issue confronting philosophers was to explain how growth, conceptual change and problem solving was or could remain a rational and objective process. That is, given two theories T1 and T2, how was one to determine if T2 was a legitimate choice over T1 and on what grounds (e.g., epistemological, political, social or otherwise). We take up this issue below in discussions of Lakatos and Laudan, two philosophers who attempted to 'save the phenomenon' of sciences' image of being rational and objective. We extend the discussion as well in the next section of model-based views of the nature of science.

Kuhn's work gave a prominent place to both history of science and psychology, both of which were absent from the received view. In science studies, the recognition of the context of theory development provides a significant place for cognitive psychology in the study of science. Kuhn relied on Gestalt psychology, but since then the new post-behaviorist cognitive revolution has produced valuable studies on learning, concept acquisition and related themes.

One very important element for our purposes is that Kuhn recognized that a scientific community shared values and *examples/exemplars* as well as explicitly articulated elements such as equations. It is partly the tacit and somewhat variable construal of the values and exemplars that makes the community essential. Just as variation among genes within a species gene pool provides the material for further adaptation, the variation in the scientific community is a valuable resource for further development of that community and the growth of knowledge within the community.

Kuhn's inclusion of the scientific community as part of the scientific process goes against Tenet 4 above, which treats the individual scientist as the basic unit for understanding scientific rationality., This, together with his rejection of Tenet 2, (inductive logic) produced negative reactions from many philosophers. Including a social dimension was seen as threatening the objectivity and rationality of scientific development. His denial of Tenet 5, that scientific development is always cumulatively progressive, and his arguments against Tenet 6 that disciplinary matrices on different temporal sides of a revolutionary change are incommensurable also produced negative reactions. Some of this was because of misunderstanding of "incommensurability" which he took to mean that the competing matrices were not comparable in any rational way. Kuhn was using the term in the old Greek sense meaning that no "ratio" exactly captures the relation, though one can approximate it as closely as one likes. And his denial that science is always progressive also produced criticism and outrage. As Kuhn confessed in 1993

> To my dismay, ... my 'purple passages' led many readers of Structure to suppose that I was attempting to undermine the cognitive authority of science rather than to suggest a different view of its nature. And even for those who understood my intent, the book had little constructive to say about how the transition between stages comes about of what its cognitive significance can be. (Kuhn, 1993, 314)

Also of considerable significance in understanding the impact of Kuhn's work was the "strong programme" in sociology of science which saw Kuhn's work as showing that science is essentially a matter of power struggles and personalities and not a matter of evidence and explanation; objectivity and rationality. Kuhn strongly disagreed with these alleged consequences of his work, in fact countered by arguing:

> Properly understood...incommensurability is far from being the threat to rational evaluation of truth claims that it has frequently seemed. Rather, ... its needed to defend notions like truth and knowledge from, for example, the excesses of postmodern movements like the strong program. (Kuhn, 1990, 91)

But because he was claimed as a progenitor by the strong programme many critics saw him as part of that movement. The assessment is complicated by the fact that while Kuhn rejects one aspect of Tenet 5, he accepts another. He believes that if we consider two scientific theories, one of which is a descendant of the other, we could tell which is the earlier and which the later by such criteria as comparative accuracy of predictions, number of problems solved, and degree of specialization. He comments that "For me, therefore, scientific development ... is unidirectional and irreversible. One scientific theory is not as good as another for doing what scientists normally do" (Kuhn, 1970, p 160).

However, while embracing increasing empirical adequacy, Kuhn denies that science is coming closer to a description of ultimate reality. He discusses philosophers who "Granting that neither theory of a historical pair is true ... seek a sense in which the later is a better approximation to the truth. I believe nothing of that sort can be found (160). (For further discussion of Kuhn's views on these matters, see Grandy, 2003a.)

A number of philosophers saw the merit in Kuhn's criticism of Tenets 1-3, but disagreed with his criticisms of Tenets 4-6 and proposed alternative frameworks. Thus a second important figure in the development of the conceptual change view of philosophy of science was Imre Lakatos. Lakatos was one of the critics of

Structure who felt that Kuhn was misrepresenting science as less rational and objective than it actually was. "...in Kuhn's view scientific revolution is irrational, a matter for mob psychology. Lakatos agreed with Kuhn in rejecting Tenets 1 and 2, but insisted on 3 and remained sympathetic to the remainder. In order to provide what he saw as rational criteria for theory choice without Tenets 1 and 2, he introduced the terminology of research programmes and distinguished between negative and positive heuristics (Lakatos, 1978). Negative heuristics signaled theory development that was deteriorating while positive heuristics represented theory development that was progressive. Among the positive heuristics are better solutions, or more accurate solutions to old problems, but more importantly the prediction of novel facts that are confirmed (e.g., *Mendeleev's prediction of elements in support of the periodic law; Huxley's prediction of two-toed horses in support of the theory of evolution, Tuzo Wilson's prediction of transform faults in support of the theory of plate tectonics*). Using that theoretical apparatus, Lakatos distinguished progressive and degenerative research programme trajectories.

A third significant figure was Larry Laudan who sought to represent the growth of scientific knowledge in terms of the progress being made among broad based communities of scientists which he called "research traditions". Laudan interpreted Kuhn as concluding that "scientific decision making is basically a political and propagandistic affair, in which prestige, power, age, and polemic decisively determine the outcome of the struggle between competing theories and theorists" (Laudan, 1977 p. 4). We disagree with that interpretation of Kuhn, interpreting him rather as rejecting a traditional conception of rationality and urging the philosophical and historical examination of the history of science to eke out a more realistic conception of rationality. (See Kuhn, 1977, 2000; Hempel, 1983a, 1983b, for further discussions.)

Although we disagree with Laudan on this central issue of Kuhn interpretation, we believe that his distinction between empirical and conceptual problems is helpful, as too is his term "research tradition" and his characterization of when a research tradition is progressing. Empirical problems are matters of fitting descriptions to the world. Galileo's search of the laws of falling bodies was an attack on an empirical problem, as was the problems involved in filling in the periodic table after Mendeleev had the basic insight. Mendeleev solved the conceptual problem of how to organize the elements – portraying them as a two dimensional array made evident important relationships that had been invisible using the older one dimensional array of elements. Thus, conceptual problems are those that arise when a theory or theoretical framework in one domain conflicts with theories from other cognate domains. Empirical problems represent the extent to which the theory or theoretical frameworks fits or accounts for object and material evidence. Copernican and Ptolemaic astronomy provided empirically equivalent predictions of the retrograde motions of the planets. However, the Copernican system provided a principled explanation of the retrograde motions while the Ptolemaic system had only an ad hoc adjustment to fit the empirical data.

Using the problem solving theoretical apparatus, Laudan (1977) attempted to characterize when a research tradition is progressing, and when it is not, in terms of increased problem solving ability in one or both of these areas. A research tradition progresses when it successfully solves conceptual and/or empirical problems.

The research tradition developing Newton's theory of universal gravitation made continual empirical progress in early stages as new phenomena such as tides and comets were brought within the scope of the theory. It made empirical progress in later stages as new planets were postulated, and discovered, on the basis of the theory. On the other hand, critics of the theory, especially Cartesians in France, continued to point to the *conceptual* problem that the theory lacked any account of how gravity produced the effects it does. It was not conceptually progressive and was eventually superceded by general relativity.

Lakatos, Laudan and others sought to accommodate what they regarded as the important insights of Kuhn's *Structure*, but *also tried* to restore a more familiar sense of rationality and progress. Kuhn disputed their claims and thought that the dialectical social processes that take place in scientific communities constituted the best example of rationality. In any event, one of the lessons from science studies is that social processes such as peer review are essential to scientific objectivity, and thus for students to understand the processes of science we must introduce argumentation discourse in the classroom. And this should be situated in the context of improving and refining perspectives, theories, data, models and knowledge claims.

Returning to the theme of "the scientific method", we can now recognize that the initial process of "observation" in Step 1 already involves many presuppositions about how the world works, that the choice of hypothesis will be guided, perhaps constrained, by the appropriate disciplinary matrix, that the deduction of consequences typically requires auxiliary assumptions, that the design of the experiment to make the subsequent observation possible depends on theoretical assumptions about intervening causes and the design and function of instruments, and finally that the evaluation of the fit between "observation" and "prediction" is less than clear-cut in many cases.

Having given up the logical positivist assumption that there is a unique algorithm for evaluating the fit between theory and data, and the assumption that there is a privileged unproblematic observation language, we recognize that the dialogic processes by which theory evaluation and data construction occur in the community are an indispensable part of the rational structure by which science unfolds. One central question is where in the trajectory of science education we should introduce these complications of simultaneous revision of theory and data. We know that six year olds come to the classroom with concepts, theories and belief in data. Can we already introduce some elements of the dialogic processes at that level?

Recent research on the persistence of false beliefs (Gilbert, 1991; Gilbert et al., 1993) indicates that the process of dispelling prior misconceptions is much more difficult than has been appreciated among advocates of conceptual change teaching methods. More investigation of the processes involved in the relevant contexts would be very helpful.

MODEL BASED SCIENCE: LANGUAGE, REPRESENTATION AND COMMUNICATION[2]

Emphasis on understanding the relations between data, theories and naïve observation stems from the work of Kuhn, Hanson and others, but has only become a focus for philosophy of science in the last 20 or 30 years. Among the major contributors are Ackermann (1985), Bechtel (Bechtel & Richardson, 1993), Cartwright (1983), Giere (1988, 1999), Longino, (1990, 2002), Mayo (1996) and Fred Suppe (1989). Surely having an erroneous view of the nature of scientific theories was an impediment to science education, and having a correct one should be a boon. But we will see that reaping these benefits will probably require more consensus, or at least more discussion and clarification than we have achieved thus far. For a start, there is no generally agreed on name for the new view, unfortunately. It has been called "semantic", "structuralist", "model-theoretic", "set theoretic", "non-statement" and more recently a favored term has been "model-based". This may just be an issue of nomenclature, but it may also reflect some important uncertainty about details.

One of the most careful statements of the philosophers' new views is probably Suppe's. His "nuanced sloganistic analysis" of the new view is that, "Scientific theories are causally-possible collections of state-transition models of data for which there is a representation theorem" (Suppe, 2000, p. S111). A representation theorem means that there is a mapping from measurable quantities of physical systems to the mathematical system in which the structure of relations among quantities is reflected by parallel relations among real numbers. Causally possible collections are those that are consistent with experiment and theory, while a 'state transition model' is a systematic description of how the states of the physical system change over time.

To take a simple example, if we are theorizing about cylinders rolling down an inclined plane, then the relevant physical quantities are the slope of the plane, the radius of the cylinder, distribution of mass in the cylinder, the location of the cylinder on the plane and its velocity. A state transition model would describe how later states of the cylinder are related to earlier states and the causally possible ones are those consistent with Newton's laws of motion and gravitation.

Developmental psychologists have also shown an interest in the new *model-based* view and used it in arguing for cognitive continuity between infant and adult science; i.e., children are fundamentally the same kind of reasoners as adults. The prospect of bringing together philosophy of science and cognitive developmental psychology in the service of science education is exciting and promising. However, to fulfill the promises, the various parties must pull in the same direction.

Gopnik is one of the most explicit of the cognitive psychologists on cognitive continuity and her conceptual change account of learning – the theory-theory – is that, "A person's theory is a system that assigns representations to inputs..." These representations are distinctive in having "...abstract, coherent, causal, ontologically committed, counterfactual supporting entities and laws...". Moreover, "The representations are operated on by rules that lead to new representations; for example, the theory generates predictions" (Gopnik, 1997, 43).

The most sophisticated attempt to bring together psychology with the new *model-based* view is Giere (1988, 1999). His account is that a theory is a family of models and that a model is an abstract mathematical entity. Giere's models include those envisioned by Suppe but could include a broader range of other mathematical structures.

We do not want to worry further here whether Suppe's models are the same kind of creature as Giere's, nor whether Gopnik's representations have anything to do with Suppe's representation theorems. We simply want to alert the reader to the discrepancies that exist in terminology, and probably in underlying views. While the recognition that "models are essential to science" is a potentially valuable discovery, it risks sowing confusion since there are many different meanings attached to "model".

We can readily distinguish at least five meanings that are prevalent in the literature in science studies and science education.
– Model as mathematical descriptions
– Model as physical analogues
– Model as iconic

In addition, models can either be "mental", i.e., within the cognitive system of a teacher or learner, or can be represented in a publicly accessible fashion.

We will begin to delve into this question by asking when an external mathematical model is required in place of a mental model. This may help to illustrate the first sense of model and the issues it raises. Here we draw on the work of Profitt and Gilden (1989) on the understanding and misunderstanding of rotating objects. The simplest problems in mechanics have only one relevant parameter, the location of the center of mass of the object. However, when rotating objects are considered, the distribution of mass sometimes becomes relevant. Given two objects sliding down a frictionless plane, their distribution of mass is irrelevant. In contrast, given two cylinders of the same size, shape and mass, on an inclined plane, one still needs to know the distributions of mass in order to predict which cylinder will roll to the bottom first. Two cylinders of may have the same diameter and mass, but different distributions of mass. One example would be a solid wooden cylinder and a hollow metal cylinder of the same diameter and mass. In the latter all of the mass is concentrated at the periphery, and the latter will roll more slowly down an inclined plane than the former.

In their studies, a variety of subjects were able to provide correct answers much more frequently to the one parameter problems than to the two-parameter problems. Most telling was that when the two-parameter problems were presented to high-school physics teachers and the teachers were forced to answer rapidly, their answers were no better than those of undergraduates without time constraints. Summarizing:
– Given problems about rolling cylinders in which mass distribution is a relevant parameter, over 80% of University of Virginia undergraduates gave incorrect responses
– Members of the University of Virginia Bicycle club did no better
– Given the same problems and required to answer immediately without calculations, University of Virginia physics faculty did no better
– Given the same problems and required to answer immediately without calculations, high school physics teachers did no better.

To quote from Profitt and Gilden's conclusion:

> ... physicists have a dual awareness of the characteristics of mechanical systems. This awareness is quite interesting to observe and it is easily elicited .... Prevent competent physicists from making explicit calculations about such events as rolling wheels, and they exhibit the basic confusions that are found in naive observers .... However, most physicists could work the problem out in a few minutes. (391)

What is required for the experts to solve the problems is the opportunity to make explicit calculations. This suggests that the most pragmatic understanding of "having a model" is "having the ability to deploy relevant equations to describe the problem."

Evidence both from the history of science and from research on cognitive and learning indicates that while people can fairly accurately represent relations between one independent variable and dependent variable which is a linear function of the first, that any step beyond that in mathematical complexity requires explicit mathematical concepts and notation. Galileo struggled for decades to come to grips with the first non-linear relationship in the history of science, the dependence of the distance a body falls in a given time on the square of the time of fall.

There is a large body of evidence (Chi, etc.) that even those who have been formally trained in mechanics revert to erroneous informal heuristics if pressed to solve problems under severe time constraints if those problems involve non-linear equations, e.g., $y=ax^2+bx+c$. Although we know of no historical struggles comparable to Galileo's, there is also ample literature testifying that humans have difficulty with even multivariate linear relations involving more than one independent variable, e.g., $y = ax +bz +c$.

The development first of algebra and then calculus were indispensable to the rise of modern science and no science curriculum, inquiry based or not, can ignore the role of mathematics. How, and how much, mathematics to integrate is a difficult question we must address. There is in mathematics education a great deal of work on introducing calculus concepts (math of motion, space, time) in early grade levels. The commitment is to locating the 'Big Ideas' in mathematics and then thinking hard about how to design adevelopmental sequence or corridor of instruction that leads to increasingly sophisticated understandings. The bottom line for us is that the integration of science and mathematics needs more attention!

In addition to these older historical developments, as indicated earlier the development of the computer has provided another indispensable tool for constructing mathematical models and provides another host of questions about teaching and integration of computers and science education.

The second sense of model, physical model, is equally familiar but quite different. Physical models have sometimes been indispensable in scientific discovery, e.g., the billiard ball model of gases and Watson and Crick's physical construction of possible models of DNA.

Fuller discussion of physical and other types of model are essential,but are beyond the scope of this paper. While we believe that our classification is useful, it is important to bear in mind that a model may have characteristics of several different kinds. For example, Maxwell used a picture (visual model) of an imagined (mental model) physical structure (physical model) to arrive at his equations (mathematical model) (see Nersessian, 1992 p. 62).

CHARACTERISTICS OF MODELS

Woody (1995) identified four properties of models: approximate, projectability, compositionality, and visual representation. A model's structure is approximate. In other words, the model is an approximation of a complete theoretical representation for a phenomenon. The model omits many details based on judgements and criteria driving its construction. In many cases the model omits many variables that we know are relevant, but whose contribution is (believed to be) minor in comparison with the factors that are included in the model (see Harte, 1988).

The second characteristic of models according to Woody is that they are productive or projectable. In other words, a model does not come with well-defined or fixed boundaries. While the domain of application of the model may be defined concretely in the same sense that we know which entities and relationships can be represented, the model does not similarly hold specification of what might be explained as a result of its application.

Woody further argues that the structure of the model explicitly includes some aspects of compositionality. There is a recursive algorithm for the proper application of the model. Thus, while the open boundaries of the model allows it potential application to new, more complex cases, its compositional structure actually provides some instruction for how a more complex case can be treated as a function of simpler cases.

Finally, in Woody's (1995) framework, a model provides some means of visual representation. This characteristic facilitates the recognition of various structural components of a given theory. Many qualitative relations of a theoretical structure can be efficiently communicated in this manner.

Our taxonomy of models and their apparently disparate nature might lead readers to wonder if anything units them other than the label. We believe that the common element to all models is that they are external aids to reasoning – they are cognitive prostheses. Moreover, just as we now conceive the scientific community as a fundamental part of the process, the models are also a fundamental part and the cognitive processes should be thought of as being distributed throughout the system of people, instruments and models (Hutchins, 1995; Giere, 2002).

Adopting a purely language-based perspective for the framing of scientific theories has proven unsuccessful, and have led to the rejection of Tenet 7, the assumption that the best way of represeneting scientific theories is as a set of sentences in a formal logical language. Closely related is the observational-theoretical (OT) distinction. (Tenet 1) As discussed earlier, language-based perspectives maintained that the observational language of science can and should be distinguished from the theoretical language of science. Hanson (1957) was one of the first philosophers to advance the distinction between 'seeing as' and 'seeing that', the first grounded in purely sense perception observation, the second grounded in observation accompanied with background knowledge. That is, what we see is based on what we know and believe. Subsequent research on children's conceptual understandings have shown how powerful prior knowledge can be in shaping and influencing the learning of science.

Using models to communicate the theoretical structures of science stand as an alternative to the communication of scientific knowledge claims solely on the basis of language. Philosophers of science refer to the problem of language as a basis for analyzing the theoretical structures of science as the observational/theoretical (OT) distinction problem. We now turn to a discussion of the OT problem and the implications it has for framing inquiry in science education.

Critical examinations of the structure of scientific theories also revealed that not only was it hard to make the OT distinction, it was equally difficult to come to terms with the observable and non-observable entities of theories. Van Fraassen (1980), among others, maintains that while one may elect to believe in the observable entities of a theory (e.g., age of earth, laws of motion, gas laws, anatomical structures), one may choose to remain agnostic with respect to the claims a theory makes about unobservable entities (e.g., alleles, cognitive schemata, neutrino capture). This view of the nature of science is labeled 'semantic realism' and it is a view that embraces the central role of mathematical languages in the representation of science (Suppe, 1998).

We would argue that a similar clarity in terms of observation language should be considered when rethinking the role of inquiry in science education. The traditional science education approach has been to get students to distinguish observation from inference – an approach rooted to Michael Faraday's infamous exploration of the burning candle, and later adopted as the initial investigation of 1960's NSF sponsored ChemStudy. While a distinction between observation and inference may be important for the design and interpretation of experiments, the role of observational language in statements of theory and construction of models is quite another matter.

Scientific inquiry has over the last 200 years become less and less dependent on direct sense perception acquisition of data. Increasingly, scientific inquiry relies on theory-driven methods of data collection, theories typically embodied in the tool and instruments employed by scientists. New scientific tools and technologies, like that used in GIS inquiries, have embedded visualization techniques as a corner stone for both the investigation and representation of scientific claims. Referred to as 'inscription' devices used by scientists, the visual representation of scientific evidence is an important language component of the sciences. Shapere (1980) in his exploration of the role theory has in technologically-rich scientific observation (e.g., neutrino capture experiments) maintains that any scientific observation naturally contains theories about the source, transmission and reception of observational information. An observation is deemed reliable and accurate, or scientific, if and when scientists have accounted for how information transmits from source to receptor without interference or distortion of the information. Galileo encountered such challenges as his claims about mountains on the moon were attributed to distortions of light caused by the lenses of his telescope.

A challenge for obtaining a proper perspective of scientific inquiry in science education is how to contend with the transition from sense perception based science at early grade levels (e.g., K-3) to theory-driven based science at later grade levels (e.g., 4-12). We can also conceptualize the transition problem within a specific domain of inquiry or instructional unit. That is, how ought we engage students in the phenomena of the natural world? Here we are calling attention to the recognition of patterns as well as to the concrete and abstract elements of science. Beyond the question of promoting valuable motivation for learning, should inquiry units always begin with or sustain a commitment to sense perception based exploration of phenomenon? There is a strong sentiment among some science educators that engagement with phenomenon is critically important for sustaining inquiry. If so, the questions we are raising concern what are the important features of scientific phenomenon and how do we conceptualize distinctions between perceptual and non-perceptual or hidden observational phenomenon in science lessons. What patterns of nature are revealed by our senses and which are hidden from or fooled by our senses? Louis Wolpert in his book *The Unnatural Nature of Science* claims that science is fundamentally a break from common sense. What we see is shaped by what we know.

What came to be recognized and understood through historical and contemporary case studies of scientific inquiry was that the move from experimental data to scientific theory was mediated by dialogic communication. What stands between data and theory are models. Model-based views of the nature of scientific inquiry allow the inclusion of psychological processes where the received view did not. Model-based views recognize that argumentation discourse processes serve to define dialogic communication. By standing between theory and data, models can be influenced by both data revisions and changes in theory commitments.

Lab investigations situated in immersion units afford such opportunities for data revision and theory change. Thus, experiment and theory structure are important elements of the nature of scientific inquiry but now must be understood in relation to the dialectical processes that establish data as evidence and then take the evidence to forge explanations. The implication for high school laboratories is that science education should provide more opportunities that lead to model-based inquiry and support the dialectical processes between data, measurement, and evidence on the one hand, and observation, explanation, and theory, on the other. In such immersion units, students can be enticed to experiment for reasons and reason about experiments.

PHILOSOPHY OF SCIENCE AND THE LEARNING SCIENCES: THE CASES OF CONSTRUCTIVISM[3]

While some philosophical disputes, such as those about the nature of models and theories, are important and relevant for science education, others are not. In particular, we believe that the debates over metaphysical constructivism are irrelevant to teaching and if we can distinguish them from other cognitive constructivism, an important step forward will have been achieved. Recent developments in the design of learning environments that support learners' acquisition of scientific reasoning, argue for active engagement of learners in the process of reasoning and for embedded assessment activities that make thinking visible.

This may be a more generally useful exercise, of course, since it seems to us that there is not sufficient clarity about the variations on constructivism, let alone their relations and implications. We also believe that an understanding of the various elements and kinds of constructivism will be helpful in evaluating what is required of teachers in implementing full inquiry/immersion units or any other curriculum that incorporates the important elements of constructivism delineated below. What views teachers hold about constructivism will effect not only the content of what they teach and the methods they use, but also what they take as the goals of science teaching.

There are a wide range of terms (Fosnot, 1993; Gianetto, 1992; Glasson, 1992; Goldin, 1992; Mathews & Galle, 1992; Mathews, 1994; O'Loughlin, 1992, 1993;

von Glasersfeld, 1990, 1992), we are sure we will offend some authors by using the following distinctions rather than theirs. We distinguish cognitive constructivism, and metaphysical constructivism, both of which trace back to the work of Kuhn.

Cognitive constructivism is the view that individual cognitive agents understand the world and make their way around in it by using mental representations that they have constructed. What they could in principle construct at a given time depends on the conceptual, linguistic, and other notational resources, e.g., mathematics and graphing, at their disposal and on their current representations of the world that they have constructed through their personal history. What they actually construct depends also on their motivations and on the resources of time and energy available to devote to this particular task. This view derives from Kuhn's insights that perception is influenced by theory, and that theory evaluation always takes place in a context that includes competing theories.

By metaphysical constructivism we mean the (collection of) views that the furniture of the world is constructed by us. This view can be subdivided into the individualistic, which postulates individual constructions of individual worlds, and the social, which postulates social constructions of shared worlds. This view typically contrasts with metaphysical realism, the view that (much of) the furniture of the world exists independently of minds and thoughts. There are some obvious issues for the social sciences that we will not explore here, since social institutions are clearly human creations; the implications for geography or psychology or human biology are unclear and will also not be explored here. This view stems from, or at least is often thought to be supported by, the misinterpretation of Kuhn's work alluded to earlier. If one interprets Kuhn as thinking that "scientific decision making is basically a political and propagandistic affair, in which prestige, power, age, and polemic decisively determine the outcome of the struggle between competing theories and theorists" (Laudan, 1977 p. 4), then teaching should not emphasize distinctions between good reasons and bad, nor good arguments and bad, but rather emphasize arguments that persuade and those that do not, for whatever reasons. There is a subtle distinction between giving approximately equal weight to experiment, interpretation of data and argumentation, and giving less weight to experiment and more to interpretation and argumentation. We view it as essential that inquiry retain the importance of experimentation as providing an access to the world and not simply as a source of rhetoric.

Metaphysical realism itself comes in a range of positions on the optimism/pessimism scale with regard to the knowability of the structure. Optimistic metaphysical realism holds that not only does the universe have an intrinsic structure, that God used a blueprint if you like, but that the structure is knowable in principle by humans – we could understand the blueprint. A less optimistic, though still guardedly hopeful, version would be that we may develop representations which are approximately correct descriptions of some aspects of the universe. How either of these positions is justified philosophically is a matter we will not linger over here, for our main point is that these issues are irrelevant for science education once we understand fully the implications of cognitive constructivism.

A metaphysical realist who accepts cognitive constructivism must recognize that whatever knowledge is attained or even attainable about the ultimate structure of the universe must be represented in the constructions of the cognizer. While accepting cognitive constructivism has very important consequences for the teacher, which we will elaborate on shortly, once you have embraced cognitive constructivism it makes very little difference what attitude one has toward the metaphysical realist issues. However independent of us the structure of the universe may be, what we can achieve by way of producing more knowledgeable students depends on the representations they can construct. This seems to us of great importance because if Constructivism is presented as a package which includes both cognitive constructivism and metaphysical anti-realism then teachers who have longstanding philosophical inclinations toward realism will find the package unpalatable. Cognitive constructivism is a relatively empirical theory which has strong evidential support from psychology, artificial intelligence and education; metaphysical realism is a venerable philosophical doctrine supported by philosophical arguments, and subject to equally venerable philosophical objections.

Accepting cognitive constructivism has very significant consequences for understanding the tasks, and the demands of the tasks, required of the science teacher (Bloom, 1992). All science teachers have, and must necessarily always have had, a philosophy of science – a set of beliefs about the nature of scientific inquiry, of scientific progress, of scientific reasoning, of scientific data, theories, and so on. Often this has been somewhat unconscious and implicit, often acquired unreflectively along with the content knowledge in science classes. And often, in the past at least, this philosophy of science incorporated beliefs in the continuous linear progress of science, of the logical empiricist inductive scientific method, in the immutability of scientific facts, in scientific realism, perhaps even metaphysical realism, and so on. It has often included the philosophy of science education which is described as direct teaching (Duschl, 1990) or, as we think of it, the modified Dragnet theory of teaching. Unlike the old Dragnet show we don't just give them the facts, but on that model we do just give them the facts, definitions and theories, and nothing but the facts, definitions and theories.

Whatever other positions one adopts in philosophy of science, to accept *cognitive* constructivism means recognizing that each student constructs a representtation based on their experience, including but by no means limited to teachers verbal input. The teacher must assess the extent to which the student's representation is isomorphic to the teacher's, but of course cognitive constructivism applies reflexively and the teachers have no direct infallible access to the students' representations but instead construct their own representations of the students' representations. Since one of the typical student's motivations, for better or worse, is to please the teacher it may also be valuable for the teacher to construct a representation of the student's representation of the teacher's representation.

The practical issues of this process need to be discussed much further, but we want to make some general pragmatic points about the process. Teachers are necessarily pursuing this process under severe time constraints and must repeatedly balance the potential value of further exploring the student's representation in all of its detailed uniqueness against categorizing the students' representation as sufficiently similar to others seen in the past to allow a particular course of further instruction to be developed without further investigation.

Of course that is only part of the task, for the teacher may well need to understand also why the student has constructed that particular representation. The divergence from the desired kind of representation may result from lack of the tools to construct an alternative, lack of accepted evidence that the current representation is insufficient or lack of motivation to construct an alternative (Ames, 1990). The next step to bringing about desired change will likely depend on which of these factors is prevalent, and this implies that the teacher must have an understanding of motivational psychology, of the evidence the student accepts, what the student counts as evidence, and on what conceptual tools the student can make use of. If this is correct, then the conceptual change movement was in the right direction, but the process of instigating conceptual change in the student is more complicated than was initially recognized (cf. Clement, 1982).

THE LEARNING SCIENCES

Conceptualizing model-based science is prompted by cognitive science research demonstrating that higher-level thinking or reasoning is domain specific and specialized (Bransford et al, 1999). Research on learning with an eye toward informing educational processes suggests that we must attend to 4 types of knowledge:
– declarative "what we know" knowledge,
– procedural "how we know" knowledge,
– schematic "why we know" knowledge, and
– strategic "thinking about thinking" knowledge.

Given this deeper understanding of learning we can better appreciate the request that an important dynamic in the science classroom, and all classrooms for that matter, is to make students' thinking visible. The application of theory change processes to science education developed a focus on conceptual change teaching (Posner, Strike, Hewson & Gertzog, 1982). A consideration for the role dialectical processes have in science and in science learning led to the development of a focus on conceptual change teaching that was embedded both in motivating and relevant curricular contexts and in learning environments that promoted meaningful learning and reasoning (Pintrich, Marx & Boyle, 1993).

Robert Glaser (1995), in a major review of how psychology can inform educational practice develops and outlines the components of a coherent learning theory that can inform instruction and illuminates how Kuhn's approach might be achieved. He identifies 7 research findings that inform us about the structure and design of learning environments – aspects of which are further elaborated in *How People Learn* (Bransford, et al, 1999).

1. *Structured Knowledge* – "Instruction should foster increasingly articulated conceptual structures that enable inference and reasoning in various domains of knowledge and skill" (Glaser 1995, 17).
2. *Use of Prior Knowledge and Cognitive Ability* – "Relevant prior knowledge and intuition of the learner is … an important source of cognitive ability that can support and scaffold new learning … the assessment and use of cognitive abilities that arise from specific knowledge can facilitate new learning in a particular domain" (p 18).
3. *Metacognition Generative Cognitive Skill* – "The use of generative self-regulatory cognitive strategies that enable individuals to reflect on, construct meaning from, and control their own activities … is a significant dimension of evolving cognitive skill in learning from childhood onward ... These cognitive skills are critical to develop in instructional situations because they enhance the acquisition of knowledge by overseeing its use and by facilitating the transfer of knowledge to new situations … These skills provide learners with a sense of agency" (p. 18).
4. *Active and Procedural Use of Knowledge in Meaningful Contexts* – "Learning activities must emphasize the acquisition of knowledge, <u>but</u> this information must be connected with the conditions of its use and procedures for it applicability … School learning activities must be contextualized and situated so that the goals of the enterprise are apparent to the participants" (p. 19, emphasis in original).
5. *Social Participation and Social Cognition* – "The social display and social modeling of cognitive competence through group participation's is a pervasive mechanism for the internalization and acquisition of knowledge and skill in individuals. Learning environments that involve dialogue with teachers and between peers provide opportunities for learners to share, critique, think with, and add to a common knowledge base" (p. 19).
6. *Holistic Situations for Learning* – "Learners understand the goals and meanings of an activity as they attain specific competencies … Competence is best developed through learning that takes place in the course of supported cognitive apprenticeship abilities within larger task contexts" (pp. 19 -20).
7. *Making Thinking Overt* – "Design situations in which the thinking of the learner is made apparent and overt to the teacher and to students. In this way, student thinking can be examined, questioned, and shaped as an active object of constructive learning" (p. 20).

Prominent in the components of effective learning environments identified by Glaser is recognition of the important role prior knowledge, context, language and social processes have on cognitive development and learning. Such components are not marginal but centrally important to the process of learning. Such understandings have guided many educational researchers to recognize that reasoning is socially driven. (Brown, 1992; Cobb, 1994; Rogoff, 1990), language dependent (Wertsch, 1991; Gee, 1994), governed by context or situation (diSessa, 2000; Brown, Collins & Durgid, 1989) and involving a variety of tool-use and cognitive strategies (Edelson, Gordin & Pea, 1999; Kuhn, 1999). Putnam and Borko (2000), examines the challenges these new ideas about knowledge and learning have for teacher education, summarize these newer conceptions of learning as cognition as social, situated and distributed. It is social in that it requires interaction with others, situated in that it is domain specific and not easily transferable and distributed in that the construction of knowledge is a communal rather than an individual activity. The various programs of research conducted and coordinated by cognitive, social, developmental and educational psychologists now present a more coherent and multi-faceted theory of learning that can inform the design of learning environments (Bransford et al., 1999). In science education, we interpret this to mean that students must have an opportunity to engage in activities which require them to use the language and reasoning of science with their fellow students and teachers – that is to engage in the construction and evaluation of scientific arguments and models through a consideration of the dialogic relationship between evidence and explanation.

Today, understanding scientific knowledge, developing abilities to do scientific inquiry and understanding the nature of scientific inquiry requires going further than simply learning the conceptual frameworks of science. Thus, the role of laboratories has an enhanced importance more than ever because science as a way of knowing is a cultural entity. That is, the claims of science and the methods of science impact our lives in very direct ways, defining, challenging and redefining our perspectives about nature and the world. The National Science Education Standards recognize this through the inclusion of social perspectives about science and technology as a distinct content standard. The game of science, not surprisingly, has become more nuanced as scientific inquiry becomes more focused on disciplinary specializations and increasingly takes on problems concerning the human condition like hunger, risk assessment, environmental quality, disease, energy, and information technology, among others. Our high school students and their parents do indeed have a perspective and partial understanding of scientific issues and frequently espouse alternative perspectives that challenge established scientific claims. The implication is that the role of the laboratory in high school science should be more than a mechanism for teaching what we know.

The emphasis on concept and process learning in science education has, in our opinion, contributed to an image of science education that emphasizes the confirmation role of laboratory work, as opposed to a model-based role of laboratory. In a confirmation lab, the concepts and processes are presented via lecture and/or text and then a demonstration and/or investigation is conducted to confirm how the evidence links to the conceptual understanding. In this way, science education becomes final form science (Duschl, 1990) or as Schwab warned a 'rhetoric of conclusions'. What the 4 types of knowledge indicate is that a model of science learning and the goal of laboratory experiences must go well beyond the acquisition of declarative knowledge.

A strong implication from cognitive research on learning and teaching is that conceptual understandings, practical reasoning and scientific investigation are three capabilities that are not mutually exclusive of one another. Thus, an educational program that partitions the teaching and learning of content from the teaching and learning of process, cognitive and manipulative, will be ineffective in helping students develop scientific reasoning skills and an understanding of science as a way of knowing.

Another development that affects thinking about the role of inquiry in science education is the learning science research on effective learning environments. Again, in very broad brushstrokes, the research informs us effective learning environments are those that scaffold or support learning through the use of effective mediation or feedback strategies. Research on learning has demonstrated the importance of social, epistemic and cultural contexts for learning (Bruer, 1993; Brown & Campione, 1994; Pea, 1993; Goldman et al., 2002; Sandoval & Morrison, 2003). One of the implications coming out of the research is the assessment of laboratory work in science education. Another conclusion coming out of such research is that language development for learning and reasoning is critical.

ASSESSING SCIENCE LAB[4]

A review of the science education assessment literature indicates that the compartmentalisation of scientific learning continues to be a dominant perspective even today as measured by the contents of standardised tests used for international and national assessments in the United States. The review of research on assessment in science by Doran, Lawrenz and Helgeson (1994) shows the prominence of the traditional separation orientation. Their review of large scale international (IAE), national (NAEP) and state and provincial government assessments is a comprehensive listing of testing programs that rely on traditional distinctions and beliefs.

Historically science education has partitioned assessment of conceptual components of science from the process, practical, inquiry and attitudinal components of science. Even new assessment procedures, such as those developed in Connecticut and California, though modifying how students report and record their responses to assessment problems, still tend to evaluate in accordance with a process/conceptual dichotomy (Baron 1990; Shavelson, Baxter & Pine, 1992). Indeed, many of the tasks are adaptations of process skill activities develop by National Science Foundation sponsored curriculum projects from the 1960s.

Many recent efforts remain strongly influenced by the tradition of laboratory practical examinations. For instance, the Doran, Lawrenz & Helgeson (1994) discussion of performance assessment focuses exclusively on laboratory and inquiry skills. For many science educators, laboratory practical examinations are alternative assessments. Hence, it is not surprising that the perspective of performance-based assessments used in science today (Doran & Tamir, 1992; Kanis, Doran & Jacobson, 1990; Shavelson, Baxter & Pine, 1992) has a great deal in common with the practical examination formats promoted years ago (Hofstein & Lunetta, 1982; Lunetta & Tamir, 1979; Lunetta, Hofstein & Giddings, 1981; Tamir, 1985). As one example, the report on alternative assessment of high school laboratory skills by Doran, Boorman, Chan and Hejaily (1993) partitions skills as they relate to planning, carrying out and analysing data from investigations as well as applying results to new contexts.

Without a doubt, there is a shift away from partitioning and toward integrating curriculum, instruction and assessment. Millar and Driver (1987), for example, argue that the processes of science cannot be restricted only to those involved in investigations. Students' prior knowledge and the context in which an inquiry is set affect the ways in which students ultimately will engage in an investigation or laboratory exercise. Hodson (1993) echoes a similar concern by reminding science educators to consider the fact that all investigative, hands-on or practical approaches in classrooms occur within an epistemological context that affects students' understandings of the tasks. This position is similar to that made about the importance guiding conceptions have in the design, implementation and evaluation of scientific inquiries (Schwab, 1962). The concern is an over emphasis in the curriculum on the methods of science. Beyond the processes involved in experimenting and testing hypotheses, they argue that science curricula need also to address the processes involved in the communication and evaluation of knowledge claims, processes they believe are fundamental to the learning of science and to learning about the nature of science.

Thus, changes in social values have given rise to a new generation of assessment items and instruments (e.g., authentic tasks, performance-based task and dynamic assessments) and a set of new strategies and formats (e.g., portfolios). Champagne and Newell (1994) argue that an expanded role of assessment in science ought to include consideration of three areas of performance capabilities: (1) conceptual understanding; (2) practical reasoning; and (3) scientific investigation. They offer some assistance toward understanding the ways in which assessments need to be expanded to take into consideration the reasoning of learners. They divide the diverse roles of performance assessments into three groups:

1. Academic performance assessments which include traditional laboratory practical exams and other closed-ended school problems;
2. Authentic tasks such as BSCS's 'Invitations to Inquiry' which involve real-world, open ended tasks that involve students in question framing, experimental design, and data analysis;
3. Dynamic or developmental assessments given over the course of a year or several years which measures students' potential for change over time as determined by students' responses to feedback on a task.

Together, changing social values of the purpose of assessment and of the nature of school science have profound implications for science assessment and for the role of inquiry in science education. A recent trend in assessment has been that of performance assessment wherein pupils are asked to complete or perform complex investigations. Moving to performance assessments that emphasize the integration of conceptual understanding, reasoning and investigatory skills has important consequences for what comes to count as inquiry and for the kinds of inferences that are to be made about science learning and achievement. Given the new views about science achievement from the AAAS Benchmarks and from the National Science Education Standards, the redefinition of learning in terms of performances situated in specific domains poses challenges with regard to providing evidence for claims that student ability in one domain transcends any one particular performance. This is the generalizability problem in measurement.

Furthermore, increased emphasis on the role of assessment in supporting instruction and educational reform forces greater attention to the consequences of assessment than has been usual. The validation of inferences is not only required about student achievement with respect to a defined domain and construct, but evidence is needed concerning the consequences of assessment practices for supporting instructional practices that lead to more successful learning for larger groups of students. This is, in assessment terminology, the consequential validity problem.

LANGUAGE DEVELOPMENT & ARGUMENTATION IN SCIENCE EDUCATION[5]

Conceived as a sociocultural process, language development in science, in mathematics, in music, or in history involves development of the syntactical, semantic and pragmatic language structures of a domain. A critical aspect to the development of reasoning in a domain is the appropriation of language in that domain (Gee, 1994; Lemke, 1990). The implication of focusing on data texts (Ackerman, 1985) and model-based science is that the language of science is extended beyond the narrow historical conception of language. The language of science includes mathematical, stochastic and epistemological elements as well, among others. The challenge for research in learning science is one of understanding how to mediate language acquisition in these various ways of communicating and representing scientific claims. Here again, immersion units have been found to be successful (Bell & Linn, 2000; Schauble, Glaser, Duschl, Schulze & John, 1996; Sandoval & Reiser, 2004).

A significant insight towards changing the role of laboratory that has developed over the last 50 years, and yet only partially realized at the level of the classroom, is the important role language plays in learning, and in the design of effective learning environments. For a prominent, if not central feature, of the language of scientific enquiry is debate and argumentation around competing theories, methodologies, evidence and aims. Such language activities are central to doing and learning science. Figure 1 (on page 57) presents a schematic representation of the evidence to explanation continuum and the dialectic opportunities for epistemic conversations across science lessons. Importantly, 4-6 week long full-inquiry units provide affordances to focus on the data texts of scientific inquiries. Thus, enabling an understanding of science and opportunities for appropriating the syntactic, semantic and pragmatic components of its language. The question with regard to the role of the laboratory or inquiry in high school science is the extent to which students need to be engaged in the collection of data, the analysis of data or both.

The issue is one of emphasis and educational aims with an eye toward engaging students in practicing and using scientific discourse in a range of structured activities. The role of the laboratory is important but a proper balance must be struck between time allocated to the collection of data and the skills of measurement and time allocated to data analysis and modeling. The role of the laboratory should NOT be to confirm conceptual frameworks presented in textbooks. At the high school level, the role of the laboratory should begin to take advantage of the skills, questions and interests students bring to the classroom. That is, if the structures that enable and support dialogical argumentation are absent from the classroom, it is hardly surprising that student learning is hindered. Or, put simply, teaching science as a process of inquiry without the opportunity to engage in argumentation, the construction of explanations and the evaluation of evidence, fails to represent a core component of the nature of science and to establish a site for developing student understanding. Herein lies the importance of immersion units situated in design, problem, and project contexts.

Teaching science as an enquiry into enquiry must address *epistemic goals* that focus on *how we know* what we know, and *why we believe* the beliefs of science to be superior or more fruitful than competing viewpoints. Osborne (2001) has argued, if science and scientists are epistemically privileged, then it is a major shortcoming of our educational programs that we offer so little to justify the accord that the scientists would wish us to render unto scientific knowledge. Hence, if the rationale for universal science education lies in its cultural pervasiveness and significance, then we must attend to explaining why science should be considered the epitome of rationality, In short, the challenge is exposing the nature of science and the values that underlie it. The role of the laboratory is vital in meeting this challenge because the authority of science lies within the process of constructing the evidence and within the reasoning linking evidence to explanation.

From such a perspective, one critically important task is establishing a context in which epistemic dialogue and epistemic activities occur (De Vries et al., 2002). Fundamentally this requires creating the conditions in which students can engage in informed argumentation; i.e., to explore critically the coordination of evidence and theory that support or refute an explanatory conclusion, model or prediction (Suppe, 1998). Situating argumentation as a critical element in the design of science laboratory learning environments not only engages learners with conceptual and epistemic goals but can also help make scientific thinking and reasoning visible for the purposes of the practice of formative assessment by teachers. Central to a view of the value of argumentation is a conception, proposed by Ohlsson (1995), of discourse as a *medium*, which stimulates the *process* of reflection through which students may acquire conceptual understanding. As De Vries states, discourse activities are important because:

> In comparison with problem solving activities, they embody a much smaller gap between performance and competence. In other words, the occurrence of explanatory and argumentative discourse (performance) about concepts effectively reveals the degree of understanding of those concepts. Epistemic activities are therefore discursive activities (e.g. text writing, verbal interaction, or presentation) that operate *primarily on knowledge and understanding*[6], rather than on procedures. (DeVries et al., 2002, 64)

Epistemic goals are not extraneous aspects of science to be marginalized to the periphery of the curriculum in single lessons (Duschl, 2000). Rather, striving for epistemic goals such as the ability to construct, evaluate and revise scientific arguments attains cognitive aims as well. Yet, as Newton, Driver and Osborne (1999) have shown, opportunities for such deliberative dialogue within science classrooms are minimal. Thus, a renewed image of the role of laboratory in science ought to embrace and then include in instruction the dialectical processes that engage learners in linking evidence to explanation, i.e., argumentation.

Argumentation has three generally recognized forms: analytical, dialectical, and rhetorical (van Eemeren et al, 1996). The application of analytical arguments (e.g., formal and informal logic) to evaluate science claims is extensive and pervasive. The capstone event of applying argumentation to the sciences is perhaps Hempel-Oppenheimer's Deductive-Nomological Explanation Model (Hempel, 1965) wherein the argumentation form is used as an account to establish the objectivity of scientific explanations. Toulmin's (1958) examination of argumentation was one of the first to challenge the 'truth'-seeking seeking role of argument and instead push us to consider the rhetorical elements of argumentation. For Toulmin, arguments are field dependent; in practice, the warrants and what counts as evidence, and the theoretical assumptions driving the interpretations of that evidence, are consensually and socially agreed by the community – an idea recognized by Schwab who saw the teaching of science as an investigation of the guiding conceptions that shape enquiry.

Likewise, case studies of scientists immersed in scientific enquiry show that the discourse of science-in-the-making involves a great number of dialectical argumentation strategies (Dunbar, 1995; Latour & Woolgar, 1979; Longino, 1994; Gross, 1996). Research in the sociology of science (Collins & Pinch, 1994; Taylor 1996) has also demonstrated the importance of rhetorical devices in arguing for or against the public acceptance of scientific discoveries. In short, the practice of science consists of a complex interaction between theory, data and evidence.

The rationality of science is founded on the ability to construct persuasive and convincing arguments that relate explanatory theories to observational data. Science requires the consideration of differing theoretical explanations for a given phenomena, deliberation about methods for conducting experiments, and the evaluation of interpretations of data. Clearly, argumentation is a genre of discourse central to doing science and thus is central to *learning* to do science (Lemke, 1990; Kuhn, 1993; Siegel, 1995; Kelly & Crawford, 1997; Kelly, Chen & Crawford, 1998; Suppe, 1998; Newton, Driver and Osborne, 1999; Driver, Newton & Osborne, 2000; Loh et al., 2001).

If students are to be persuaded of the validity and rationality of the scientific world-view then the grounds for belief must be presented and explored in the context of the science classroom. In short:

> the claim 'to know' science is a statement that one knows not only *what* a phenomenon is, but also *how* it relates to other events, *why* it is important and *how*[7] this particular view of the world came to be. Knowing any of these aspects in isolation misses the point. (Driver, Newton & Osborne, 2000)

Such an aim requires opportunities to consider *plural* theoretical accounts and opportunities to construct and evaluate arguments relating ideas and their evidence. For as D. Kuhn (1993: 164) argues, 'only by considering alternatives – by seeking to identify what is not – can one begin to achieve any certainty about what is.' To do otherwise leaves the student reliant on, or skeptical about, the authority of the teacher as the epistemic basis of belief. This leaves the dependence on evidence and argument – a central feature of science – veiled from inspection. Or, in the words of Gaston Bachelard (1940), the essential function of argument is that, 'two people must first contradict each other if they really wish to understand each other. Truth is the child of argument, not of fond affinity.' Indeed, Ogborn et al. (1996) show elegantly how one of the fundamental strategies of all science teachers is the creation of difference between their view and their students' view of phenomena. For without difference, there can be no argument, and without argument, there can be no explanation. Within the context of science, dedicated as it is to achieving consensus, it is argument, then, that is a core discursive strategy, and a *sine qua non* for the introduction of argument is the establishment of differing (i.e. plural) theoretical accounts of the world. This is not to suggest that argument is something, which is unique to science. Argument plays a similar function in many other disciplines. Rather, the intent is to show that argument is as central to science as it is to other forms of knowledge and, therefore, cannot be ignored in any science education.

Mitchell (1996) is helpful in this matter by distinguishing between two types of argument – regular and critical arguments. Regular arguments, she states, are rule-applying arguments that put forward applications of theories that are not in themselves being challenged. Such arguments are generally predictive and a central feature of the work of scientists. These arguments are of the kind recognized by the logical positivists, and also appear in the Kuhnian framework of normal science. In contrast, critical arguments do challenge the theories and ideas but have as a fundamental goal the refinement of existing theories or introduction of alternative ideas and not the defeat of another. In the moral community that is science, personal conflicts and aspirations are always secondary to the advancement of knowledge.

We must remember, therefore, that initial efforts with engaging children in argumentation will require setting ground rules to avoid, for instance, ad hominem arguments that attack the person and not the ideas (Dillon, 1994). Such preliminary attempts to initiate argumentation practices will also require modeling and practicing the standard inductive (argument by example, argument by analogy, argument by causal correlation) and deductive (argument from causal generalization, argument from sign, syllogisms) forms of argument. Worth mentioning at this point is the research (Duschl, Ellenbogen & Erduran, 1999) that shows children do seem to have a natural tendency to engage in such inductive and deductive forms of argument when a sound context is provided.

Thus, Cohen (1994) seeing argumentation as war is an ineffective and inappropriate metaphor for promoting dialogic discourse – a metaphor that must be explicitly countered when initiating the contexts for argument in the classroom. The alternative is to envision argumentation as a process that *furthers* inquiry and not as a process that ends inquiry. Thus, alternative and more apposite metaphors for Cohen (1995) include argumentation as diplomatic negotiation, argumentation as growth or adaptation, metamorphosis, brainstorming, barn raising, mental exercises for the intellect or roundabouts on the streets of discourse. Science as a way of knowing does seek consensus on matters but, more often than not, progression in scientific thinking involves the use of critical arguments and processes that are more akin to diplomatic negotiation than to conflict. In this way, lab based science can be a dialogic process.

Harvey Siegel (1995) in "Why should educators care about argumentation?" takes the position that if one ideal of education is the development of students' rationality, then we must be concerned not only with how students reason and present their arguments but also with what students come to consider as criteria for good reasons. Siegel sees argumentation as the way forward because of the correlation between the ideal of rationality and the normative concerns and dimensions of argumentation and argumentation theory. He writes, "Argumentation ... is aimed at the *rational* resolution of questions, issues, and disputes. When we engage in argumentation, we do not seek simply to resolve disagreements or outstanding questions in any old way .... Argumentation ... is concerned with/dependent upon the *goodness,* the *normative status*, or *epistemic forcefulness*, of candidate reasoning for belief, judgment, and action" (p. 162, emphasis in original). Thus the second concern with the introduction of argumentation is the necessity to model effective arguments in science, to expose the criteria which are used for judgment such as parsimony, comprehensiveness and coherence, and why some arguments are considered better than another. For instance, given two arguments to explain the 24-hour rotation of the Sun and stars why do we pick the argument that it is the Earth that is moving rather than the Sun and stars?

Critically important to using argument constructively is allowing learners to have the time to understand the central concepts and underlying principles (e.g. the "facts") important to the particular domain (Goldman et al., 2002). In other words, a necessary condition for good arguments is a knowledge of the "facts" of a field as otherwise there is no evidence which forms the foundation of a scientific argument. Alternatively, students must be provided with a body of 'facts' as a resource with which to argue (Osborne, Erduran, Simon & Monk, 2001). However, argumentation does not necessarily follow from merely knowing the "facts" of a field. Equally important is an understanding of how to deploy the "facts" to propose convincing and sound arguments relating evidence and explanation. Herein lies the need for learners to develop strategic and procedural knowledge skills that underpin the construction of argument. Herein lies the need for the laboratory to be situated into immersion units in high school science programs.

In summary, the challenge is providing teachers and students with tools that help them build on nascent forms of student argumentation to develop more sophisticated forms of scientific discourse (Duschl et al., 1999; Osborne, Erduran, Simon & Monk, 2001). Such tools need to address the construction, coordination, *and* evaluation of scientific knowledge claims. Equally important, as Siegel (1995) argues, is the need to address the development of criteria that students can employ to determine the goodness, the normative status, or epistemic forcefulness of reasons for belief, judgment, and action. What has been presented is that the central role of argumentation in doing science is supported by both psychologists (Kuhn, 1993) and philosophers of science (Siegel, 1995; Suppe, 1998) as well as science education researchers studying the discourse patterns of reasoning in science contexts (Sandoval & Reiser, 2004; Bell & Linn, 2000; Driver, Newton & Osborne, 2000; Hogan, Nastasi & Pressley, 2000; Kelly, Chen & Crawford, 1998; Kelly & Crawford, 1997; Lemke, 1990).

Designing learning environments to both facilitate and promote students' argumentation is, however, a complex problem, for the central project of the science teacher is to persuade students of the validity of the scientific world-view. Conceived of in this manner – as a rhetorical project – the consideration of plural enterprises may appear to undermine the teacher's task and threaten the learner's knowledge of 'the right answer'. Moreover, normal classroom discourse is predominantly monologic and it is difficult for teachers to transcend such normal modes of discourse. Therefore, changing the pattern and nature of classroom discourse requires a change both in the structure of classroom activities and the aims that underlie them. Laboratory style investigations embedded in immersion units are the way forward.

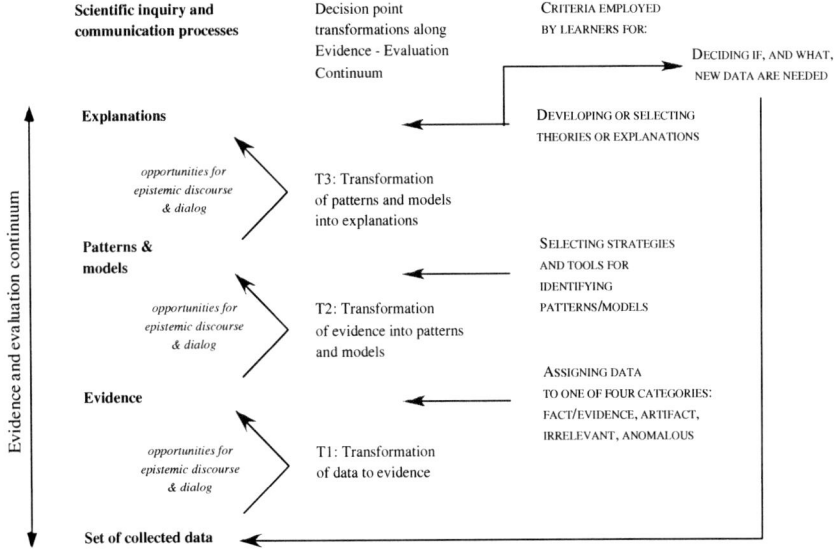

Figure 1. Schematic of Evidence-Evaluation continuum model for consideration of epistemic dialog opportunities

NOTES

[1] Kurt Godel, a member of the Vienna Circle, demonstrated in 1931 that there could not be a complete axiomatization of mathematics – this discovery seems to have had less impact on the development of logical empiricism than, with hindsight, it should have.
[2] Some material in this section derives from (Grandy, 2003a))
[3] Some of the material in this section is derived from (Grandy, 1993)
[4] This section is derived from Gitomer and Duschl (1998).
[5] This section is derived from Duschl & Osborne (2002).
[6] Emphasis added.
[7] Emphasis in the original.

Richard A. Duschl
Graduate School of Education, Rutgers University

Richard E. Grandy
Department of Philosophy, Rice University

WILLIAM F. BREWER

IN WHAT SENSE CAN THE CHILD BE CONSIDERED TO BE A "LITTLE SCIENTIST"?

In the area of developmental psychology many recent writers have proposed an analogy between children and scientists (Brewer & Samarapungavan, 1991; Carey, 1985b; Gopnik & Wellman, 1992; Gopnik, 1996b; Gopnik & Meltzoff, 1997; Karmiloff-Smith, 1988; Karmiloff-Smith & Inhelder; 1975; Wellman, 1990). Within the area of cognitive development this approach can be seen as part of a larger shift away from Piagetian stage theories toward knowledge-based theories (Carey, 1985a; Wellman & Gelman, 1992). The child scientist proposal has met much opposition. Gopnik characterized the typical response of many scholars as one of "shocked incredulity" (1996, p. 486). She is certainly correct about this. Opponents of the child scientist analogy have written articles with titles such as "Why the young child has neither a theory of mind nor a theory of anything else" (Gellatly, 1997) and "The baby in the lab-coat: Why child development is not an adequate model for understanding the development of science" (Faucher, Mallon, Nazer, Nichols, Ruby, Stich & Weinberg, 2002).

In the area of science education there was a similar and independent move toward the view of the child as a young scientist (Driver & Easley, 1978; Helm & Novak, 1983). The major impetus in science education for the analogy between the child and the scientist was the realization that school children had strongly held "misconceptions" or "alternative frameworks" about many aspects of the physical world. These beliefs could easily be seen as false theories of the physical world, thus implying that children must be constructing theories just as scientists do. More recently Metz (1995) has reviewed the arguments for the child scientist position in science education and was met with a counterattack from an opponent of the child scientist view (Kuhn, 1997).

The rest of this paper will be an attempt to understand in what sense the child as scientist analogy can be considered to be correct.

THE CHILD AND THE SOCIAL/CULTURAL INSTITUTION OF SCIENCE

Brewer and Samarapungavan (1991, pp. 220-223) point out that in making the child scientist analogy one must distinguish between the *individual child* constructing theories, the *adult scientist* constructing theories, and the *cultural institutions* of science (cf. Harris, 1994, for a similar point).

Clearly science is one of the major social constructs of modern culture. It has many aspects: (1) It is a profession with entry requirements, professional organizations, professional journals. (2) The members of the social group have norms such as organized skepticism, beliefs about the public nature of science (see Merton 1942/1973; Ziman, 1984). (3) The group has a specialized technology (e.g., particle accelerators, cloning techniques). (4) The social institution has a specialized set of methodologies (cf. Laudan, 1987). (5) The social organization has an enormous set of accumulated knowledge about the physical world. (6) The group has procedures for passing down the accumulated knowledge (e.g., graduate school, science textbooks, journals). (7) The social group has procedures for evaluating the knowledge produced by members of the group (e.g., the journal review process) (8) Most of the members of the social group earn their living by teaching or doing science.

It is obvious that preschool and elementary school children are not scientists in the sense of being members of the cultural institution of science. If one considers science to be essentially a social construction (Bloor, 1976; Shapin & Schaffer, 1985) then the debate is over. The child is not a member of the social institution of science and thus can not be a little scientist. A number of critics of the child as scientist view have used this logic to argue that the child as little scientist analogy is false (e.g., Gellatly, 1997).

Another way to view the same issue is to focus on the relatively recent historical development of the institution of science. The position that the child generates theories of the natural world has to assume that this process has taken place in generation after generation of children in the history of the human species. Yet the social institution of science has only come into being in the last 400-500 years. Alison Gopnik, who takes a very strong form of the child scientist view, worries about this problem (1996a, 1996b), but does not appear to have a good solution (cf. Faucher et al., 2002; Giere, 1996; Stich & Nichols, 1998).

Thus, it seems to me that if one takes an extreme social constructive view of science, the game is over. However, if one adopts a view that considers science to include both social and cognitive aspects (e.g., Nersessian, 2005), then the question of the validity of the child scientist analogy is still open. Brewer and Samarapungavan (1991) argue that attempts to compare the child to the social institution of science are simply inappropriate and that the analogy must be made between the child and the scientist.

THE INDIVIDUAL CHILD AND THE INDIVIDUAL SCIENTIST

Brewer and Mishra (1998) proposed that for the purpose of a cognitive science evaluation of science one could break down the work of scientists into three major:

1. Understanding and evaluating scientific information.

We pointed out that scientists spend much of their time reading the scientific literature and attending scientific conferences (such as this one!). Clearly through both formal and informal means children learn aspects of science content from the adult culture so there is no *qualitative* difference here between the child and the scientist. However, one of the major insights in knowledge-based approaches to cognitive development (e.g., Carey, 1985a) is the enormous gulf between the knowledge of the child and the knowledge of the adult (scientist and nonscientists) and the profound implications this has for the child's understanding of the natural world. However, this particular contrast between children and scientists does not seem to be one that we should focus on in looking at the child scientist analogy because the same difference between the child and the adult is true for other forms of organized knowledge (e.g., music, car repair). As Brown and DeLoache (1978) put it, the child is a "universal novice."

Many working scientists and scholars in the philosophy of science have argued that in evaluating scientific theories scientists apply criteria such as empirical accuracy, consistency, scope, and simplicity. Samarapungavan (1992) has carried out an ingenious series of experiments that found that young elementary school children can be shown to evaluate theories using criteria such as empirical accuracy, logical consistency, and scope. This is an important area for exploring the similarities and differences between children and scientists and needs additional empirical study.

2. Disseminating scientific knowledge.

Scientists spend another large part of their lives writing papers (the fact that I am sitting here at my computer late in the night writing this paper seems good evidence for this aspect of scientific life). We also give talks at professional meetings (more evidence to come in February) This component of what scientists do does not seem to me to be shared with young children, so this is another aspect of science where the child as scientist analogy does not hold.

3. Generating new scientific knowledge.

Scientists design and carry out experiments. Gopnik and Meltzoff (1997) argue that very young children carry out activities that are analogous to carrying out experiments (e.g., repetitive hiding and uncovering an object to explore the nature of object permanence). Karmiloff-Smith and Inhelder (1975, pp. 207-208) make a similar argument for somewhat older children.

However, for older children it seems to me that this is another area of disanalogy between children and scientists. Children are very good at trying to make sense of the phenomena of the everyday world, but they typically do not go out of their way to systematically collect data relevant to a belief or theory that they have developed. Wellman (1990) has made a similar point. He concluded "children do not craft their theories on the basis of explicit, rigorous activities akin to scientific formulation and test" (p. 130). Reif and Larkin (1991) have come to the same conclusion. They state "knowledge in everyday life is ordinarily not deliberately sought, but spontaneously acquired through interaction with the world and other people (p. 741). As I will argue later, it appears that the children's approach to understanding the natural world is more like a PreSocratic Greek philosopher than it is like that of Galileo.

Another part of generating new scientific knowledge is the construction of new scientific theories. This is the aspect of the child as scientist analogy that seems to me to be the strongest and the one that has attracted the most attention. Most of the rest of this paper will be devoted to exploring theory construction by children and by scientists. However, before addressing those issues I will turn to another aspect of the child as scientist analogy that has found its way into the recent debates.

THE THEORY-LADENNESS OF DATA AND RESPONSES TO ANOMALOUS DATA

Issues dealing with the theory-ladenness of data have been used both to support and to criticize the child as scientist analogy. One of the major advances in the philosophy of science in the last century was the proposal of Hanson (1958) and Kuhn (1962) that scientists' theories influenced their observations. Brewer and Lambert (2001) argued that this debate within the philosophy of science continued a strong concern with visual perception and scientific observation in philosophy that began with the British Empiricists, became very powerful with the work of the Logical Positivists, and continued into the work of the Post-Positivists. We proposed that a scientist's theories play a role across essentially all of the activities that scientists engage in. More specifically we argued that, in addition to their role in perception and attention, top-down processes play a major role in: (a) data evaluation, (b) data production, (c) memory, and (d) communication; and we provided cases from the history of science to support these claims.

One important class of theory-driven phenomena are cases where theory overrides evidence in data evaluation and data production. Chinn and Brewer (Brewer & Chinn 1994; Chinn & Brewer, 1993) provide evidence from the history of science and from studies of nonscientists that both scientists and nonscientists frequently allow their theoretical beliefs to override the data. We argue that there are seven basic responses to a situation where an individual holds theory A, but obtains data supporting theory B. The seven responses are: (1) *ignore* the data and retain theory A; (2) *reject* the data and retain theory A; (3) *exclude* the data from the domain of theory A; (4) hold the data in *abeyance* and retain theory A; (5) *reinterpret* the data and retain theory A; (6) reinterpret the data and make *peripheral changes* to theory A; and (7) *change* theory A, possibly in favor of theory B. Note that only the last response involves completely giving up the individual's initial theory.

In discussing the child scientist issue Deanna Kuhn (1989) implicitly adopts the Pre-Kuhnian positivist view (e.g., Hempel, 1966) that there is a clean separation of data and theory. She assumes that scientists make the distinction and so when she obtained evidence that the children in her experiments "either failed to acknowledge discrepant evidence or attended to it in a selective distorting manner" (p. 677) she concludes that children show a fundamental inability to coordinate theory and evidence. Thus she argued that, at least on this feature, children are not like scientists (see also D. Kuhn, 1997).

Ironically some of the proponents of the child scientist view (e.g., Carey, 1985a; Karmiloff-Smith, 1988) have used evidence of children's theories overriding data to support *their* side of the argument. Karmiloff-Smith and Inhelder (1975) carried out an experiment in which children attempted to balance blocks on a beam. They found that the children often excluded blocks that did not balance according to their current theory. In a later paper Karmiloff-Smith (1988) used this data to conclude that "these children do exactly what Kuhn (1962) has argued scientists do. Children do not abandon their theory despite glaring counter-examples!" (p. 186-187). It seems to me that while Karmiloff-Smith's argument carries considerable weight it also has some difficulties. Other molar forms of knowledge representation such as schemas can cause individuals to make errors (cf. Brewer, 1999b), and so for the evidence to have maximum force in the child scientist debate it is necessary to focus on arguments that the forms of knowledge involved in generating the errors are (scientific) theory-like in character. In addition one has to worry about the degree to which theory-based distortion are specific to scientists (Gellatly, 1997).

THEORIES AND MODELS IN SCIENCE

The core intuition that leads many researchers to adopt the child as scientist view is that young children often appear to hold beliefs that they have developed about the world that if held by a scientist we would call a scientific theory. To unpack this core intuition we need to specify: (a) what theories are in the institution of science; (b) what representations are in the heads of scientists, and (c) what representations are in the heads of children.

Theories as Formal Objects

One of the reasons that some individuals show "shocked incredulity" when confronted with the proposal that children generate theories is that they have a view of theories taken from the Logical Positivists. The positivists (e.g., Carnap, 1956) conceptualized scientific theories as hypothetico-deductive structures. I too find it hard to believe that young children are operating with complex formal systems of the kinds proposed by the Logical Positivists. However, if one does not accept the Positivists' particular proposals about the nature of theories then the position that the child constructs theories that are like those of scientists is much more plausible.

Theories as Models

Within the philosophy of science there has been a minority view (Black, 1962; Campbell, 1920/1957) which held that theories are better interpreted as models. These theorists asserted that model-based theories are theories that contain entities that have a structural relationship with other real or hypothetical entities. A typical example they used to make their point was the account of the gas laws in terms of moving molecules. The model-based theorists claimed that it was the model component of the theory that gave it the power to explain phenomena, to cover new phenomena, and to provide an intellectually satisfying account of the world. In recent times philosophers of science have shown a renewed interest in models (e.g., Giere, 1988), but it is not clear to me that the current approaches to models are quite as different from the Positivists' views as were the earlier model-based approaches.

Many working scientists have also favored model-based approaches to scientific theory. This view was particularly strong among British physicists at the end of the 19th century. For example, Lord Kelvin once stated, "It seems to me that the test of 'Do we or not understand a particular subject in physics?' is 'Can we make a mechanical model of it?'" (Thomson, 1884, p. 132). Boltzmann (1899/1974) characterized proponents of the model-based approach in the following terms: "physical theory is merely a mental construction of mechanical models, the working of which we make plain to ourselves by the analogy of mechanisms we hold in our hands, and which have so much in common with natural phenomena as to help our comprehension of the latter" (p. 790).

In several recent papers (Brewer, 1999c, 2001) I have bitten the bullet on this issue and argued that in most domains (the double helix, plate tectonics) the models are the theories. Note that a model-based approach to theories makes the proposal that children develop theories that are like scientific theories much more plausible. In fact, Barsalou (1999) has proposed a form of mental representation he calls "perceptual symbols" which seems just the right type of representation for the sub-class of theories that are model-based. The perceptual symbols in Barsalou's approach combine the nonarbitary representation of images with the generative power of symbolic systems to give a form of representation that is well suited for representing model-based theories (cf. Brewer, 1999a).

THEORIES AS A FORM OF MENTAL REPRESENTATION

Schwitzgebel (1999) has argued that neither the Logical Positivist nor the more recent semantic account of theories seems appropriate for general psychological accounts of theories. The richest general account of theories as a psychological form of representation has developed out the literature in developmental psychology directed at understanding what children know about the minds of other human beings. Many of the researchers in this area have adopted the view that children develop theories about the mental states of other individuals. This view is frequently referred to as the "theory theory" (Gopnik & Wellman, 1994). These researchers (Gopnik & Wellman, 1992; Wellman, 1990) argue that in their most general sense theories are mental structures that contain theoretical entities (sometimes nonobservable), relations among the theoretical entities, and relationships between the theoretical entities and some phenomena. I believe that this account can be enriched by a more explicit treatment of explanation. In a recent chapter we (Brewer, Chinn & Samarapungavan, 2000) have tried to do this and state that in theories "an explanation is an account that provides a conceptual framework for a phenomenon (e.g., fact, law, theory) that leads to a feeling of understanding in the reader/hearer. The explanatory conceptual framework goes beyond the original phenomenon, integrates diverse aspects of the world, and shows how the original phenomenon follows from the framework" (p. 120).

CHILDREN'S THEORIES ABOUT THE NATURAL WORLD

It seems to me that when a child develops a theory that heat and cold are substances that move through objects (Clough & Driver, 1985) or that when the sun vanishes at night it goes behind hills or to the other side of the earth (Vosniadou & Brewer, 1994), the child has a theory of the type specified by the "theory theory" approach. Or as Karmiloff-Smith (1988) phrased it, "the child is a spontaneous theoretician" (p. 183).

CHILDREN'S EXPLANATIONS OF PHENOMENA IN THE NATURAL WORLD

Many of the proponents of the view that the child is a little scientist have pointed to explanation as a crucial aspect of child theories. For example Carey (1985b) stated, "Explanation is at the core of theories. It is explanatory mechanisms that distinguish theories from other types of conceptual structures" (p. 201). Gopnik and Wellman (1992) stated that theories "are designed to explain (not merely type of generalize)" (p. 146). We (Brewer, Chinn & Samarapungavan, 2000) have argued that while child explanations differ from those of scientists in several respects (e.g., precision, using formalisms) they nevertheless have the same essential form as those used by scientists.

THE CHILD AS LITTLE SCIENTIST

It also seems to me that the account of theories given in the "theory theory" approach and the account given for the sub-class of theories that are model-based are reasonable representations for many historical and current scientific theories. Thus, I think we have finally reached the core aspect of the child as scientist analogy that is, in fact, true: *Children generate theories that are not qualitatively different from those generated by scientists.*

INDIVIDUALLY GENERATED CHILD THEORIES

One of the striking aspects of children's theories of the world is the evidence that children are little Descartian thinkers trying, by themselves, to understand the natural world. Karmiloff-Smith (1988) noted "Although plunged from the start into a social context, the child is also an individual cognitive organism and much of her theory building is endogenously provoked rather than socially mediated" (p. 184). Harris (1994) asserted that children "rarely challenge each other's conceptualization...children are mostly left to get on with their own cognitive development" (pp. 306-307). Our studies (Vosniadou & Brewer, 1994) of children's theories of observational astronomy show that young Western children believe the earth to be flat in spite of the almost universal belief in the adult culture that the earth is a sphere.

THEORY CHANGE

The hypothesis that children spontaneously generate theories about the natural world provides a natural account about a wide range of data from cognitive development. It is known that children frequently show initial beliefs about the natural world that are different from the beliefs of the adults around them and that over time they often shift to quite different beliefs (e.g., Karmiloff-Smith and Inhelder, 1975; Vosniadou & Brewer, 1994). The child scientist approach provides a very natural account of these changes – they can be understood as examples of theory change (Carey, 1985b, 1986; Karmiloff-Smith and Inhelder, 1975).

While the child scientist approach gives a very powerful account of the endpoint of the observed changes it has not been as successful in providing a detailed account of what mechanisms lead to the theory changes (cf. Harris, 1994; Russell, 1992).

DOMAINS OF APPLICATION

At this stage in the development of our understanding it is not clear how far the child scientist approach can be extended. I think the clearest examples of spontaneously developed child theories are those generated by early elementary school children dealing with easily observed phenomena of the natural world (e.g., moving objects, chemical change, biological change, observational astronomy, etc.). In these cases one frequently finds examples of theories held in opposition to the adult culture and the child theories sometimes show strong similarities with the earliest scientific theories in the same domains. These types of evidence provide strong support for the view of the child as a generator of scientific theories.

In the field of science education the classic literature on children's misconceptions or alternate conceptions (e.g., Driver & Easley, 1978; Helm & Novak, 1983) has focused on just the domain outlined above – elementary school children's beliefs about the natural world.

However, in the area of developmental psychology the locus of the child scientist approach has been quite different. Initially the "theory theory" approach (Wellman, 1990; Gopnik & Wellman, 1992) as it has come to be called was applied to the development of young (age 3-5) children's beliefs about the mental states of others ("theory of mind"). More recently Gopnik and Meltzoff (1997) have extended the approach down to infancy and applied it to domains such as infant understanding of object appearances and infants' beliefs about the conceptual grouping of objects. I remain agnostic about the application of the child scientist approach to the domain of folk psychology and to the beliefs of infants. While there are some interesting analogies that can be made it seems to me that the case for the child as little scientist is much harder to make in these domains.

LIMITATIONS ON THE HYPOTHESIS OF THE CHILD AS SCIENTIFIC THEORY GENERATOR

In our paper on children's theories versus scientific theories (Brewer & Samarapungavan, 1991) we point out that comparing the individual child's self-generated theories to theories that have been generated through the institutions of science is inappropriate. Clearly the shared knowledge developed over hundreds of years and the regulatory mechanisms (e.g., organized skepticism in the form of the review process) of the institution of science will produce theories that are more internally consistent and more precise fits to the data than are child theories. In addition the child does not, for the most part, have access to the general-purpose tools of mathematics. Thus child theories must be qualitative in nature or only loosely quantitative (e.g., bigger objects are more likely to knock you over than smaller objects). Brewer and Samarapungavan argued that child theories are more appropriately compared with a recently developed theoretical insight of the individual scientist before it has been written down, exposed to review by other scientists, and put to experimental test. With this more even-handed comparison the similarity of the theories of the child and those of the scientist seem very strong.

Meta-Knowledge about Theories

However, there remains one difference between child theories and scientific theories that I do not want to discount. There is a fundamental way in which child theories and scientist theories differ. Scientists are usually aware of the hypothetical state of their theories, and children typically are not. Karmiloff-Smith and Inhelder (1975) note that children's "theories remain implicit, since the younger child clearly cannot reflect on hypothetical situations which might confirm or refute his theory" (pp. 208-209). Wellman (1990) made a similar point, "these early strategies lack self-conscious attempts to think about a theory (rather than merely hold one and use it)" (p. 131). I realize there are some occasions when the scientist acts like the child. We all know some scientists who are so dogmatic in holding to the truth of their theories that one sometimes wonders if they *do* understand that there are other alternatives. Also many of the major revolutions in science have occurred when someone overthrew an *implicit* assumption in previous theories (e.g., Kepler rejecting the implicit constraint that all orbits must be circular). Even though these examples of lack of awareness in scientists may blunt some of the force of the difference in meta-awareness of theories, I believe this is a real difference between the initial self-generated theories of children and the theories of scientists.

THE SCIENTIST AS CHILD

Gopnik (1996a, 1996b; Gopnik & Meltzoff, 1997) has proposed a more extreme form of the analogy between children and scientists. She has reversed the logic and argued that it "is not that children are little scientists but that scientists are big children" (p. 486). She moved the child scientist analogy down to children age 1-2 years and argued that infants show curiosity, rapid development and revision of theories, and systematic "experimentation." She proposed that it is these child characteristics that are retained in that small percentage of the population who become scientists. While this approach certainly does place an interesting focus on early childhood curiosity and exploration, it probably stretches the child scientist analogy too far (see Carey & Spelke, 1996 for a similar evaluation).

THE CHILD AS SCIENTIST

In this paper I have argued that the component of the child scientist analogy that is strongest is the argument that children develop theories of the world that are qualitatively like those that individual scientists develop. So does this make the child a little scientist? Here I think we face a matter of semantics. Most scholars will probably want the term science to include systematic gathering of data. However, if one takes the view that the crucial component of science is the construction and critical evaluation of theories, as I do, and as Karl Popper has argued (1959), then the first scientists were the PreSocratics. If you believe that Galileo was the first scientist, then children are probably not little scientists; but if you believe that the PreSocratics were the first scientists, then children are little scientists.

THE INDIVIDUAL CHILD AND THE SOCIAL CONSTRUCTION OF SCIENCE

For me the view of the child as little scientist is an extraordinary insight into human nature. Generation after generation of little human beings are, on their own, constructing theories of the natural world often in direct opposition to those of the adult culture around them. This finding stands in conflict with a strong form of the view that scientific theories are socially constructed entities.

On the other hand one might want to argue that an individual's insight into the working of the natural world is not a "real" scientific theory until it has found its way into the social fabric of public science (was Leonardo da Vinci a scientist?). If one requires that a theory is not a scientific theory until it has entered the social institutions of science then the child's theories are not scientific theories.

A TEST OF THE CHILD AS SCIENTIST HYPOTHESIS

In earlier work with Stella Vosniadou (Vosniadou & Brewer, 1992, 1994) we investigated children's acquisition of knowledge about observational astronomy (e.g., the shape of the earth, the day/night cycle). We found that across children there was a small set of initial theories about these phenomena. For example, their explanations of where the sun is at night included theories that involved occlusion mechanisms, such as clouds coming in front of the sun, or movement mechanisms such as the sun going out into space or moving down behind mountains or trees.

Since the adult culture that these children were exposed to accepted a Copernican account of these phenomena we argued that the children's theories did not derive from the adult culture but must have come from the children's basic human cognitive equipment interacting with aspects of the perceived natural world. This is clearly a strong claim of universality. Data on children's theories of observational astronomy show strong similarities across cultures (e.g., Samarapungavan, Vosniadou & Brewer, 1996) and thus support the claim.

Recently I have explored another implication of the claim of universality. The constraints that lead the children to a limited set of theories should also have been operating in the adults who produced the very earliest cosmologies (before the institutions of science began to have such a large impact). Therefore we should expect very early cosmological models to show strong similarities to those we have uncovered in young children. Examination of accounts of the shape of earth and the day/night cycle found in the earliest Greek cosmologies, early Egyptian cosmology, early Hebrew cosmology, and early Chinese cosmology appear to show strong similarities to the child data and thus provide additional support for the universality hypothesis and for the child as (very early!) scientist.

IMPLICATIONS FOR EDUCATION

I think of myself as a basic researcher interested in how the mind works. I do not think of myself as having expertise in the area of education so I prefer to leave it to others to draw the educational implications from my work and that of others in the area of cognitive science.

However, given the goals of this Conference, I will make a few short comments. First, it seems to me that the distinction between the cultural institution of science and the child as generator of scientific theories has obvious implications (cf. Metz, 1995, 1997 for some similar suggestions). Presumably one of the goals of science education ought to be to teach the child some selected parts of the culture of science. (Perhaps we can pass on having them be required on occasion to dress up in medieval gowns and hats!) I, as a citizen, think it would be good if children learned something about scientific methods and about the norms of science. Apparently this may be easier said than done (cf. Abd-El-Khalick & Lederman, 2000). Most children in science classes are, as adults, going to be consumers of science information. Thus developing some understanding of how to evaluate scientific claims would seem to be an important life skill. One would also like children to know something about how scientists carry out experiments. For example, the culture of science has developed the norm that one should record data in laboratory notebooks and not trust memory. Schauble (1990) has shown that children do not understand this piece of scientific culture. I also think that learning some of the *content* of scientific knowledge is a Good Thing.

What are the implications of understanding that children are spontaneous theory constructors? One aspect of this is well known to the science education community – children are not blank slates, but come to the science classroom with strongly held theories. Clearly one would like to find better techniques for theory change. It also would appear that trying to help the children be more meta-aware of the hypothetical nature of their theories could be a promising line of research.

So all-in-all, I recommend that the educational community do all it can do to steep the little PreSocratics in the modern cultural institution of science.

William F. Brewer
Department of Psychology
University of Illinois at Urbana-Champaign

LEONA SCHAUBLE

THREE QUESTIONS ABOUT DEVELOPMENT

Dr Brewer's remarks remind us that development is not like a toggle switch. It is rarely – probably never – the case that a person shifts abruptly from the state of not-having some form of thinking to having it. Almost any kind of thinking we would care about has early origins and develops gradually over an extended period. In some cases, development is life-long, if supporting conditions are in place. Dr. Brewer points out that scientific thinking is complex, there is no one essence that characterizes it, and there is no reason to believe that it springs forth intact, like Venus from the head of Zeus, at some particular age.

As psychologists and educators, we want to understand these trajectories of growth and change. However, I don't think we will get much purchase on understanding by settling the question, "Do children think like scientists?" As Dr. Brewer demonstrates, the answer is clearly yes and no. Why, then, does the question keep coming up?

The field's devotion to this particular form of argument betrays, I think, deep-seated disagreement about the nature of development. Developmental psychology has a lot of maturational assumptions buried deep in its history; Arnold Gesell (1926) was one of its founders back in the early and mid-1900s. Gesell, for those whose professional affiliation is not developmental psychology, conducted careful observations of infant motor behavior and was the first to identify age norms for benchmarks: the average age when babies first sit up unsupported, when they first use a "pincer grasp" to pick a small object, when they first begin to crawl, and the like. Gesell was impressed by the regularity of these infant behaviors, which seemed to unfold in very predictable and patterned ways across the children in his studies. Although he observed some variability in timing, the sequence was notably consistent. He also was struck with the resilience to environmental influences that these developmental timetables appeared to demonstrate. What characterized *development,* in his view, were its genetic roots and its universal form.

Within the field of developmental psychology, we have had difficulty shaking off some of these underlying assumptions, so that if maturation or biology cannot somehow be invoked, a behavior just doesn't seem to qualify for the label "development." This may be why, within the field of cognitive development, there has been a trend toward studying progressively younger and younger children. I sometimes wonder if we are embarked on a Don Quixote-like quest for the environmentally untainted essence of human thinking, Alas, of course, there is no such thing as an experience-free human, which is why Gesell's theory is no longer considered viable. Moreover, right from the start, human experience is *human*: it is constructed by and filtered through the intentions and interpretations of others. From the first days of life, babies and their caretakers are involved in a sophisticated dance of bi-directional influence.

Richard A. Duschl and Richard E. Grandy (eds.), Teaching Scientific Inquiry: Recommendations for Research and Implementation, 50–56.
© *2008 Sense Publishers. All rights reserved.*

Children – and adults, too, for that matter – live in worlds that they themselves shape, worlds that in turn, shape them. These worlds are inherently both social and physical. Of course brains and perceptual apparatus are involved in the forms of thinking that we produce, and consequently, there are real constraints in the ways we perceive and think about the world. But just as there is no environmentally untainted mind, there is no socially untainted mind. Under this interpretation, science is *indeed* socially constructed. That does not mean that the thoughts of an individual are determined by others, only that whether and how development unfolds over the long term depends upon much more than one's "basic cognitive apparatus." Unless, of course, one considers basic cognitive apparatus to *include* tools and instruments, forms of argument and explanation, symbols and representational devices and forms, learning histories, norms for valued epistemologies, and ways of talking.

Therefore, my role will be to invite us to take the next step beyond the question, "Do children or don't they think like scientists?" Please understand – I am not arguing that the *origins* question is not a valuable one. For any valued form of thinking, we would certainly want to know, "Where do its various aspects come from?" or, "Out of what precursors do they develop?" Equally important, however, are the questions that follow: "Where is thinking going?" and, "What supports productive change?" Answers to the first question help us better understand the foundation on which further development can build. Answers to the second provide a sense of developmental trajectory, or more likely, trajectories. What characteristic changes are coming up? What pathways of change are usually observed? And answers to the third question focus on how those changes can get supported in a productive way. By the way, I don't see these as issues that differentiate basic research from applied. To me, an education researcher is nothing more or less than a developmental psychologist engaged in basic research that includes a critical question usually set aside, namely, "Under what conditions?"

So, back to the three questions. We are learning a great deal about origins from many exciting investigations of the young infant's mind. Infants seem to have expectations about the world that appear very reliably at surprisingly early ages. For example, even very young babies know that two objects cannot simultaneously occupy the same place (Baillargeon, 1987). Like other species, we are apparently wired so that we find some things very easy to learn; others, for which we are not prepared in that way, may be very difficult. It may be that some of these forms of knowledge are innate. Or, it is possible that the way our perceptual and interpretive equipment operates tunes us to noticing certain kinds of events and interpreting them in particular ways. For example, when one billiard ball collides with a second, we cannot help perceiving the second ball as having been propelled by the first (Michotte, 1963). In contrast, if a brief delay occurs between the time when the first ball hits and the second ball flies off, we do not perceive the event as a causal event. Leslie (1984) demonstrated that infants see the world in much the same way. Simple causal relationships of this kind seem to be directly perceived. Moreover, by the preschool years, children are reasonably good at making causal judgments on the basis of patterns of covariation between effects and their potential causes (Gopnik & Sobel, 2000; Schultz, 1982), even when the relationships are probabilistic rather than deterministic (Gopnik, Sobel, Schulz & Glymour, 2001).

As with the research on objects and causality, researchers have also investigated children's early understanding of quantity, living things, and mental life. On the basis of the findings, some researchers have claimed − although as Dr. Brewer explains, this remains controversial − that these core beliefs are theory-like. This so-called "theory theory" asserts that children develop coherent frameworks about the world, and in fundamental ways, their theory building is much like that of adult scientists (Gopnik & Meltzoff, 1998). Moreover, at least in some instances, these theories seem to be both extensible and revisable by data, and data are not only passively observed, but in some cases, actively sought (Gopnik et al., 2001).

We are learning, then, about the beginnings of children's understanding of objects and their interactions, including attempts to understand both the early origins of this knowledge and how it expresses itself through the preschool years. These early ideas about causality seem to be universal, and you can imagine how they play out in Dr. Brewer's example of children developing notions of where the sun might go when night comes. We know a lot about objects − that they don't usually spontaneously disappear when they move out of our sight, that they maintain their boundaries, that they tend to remain on a constant trajectory unless something interferes, that the places they might occupy are limited logically by their displacements. However, this kind of knowledge does not get one all that far in reasoning about science. Indeed, in many ways, these early causal schemas can serve as obstacles for understanding scientific principles. Therefore, what is most important for science learning is not just what we come with, but what happens down the road.

Unfortunately, the research on origins of causal knowledge has little to say about what happens after children enter school. At that point, for many developmental psychologists the issue becomes one of *learning* and is no longer of much interest. Research on early causal reasoning is not conceptually connected to the education research about student's alternative conceptions of the world, including the way that objects move and interact. Strangely, most of the education research is characterized by a complementary weakness: it typically lacks any extended developmental focus. Moreover, a lot of it is purely descriptive and, with a few important exceptions, does not address questions about mechanism.

Once we get beyond the preschool years, it becomes essential to ask not just what conceptions we observe in children, but also, whether and how they continue to develop, and under what conditions. Take, for example, the development of theories as models. This is one way, as Dr. Brewer points out, that children's thinking seems like the kind of thinking we observe in scientists. There is no doubt that children generate mental representations of the way the world works, that those representations sometimes include explanatory entities that are not observable, that data can reinforce or change them, and so forth. However, when we problematize the idea of what it means to generate and reason with models of the world, we see that this mental capacity turns out to be reasonably complex. There remains plenty of room for development, because even with respect to model generation, children both do and do not think like scientists. Consistent with the three questions that I posed earlier, I will have a little to say about the potential origins of this way of thinking, a little about next developmental accomplishments that might be encouraged, and the kinds of conditions that support those next developmental steps.

First, consider potential origins. One source is almost certainly the early forms of causal reasoning that I discussed earlier. Another is the variety of symbolic functions that young children are appropriating, that is, the variety of ways in which one thing stands for another. Models and theories, after all, are representing tools. Some of the early origins of this representing function are language, pretend play analogy, drawing, and imitation. For example, in the context of pretend play, even two-year-olds intentionally use one thing to stand in for another. A child who picks up a wooden block and pretends to sip from it knows very well that it remains a block, even as he acts "as if" it is a cup. Alan Leslie has written about the development of this early capability to treat things "as if," and he points out that it requires simultaneously generating, maintaining, and coordinating two separate mental representations of the pretend play object – one in which the block remains a block; another in which it becomes a cup (Leslie, 1987). Conceptually, this is similar to what adults do when they use water flow as an analogy for understanding heat flow (Gentner, 1989). Early representational capacities like these may be involved in the intentional generation and deliberate exploration of models as a way to better understand the modeled world. This intentional use of a "stand-in" seems more self-aware and deliberate than the example of a model of where the sun might go at night. But the idea of intentionally investigating the qualities of the representing system is central to scientific thinking.

What else might be involved developmentally in understanding what Hestenes (1992) called "the modeling game"? My colleague, Rich Lehrer, and I have identified the importance of focusing on attributes rather than intact objects and events. Focusing on attributes that are critical to one's modeling intent and discarding those that are not appears to be particularly challenging, probably at all ages. As Grosslight and her associates showed (Grosslight, Unger, Jay & Smith, 1991), children frequently maintain "copy" theories of models, perhaps because they are reluctant to leave out information contained in the original phenomenon. Consider, for example, the observations that my colleague, Rich Lehrer, and I made of a group of first- and second-graders studying the growth of flowering bulbs. To represent the changing heights of the plants over time, the children cut out paper strips on each day of measurement. They then mounted the strips side by side to display a display of growth, a record that they hoped to use to identify times when the plants grew faster or slower and also to compare the growth of one plant to that of another. At the beginning of the investigation, the children insisted that all the strips had to be green like the stems of the plants, and moreover, every strip had to be adorned with a flower. Only extended efforts to compare the growth of different amaryllis plants eventually convinced the children that some attributes they had originally considered essential because they were present in the flowers they could observe – namely, color and petals – were competing with the goal of clarifying and comparing plant heights. Over weeks they invented a new solution. Strips of different color were chosen to represent and compare different plants, and flowers were no longer drawn on the top of each strip. Students found that the petals made it difficult to measure the strips with any precision. Rich and I described this shift as one in which students went from representing *plants* to representing *heights* (for more detail about this example, see Lehrer & Schauble, 2002). It is unclear whether this reluctance to focus on attributes and their measure qualities is due primarily to the "universal novice" status of children (i.e., they do not know which of the attributes are relevant to the current question) or whether it is due to some more general age-related mechanism. We suspect both factors (for example, see Lehrer & Schauble, 2000).

Bazerman's (1988) account of historical shifts in the genre of the scientific article suggests that these ideas are difficult for adults, too. Bazerman explained that early scientific articles tended to be narrative-like accounts of the general form, "This is the interesting phenomenon I witnessed, and if you go over there, you can see it, too." Apparently, not everyone saw the promised event, and over time, as readers raised objections and questions, authors included more precise descriptions of the conditions under which the target phenomenon would occur. The idea of an experiment – in which a simplification of the world, a model of it, if you like, is investigated to learn about the more complex world itself – was a relatively late invention. Indeed, it is admittedly a rather strange form of epistemology that sets aside the intact phenomenon of interest to concentrate attention instead on a simplified stand-in composed of relevant, manipulable attributes. Bazerman credits Newton with perfecting a textual genre in which the experimental report became a form of argument. Newton accomplished this transition in the scientific genre, Bazerman explains, by trying to foresee and address in advance any potential objections or alternative interpretations that a reader might conceive while reading the report. For example, something that all of us teach our students (and one that never fails to strike some of them as a "cheat") is that multi-experiment research articles need not report studies in the order in which they were actually performed. Understanding the attributes that are relevant to a theory or a question, knowing how to manipulate them, either in a physical or thought-experiment, and drawing conclusions that relate back to the world is not a form of thinking that we would expect children to spontaneously invent. Yet, it is part of taking a modeling stance toward the world.

The very idea that alternative models can be conceived is not a simple one. As our colleagues who study the "theory of mind" tell us, thinking of one's beliefs as separate from "the way the world is" and as potentially disconfirmable is a big conceptual achievement, one related to the idea of model evaluation. Gopnik and her colleagues (2001) demonstrated that even young children's causal models are modifiable by data, but many patterns of data that might be relevant to our theories fail to "speak" to us unless we have conceptual tools for structuring and interpreting data. These tools, including mathematical and representational forms, make it possible to evaluate mismatches between a model and the modeled phenomenon.

Children develop theories or models of the way that the world works, and in some ways, they do think like scientists. However, there is also a lot of variability, both in the situations that elicit these model-like representations and in the sophistication and qualities of the models. There appears to be no one point when model construction begins, and no point when we would agree that it is completed. Instead, it is a very complex form of thinking, one that includes aspects that develop only over the long term and only in situations where they are valued and supported. Under this view, development is not simply about getting older.

What else might it be about? This brings us to the third question, "What supports developmental change?" For these purposes, variability can be a valuable resource for instruction. Students who are asked to explain and justify their explanations of phenomena, and to be accountable for restating, elaborating, and constructively critiquing the contributions of their classmates, are likely to understand over time that not everyone interprets the world in the same way. Classroom norms that systematically require reasons and evidence, supplemented by instruction on ways to structure, represent, and interpret data, make it more likely that students will spontaneously seek data to support their theories. These norms also help children value communication and audience, values that are reinforced when students are asked to maintain a history of their learning. The need to be aware of what they have come to know and how they came to know it arises naturally in classrooms where knowledge builds cumulatively, so that students are always seeking to connect new learning to accomplishments that came before.

There are now several examples of elementary school classrooms where students are routinely engaged in the forms of thinking that Dr. Brewer points out they do not typically produce. In these cases, what is typical depends both on one's history of learning and the norms operating in the present context. It is certainly true that children do not typically go out of their way to systematically collect data relevant to a belief they have developed. (Neither do adults, by the way.) Yet, Kathleen Metz's (2000, 2004) first graders enthusiastically generate conjectures about their crickets' social behaviors, develop questions that follow from their theories, and design empirical studies to evaluate their initial conjectures. Everyone in the classroom is studying the same organism, but different questions are being pursued by different research teams. Therefore, disseminating knowledge publicly is a widely accepted practice. It is important to point out that students do not engage in these behaviors without thoughtful teaching assistance, the development of shared epistemological norms, well-designed instructional tasks, and a consistent focus on symbolic representations and communication.

Similarly, Dr. Brewer points out that children are not usually explicitly aware of or critical about their own theories. However, they can be. Sr. Gertrude Hennessey was the only science teacher in a small elementary school, and therefore, she could systematically focus her instruction on core ideas built cumulatively across the years of each child's education. She chose to emphasize generating, communicating, and evaluating theories via the intelligibility, plausibility, fruitfulness, and conceptual coherence of the alternatives. Research on her graduating sixth-grade students' understanding of the nature of science suggested that they had a better sense of the enterprise than did twelfth-graders from a comparable school (Smith, Maclin, Houghton & Hennessey, 2000).

In these classrooms, the teachers were certainly knowledgeable about what to expect of their students' incoming resources and competencies. Equally important, however, is that they maintained a systematic and extended press on these competencies. These are forms of thinking that require a long time to develop. Indeed, although much can be accomplished within a year, values and epistemologies need to be in place across grades if development is to be fostered. Because intelligent people adapt to the situations in which they find themselves, if students are in contexts where scientific thinking is neither needed nor valued, it is highly unlikely to appear.

In sum, the early roots of scientific thinking may be both more numerous and more variable than we have considered. They include apparently universal forms of knowledge about the world, but also knowledge and epistemologies constructed over each child's particular history. The glass is half empty and half full. Perhaps, in the end, the half that a person focuses on is a function of whether or not he is responsible for trying to get the cup to runneth over. Now *that* is a dreadfully mixed metaphor.

Leona Schauble
Department of Teaching and Learning
Peabody College
Vanderbilt University

NANCY J. NERSESSIAN

MODEL-BASED REASONING IN SCIENTIFIC PRACTICE

INTRODUCTION

The opening sentence of Thomas Kuhn's *The Structure of Scientific Revolutions* is perhaps one of the most familiar statements in 20th century science studies: "History, if viewed as a repository for more than anecdote or chronology, could produce a decisive transformation in the image of science by which we are now possessed" (Kuhn, 1970, p.1). Many have interpreted this statement to mean that if we are to understand the nature and development of scientific knowledge, we need to start from examinations of the actual practices of scientists. If we take "history" in the broad sense – as encompassing past and present science – this view forms the premise of the various approaches to the study of science that constitute the field of science and technology studies (STS): historical, sociological, anthropological, philosophical, and cognitive studies of science and technology. We are now in possession of numerous detailed accounts of practices in nearly every field and every era of science. The result being, that, unlike the situation in which Kuhn was writing in the early 1960s, we are *not* "possessed" of a single "image of science" but of multiple images produced by various perspectives. "Science" is not a monolithic practice. Just as there is domain-specific knowledge, there are domain-specific knowledge-producing practices. There are also practices that have a cross or domain-general aspect – such as the one I will focus on "model-based reasoning" – but these also can look quite different when enacted in a domain, for example, modeling in psychology will look quite different from modeling in physics or biology – and physical modeling is different from mathematical or computational modeling in physics. What the images share is that they derive from a basic philosophical stance that understanding "how science works" starts with examining "how scientists work," rather than from purely a priori considerations.

Given my expertise, the perspective on inquiry developed here focuses primarily on the cognitive practices of scientists. Studies of cognitive practices join historical, philosophical, and cognitive science methods, data, and interpretations in constructing both specific and general accounts of the cognitive practices of scientists in creating and using knowledge. The STS fields have developed diverse accounts, ranging from domain specific to domain general, from thick descriptions to abstract analysis. To date, however, these accounts tend to line up on either side of a perceived divide between cultural[1] factors and cognitive factors in knowledge construction, evaluation, and transmission. Cultural accounts have tended to claim that cognitive factors are inconsequential to explaining how

Richard A. Duschl and Richard E. Grandy (eds.), Teaching Scientific Inquiry: Recommendations for Research and Implementation, 57–79.
© 2008 *Sense Publishers. All rights reserved.*

science works. Cognitive accounts, for their part, while paying deference to the importance of the cultural dimensions of practice, have not, by and large, made cultural factors an integral part of the analysis. The situation has fostered a perception of incompatibility between cognitive and cultural accounts of science. Perceptions to the contrary, I contend that combined research on the cognitive and cultural sides shows the divide to be artificial.

A significant body of research on the cognitive practices of scientists has led to explicit recognition that the social, cultural, and material environments in which science is practiced are critical to understanding cognition (see, e.g. Dunbar, 1995, 1999; Giere, 1988, 1992; Gooding, 1990; Gorman, 1997; Gorman & Carlson, 1990; Kurz & Tweney, 1998; Kurz-Milcke, Nersessian & Newstetter, 2004; Nersessian, 1984, 2002, 2002, 2005; Thagard, 2000; Tweney, 1985, 2002). On the other hand, recent research in cultural studies can be read as recognizing the significance of cognitive dimensions, such as Peter Galison's (1997) concern with the "image" and "logic" traditions in the material culture of particle physicists, Karin Knor Cetina's (Knor Cetina, 1999) recent analysis of scientific practices as part of "epistemic cultures," and Hans-Jörg Rheinberger's (1997) analysis of experimentation in molecular biology as producing "epistemic things." Scientific inquiry requires the kind of sophisticated cognition that only rich social, cultural, and material environments can enable. All these factors contribute to inquiry, for example, in formulating research problems and pursing their solutions, which are central both in advancing and using scientific knowledge. Thus, the major challenge for understanding how science works is to develop accounts of how scientists work that capture the integrated nature of the cultural and cognitive factors. Accommodating this insight requires theorizing cognition in relation to context, but this does not rule out generalizations. Placing specific practices within the broader framework of research on human cognition aids in establishing that the fragments of scientific research practices examined are broadly representative of scientific practices – both those that are domain specific and those that are domain general. As Edwin Hutchins has said of studies of situated cognitive practices generally

> There are powerful regularities to be described at a level of analysis that transcends the details of the specific domain. It is not possible to discover these regularities without understanding the details of the domain, but the regularities are not about the domain specific details, they are about the nature of cognition in human activity. (Woods, 1997, p. 117)

An interdisciplinary approach

Much of my focus has been on conceptual or representational change – a central problem of both positivist and post-positivist philosophy, though framed and answered differently for each. Attacking the problem of conceptual change in science requires deep understanding of both human cognition and the history of its achievements in the domains of the sciences. The nature of creative scientific practices has been addressed in history, philosophy, and the cognitive sciences. Although each of these fields brings indispensable insights and methods of its own to bear on the core issues, deep understanding requires an interdisciplinary approach that uses methods, analytical tools, concepts, and theories from all these areas.

Conceptual innovation in science is rare and does not occur in an instant, but often spans the lifetimes of one or more scientists, and thus has little chance of being caught "on-line". To interpret from historical records how the scientific practices created new conceptual representations, one needs to employ a method that is "cognitive-historical" (Nersessian, 1992, 1995); that is, a method that draws on research in the cognitive sciences to help interpret the historical records. The goal of such investigations is not to create an historical narrative per se, but to contribute to an understanding of creative cognition. What is striking is that in conceptual change across the sciences, the practices of analogical, visual, and thought experimental modeling are widely employed in the problem solving leading to the representational change.

The historical data also lead to the conclusion that the cognitive practices of science should not be divorced from the cultural factors that both enable and contribute to their development. To address these contributions to creative outcomes, I have begun to widen my viewpoint to include carrying out empirical investigations of current scientific and engineering research practices in order to determine the range and kinds of factors relevant to creating novel understandings, construed more broadly than conceptual change. Over the last three years I have been conducting a complementary line of research that examines contemporary science "in action," specifically, in interdisciplinary research laboratories in university settings in the biosciences and bioengineering. I have chosen these kinds of laboratories for several reasons. First, much contemporary scientific practice is interdisciplinary. Interdisciplinary research laboratories provide settings where novel practices and achievements can arise from approaching problems in ways that force together concepts, methods, discourses, artifacts, and epistemologies of two or more disciplines. Second, university research laboratories are sites of learning – populated with graduate and undergraduate students. Indeed, learning in laboratories – instructional and research – is an essential part of becoming a scientist today, yet it scarcely has been studied. These particular laboratories are located in an educational environment that aims to create researchers who are in themselves interdisciplinary individuals, rather than disciplinary contributors to an interdisciplinary team. Third, numerous studies of research laboratories conducted by sociologists and anthropologists have shown the fruitfulness of studying research laboratories and groups for thinking about the cultures of science, but scarcely any research has looked from a cognitive perspective[2] or with the problem of integration in mind. My research group approaches the laboratory studies from the developing perspective in cognitive science that casts human cognition, generally, as occurring in complex systems situated in social, cultural, and material environments.

To carry out this research ethnographic methods needed to be added to the mix of data collection and analysis, and thus conducting the laboratory studies required putting together an interdisciplinary research group.[3]   Indeed, the standard methods for investigating practices of communities are ethnographic interviews and observations. However, we have found that in research laboratory environments these do not suffice. History figures into the practices of the laboratories in ways that are central to understanding the interplay of cognition and culture in current practices. Ethnography, customarily, does not seek to capture this critical historical dimension of these settings: the evolution of technologies, researchers, and problem situations over time. Thus we both collect and analyze ethnographic interviews and observations and a range of historical records (spanning 20 years), including laboratory notebooks and engineered artifacts, with an eye to understanding how laboratory history is appropriated in current practices. Ethnography is providing data on the transient and stabilized physical and social arrangements that ground research activity in lab routines, social organization, construction, and manipulation of artifacts, as well as data on the appropriation of history and on the nature of current cognitive practices. It has enabled us to delineate the dimensions along which to conduct the historical research. Cognitive-historical analysis is providing data on historical trajectories relating problems, technology, models, and researchers as well as evidence for how cognitive practices originate, develop and are used.

The major problems addressed in my research are related to what I have called "model-based reasoning practices". In going beyond still customary notions that equate reasoning with logical inference, my previous research focused on practices of physicists: analogical modeling, visual modeling, and thought experimenting, and developed an account of how these constitute the reasoning that generates representational change. In the research laboratories these play a central role, too, but so do physical models constructed for simulation purposes, and the new research I will discuss briefly here is beginning to analyze their role in creative outcomes.

Cognitive Inquiry Practices and Learning

Research on learning in science education and in cognitive development are areas of cognitive science where Kuhn's early views on scientific revolutions, conceptual change, and incommensurability have had significant influence. Many researchers in these areas have proposed an analogy between the kinds of changes in conceptual structure observed in learning and development and those that have taken place in science. The main support for hypotheses drawn from the analogy comes from research that describes the initial states of the learner's conceptual structure and compares these with the desired final states. These end-state comparisons do indicate that the kinds of conceptual changes students need to experience to learn science concepts are similar to those that take place in what philosophers and historians of science have characterized as "scientific revolutions". The high degree of similarity in the nature of the *kinds* of changes indicates a potential relationship in the cognitive tasks. From a cognitive constructivist perspective extending the analogy to the *mechanisms* of conceptual change (Nersessian 1992, 1995) in science learning is warranted in that the cognitive tasks in learning and in the initial constructive process are similar in salient ways. Extending the analogy is not based on the view that the scientists are "experts" and therefore they are automatically relevant to training "novices." Rather, it is based on the hypothesis that in both cases the cognitive tasks require constructing new representations and forming representational structures that integrate the new concepts for coherent, systematic use. The constructive practices of scientists provide a model of success in achieving these tasks. Further, neither the "child as scientist" or "scientist as child" analogies are required for the cognitive practices to be relevant to learning. Rather, all that is needed is a "continuum hypothesis": the cognitive resources used in problem solving in ordinary cases and in science are not different in kind – scientific practices extend and elaborate on mundane cognitive resources. From these kinds of considerations, the broader suggestion I am making is that the cognitive practices of scientists can provide a model for cognitive aspects of the learning activity itself and that studying scientific work in action – historical and contemporary – provides a way to discern these and develop pedagogical approaches that promise to be more successful than the traditional textbook and "canned" instructional laboratory.

Since many readers misinterpret him on this point, it needs pointing out that Kuhn, himself, characterized science education as a process of indoctrination in which textbook distillations and reformulations of current knowledge are the chief pedagogical tools. Repeatedly working problem sets and assimilating the similarities and differences among problem exemplars enables the learner to acquire the paradigm and thus the means for resolving outstanding problems during the course of normal science. On Kuhn's view, working problem exemplars also provides the basis for learning the concepts and other aspects of scientific practice such as methods, analytical techniques, and how mathematical relations map to physical situations. Finally, he believed this pedagogical approach to be highly successful and one that should be continued since "[n]othing could be better calculated to produce 'mental sets' or *Einstellungen*" (Kuhn, 1977, p. 229). Perhaps this is how things looked from the optimistic vantage point of 1962, as the US moved forward on a national agenda to win the science race with the Soviet Union. In the meantime, however, cognitive science research on conceptual change indicates that the traditional textbook-type science education has *not* been successful in producing understanding of science, let alone practitioners. Very few students learn the subject sufficiently well even to provide explanations and predictions of simple physics phenomena, never mind to go on to graduate school and become practitioners. Further, documented high drop out and failure rates along the way establish that traditional forms of science education do not produce many practitioners as a percentage of the population taking classes – or even a scientifically literate populace. Although he remained convinced of the success of the traditional textbook approach, his critique of using that image for doing history and philosophy of science can, on my view, be transferred to science education: the "concept of science drawn from them is no more likely to fit the enterprise that produced them than an image of a national culture drawn from a tourist brochure or a language text" (Kuhn, 1970, p. 1).

The proposal that the cognitive practices of scientists can provide a model for cognitive aspects of the learning activity itself also gains support from the practice of apprenticeship training in science. In point of fact, much of the credit for what success there is in creating practitioners goes to the apprenticeships – until recently experienced first in graduate school – during which practices are learned in authentic situations. This is something I think Kuhn failed to appreciate sufficiently. In apprenticeships students learn the tacit as well as the explicit practices of the discipline. Within cognitive science such situated learning experiences have been called "cognitive apprenticeships" (Brown, Collins & Duguid, 1989; Collins, Brown & Duguid, 1989). In this form of training, science students learn, among other things, how to adapt problem exemplars to current research problems through observing how practitioners tackle these and through participating in the research. In this process they are exposed directly to the formal and informal methodological practices and situated conceptual understandings and interpretive structures that constitute the practice of the science.

HOW DO SCIENTISTS WORK?

Scientific practice is inherently a problem-solving activity. In contrast to problem-solving in traditional school science, however, much of scientific inquiry involves 1) formulating problems and 2) determining or devising means of working towards their solution. A growing body of work in science studies establishes that, in contrast to the positivist and textbook "image of science" Kuhn was criticizing, much of scientific inquiry is not collecting observations and constructing theories. Rather it involves creating and using models of all kinds in experimental and theoretical work. Contrary to the standard view of scientific reasoning as hypothetico-deductive or logic-based in nature, for example, models and "model-based reasoning" practices are the *signature* of much research in the sciences, both in discovery and application (Cartwright, 1983; Giere, 1988; Hesse, 1963; Magnani, Nersessian & Thagard, 1999; Morgan & Morrison, 1999; Nersessian, 1984, 1992, 1999, 2002, 2002). Studies of historical science provide evidence that although we have developed new tools for enhancing and expanding these practices, since the inception of science creating and using models has been a standard part of practice. *A priori* philosophical notions about the nature of reasoning that informed the standard view of The Scientific Method, for one thing, stood in the way of acknowledging – or even recognizing – the generative role of modeling in scientific knowledge. Consequently, assisting students in developing versatile and informed understanding of models and model-related practices needs to be a primary goal of any science curriculum.

A model can loosely be characterized as a representation of a system with interactive parts with representations of those interactions. Models can be qualitative, quantitative, and/or simulative (mental, physical, computational). They are often multifaceted and these can come together in a problem-solving episode. For example, in the cases I discussed below of Faraday and Maxwell, they constructed visual representations of imaginary physical models representing a new conceptual system – 'the field' – conceived as animated (mental simulation) from which they derived mathematical representations, theoretical hypothesis, and experimental predictions.

What the historical and contemporary cases show is scientists engage in modeling prior to developing the laws and axioms of theories and in applying these in research. Models are also used in communicating novel results which can be thought of as a form of instructing peers within the community. Although modeling is ubiquitous and significant, it is, of course, not exhaustive of scientific inquiry practices. But given my own research focus, I will concentrate mainly on "model-based reasoning," by which I mean making inferences from and through model construction and manipulation.

An effective methodological approach for determining how scientists create and use models includes both historical research on past practices and ethnographic investigations of contemporary practices. These provide the focal points for examining existing cognitive science research in search of findings that help to explain the cognitive underpinnings of the scientific practices, to formulate hypotheses about why these practices are effective, and to discern ways in which the cognitive research might be challenged by the findings derived from examining scientific cognition.

Modeling Practices in Conceptual Change

Conceptual change in science comes about through extended periods of problem solving that often span the life of a scientist or generations of scientists, and is thus best studied through historical cases. By now there are numerous studies of instances of conceptual change across the sciences. The nature of the specific conceptual, analytical, and material resources and constraints provided by the environments in which conceptual changes have taken place have been examined for many episodes. What stands out from this research is that in numerous instances of conceptual innovation and change across the sciences, the practices of anagogical, visual, and thought experimental (simulative) modeling are used. My own historical investigations have centered on physics, but those of other sciences establish the prominent role of these practices across the sciences, making them candidates for domain general inquiry modes in conceptual change (for a sampling in addition to previous references, see Darden, 1991; Gentner et al., 1997; Giere, 1992, 1994; Griesemer, 1991; Holmes, 1985; Latour, 1986; Lynch & Woolgar, 1990; Rudwick, 1976; Shelley, 1996, 1999; Trumpler, 1997). Although not presented in this paper, cognitive science studies of analogy, mental modeling, mental imagery, imagistic and diagrammatic reasoning, problem solving, and conceptual change assisted developing the argument for the generative, domain independent nature of "model-based reasoning" in problem solving generally, and in conceptual change in specific (see, e.g. Nersessian, 2002). It is important to underscore that the complexity of scientific cognition requires integration and unification of a wide range of data and interpretive results in what are traditionally considered separate areas of cognitive science research.

This research on scientists modeling practices has been making some impact on science education. Indeed there are nation-wide and international efforts underway to build "model-based" K-16 classroom curricula in the sciences (Clement & Steinberg, 2002; Clement, 2000; Gilbert & Boulter, 2000; Gobert & Buckley, 2000; Justi & Gilbert, 1999; Van Driel & Verloop, 1999; Smith, Maclin, Grosslight & Davis, 1997; Snir & Smith, 1995; Wiser, 1995; Nersessian, 1995, 1989; Wells, Hestenes & Swackhamer, 1995; Grosslight, Unger, Jay & Smith, 1991; Hestenes, 1987). Even the 1996 *National Science Education Standards* developed by the *National Committee on Science Education Standards and Assessment, National Research Council*, specify model-based understanding and reasoning competence as a goal even for science education in grades K-12. The *Standards* emphasize how models can take many forms, including physical objects, plans, mental constructs, mathematical equations, and computer simulations and emphasize their explanatory power. What needs to be added to their emphasis are the ideas that problem solving involves not only having models of phenomena, but also such things as understanding, for example, how making a physical representation of a model provides its own representational constraints; the role of models in the context of problem finding and problem formulation; and the role of models as mediators between and integrators of research practices.

I will illustrate what is meant by "model-based reasoning" with two historical examples instrumental in conceptual changes in physics, specifically, creating the concept of electromagnetic field. There are many salient dimensions of these cases that are not developed in this brief illustrative presentation. For example, the cultural context is crucial to understanding both cases. Maxwell's location in Cambridge, for example, led to his training as mathematical physicist, proficient in the emerging field of continuum mechanics. This shaped the nature of the theoretical, experimental, and mathematical knowledge, conceptual representations, and methodological practices with which he formulated the problem and approached its solution. The work of Michael Faraday and William Thomson (later, Lord Kelvin) contributed significantly to these. Continental physicists working on electromagnetism at the same time employed quite different practices and drew from fundamentally different mathematical and physical representational structures.

In the illustration from Faraday, the model-based reasoning is in a context where experimental practice is paramount and in the illustration from Maxwell, theory and mathematical analysis are paramount, though there are elements of all in each case. Many instances of model-based reasoning in science and engineering employ physical representations that are constructed during the reasoning process, such as diagrams, sketches, and physical models. These can be viewed as providing constraints and affordances essential to problem solving that are used in conjunction with those provided by whatever mental representations are used by the reasoner during the process. In each case below, model-based reasoning makes use of *a visual representation of an imagined physical model – conceived as animated – to derive mathematical results, make theoretical hypotheses, and make experimental predictions.*

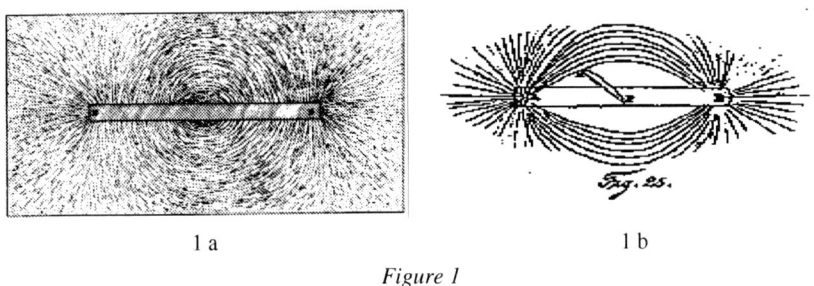

1 a   1 b

Figure 1

Faraday's lines of force model. Faraday had hypothesized that the lines of force that form when iron filings are sprinkled around magnets and charged matter indicate that some real physical process is going on in the space surrounding these objects and that this process is part of the transmission of the actions (Faraday 1831). That visual display of lines of force in geometrical configuration had a profound influence on Faraday's conceptualization of electric and magnetic actions, and of forces and matter in general. *Figure 1a* shows the actual lines as they form around a magnet. *Figure 1b* shows Faraday's drawing of the model that represents them in geometrical and dynamical form on his analysis. That is, although the visual representation is of a static geometrical pattern, Faraday reasoned with it as representing his theoretical model of dynamic processes. He envisioned the various forces of nature as motions in the lines, such as waving, bending, stretching, and vibrating. Thus, he reasoned with the static visual representation of the lines as representing a qualitative dynamical model for the transmission and interconversion of electric and magnetic forces, and, ultimately, for all the forces of nature and matter. As Maxwell (1890b) remarked, although this model is qualitative, it embodies within it a great deal of mathematical understanding. The only quantitative relationship that Faraday, himself, derived can be traced directly to the model. As he formulated it, the intensity of the induced force is related to the *number* of lines cut. However, even for Faraday this relationship is expressed incorrectly because "number of" is an integer, while "field intensity" is a continuous function. This "error" is directly traceable to Faraday's visual representation of the model, where the lines representing the configuration of the filings are discrete entities, whereas, except in rare cases the lines of force spiral indefinitely in a closed volume.

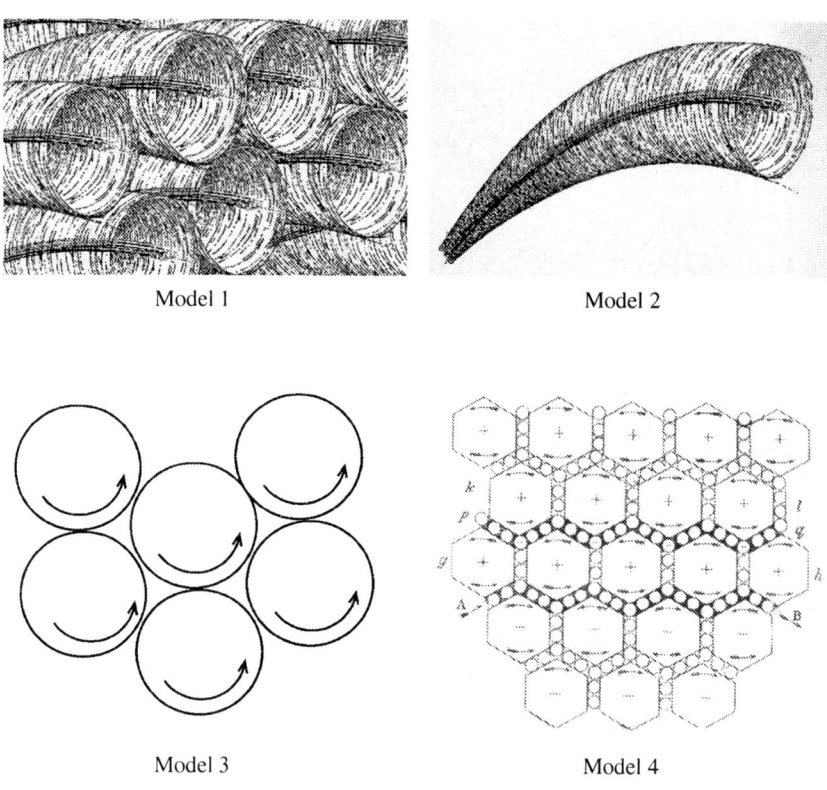

Figure 2

Maxwell's idle-wheel and vortex model. Maxwell's (Maxwell 1861-2) processes of model construction in creating a mathematical representation of the electromagnetic field concept are as follows. First he constructed a primitive imaginary physical model (Figure 2, Model 1 is my representation from his description) consistent with an initial set of constraints from the electromagnetic domain. That model was a fluid medium composed of elastic vortices and under stress. By focusing on a single vortex in this model (Model 2, my representation), he derived mathematical representations for various magnetic phenomena. Analyzing the relationships between current and magnetism led to his altering the model. Here the relationships between vortices need to be considered. If one imagines the model (Model 3, my representation of a cross-section) as animated, all the vortices are rotating in the same direction. Since they touch, friction is produced and so one can infer they will eventually stop. Mechanical consistency, thus, requires the introduction of what Maxwell called "idle wheels", such as those used in machine gears, in this case to surround the vortices. He argued that their translational motion could be taken to represent electricity. Model 4 is Maxwell's own drawing of the idle-wheel and vortex model, which he accompanied by text for how to imagine animating the model. The imaginary physical system provides a mechanical interpretation for electromagnetism: motion of the particles creates motion of the vortices and vice versa. Thus, as is known experimentally, electric current produces magnetic effects and changes in magnetic effects produce current. Using the model Maxwell was able to derive mathematical equations to represent these causal relationships. Our illustration ends here, though it then took him nine months to figure out how incorporate electrostatics into the model, which enabled him to derive a time delay for the propagation of the electromagnetic actions, thus establishing – subject to empirical testing – a mathematical representation for electromagnetic field actions, as opposed to actions at a distance. Maxwell's model construction processes involved abstracting physical and mathematical constraints that hold in common among certain kinds of continuum-mechanical systems, machine mechanisms, and electromagnetic systems and integrating these into a series of imaginary physical systems. These models represent the production and transmission of electric and magnetic forces in a mechanical aether.

A key feature of Maxwell's model-based reasoning that although his particular instantiations of the models provide specific mechanisms, he reasoned with them as representing *generic* processes. In the model representing electromagnetic induction (*Model 4*), for example, the causal relational structure between the vortices and the idle wheel particles is maintained but not the specific causal mechanisms. The idle-wheel and vortex mechanism is not the cause of electromagnetic induction; it represents only the causal structure of that unspecified process. Any mechanism that could represent the *causal structure* would suffice. As a reasoning process generic abstraction separates dynamical properties and relationships common to mechanical and electromagnetic systems – from the specifics of the models by means of which they had been made concrete.

During the modeling process, Maxwell continually evaluated the models and the inferences he drew from them, and integrated the solutions to the sub-problems into a mathematical representation. There are several ways of seeing his reasoning as taking place through model manipulation. For example, in the initial analysis of magnetic phenomena, Maxwell focused on a single rotating vortex. In the next phase of the analysis he considered the behavior of a medium full of rotating vortices, which, as discussed above, led directly to the introduction of the idle wheel particles as a way of representing the causal structure of electromagnetic induction. There are, also, significant sign "errors" in this part of Maxwell's analysis, but as I have argued in previous work, all but one (a minor substitution error) are not errors if we understand him as deriving the equations from reasoning through the models he created. That is, the signs are correct on the model, though incorrect from our point of view which no longer employs mechanical models.

Model-based reasoning in conceptual change. The hypothesis of reasoning through constructing and manipulating models helps to make sense of much of scientific practice that has been ignored by historians or ruled as insignificant to scientific reasoning by philosophers. Traditional philosophical accounts view reasoning as carrying out logical operations on propositional representations. Yet clearly modeling is a creative engine driving much of scientific change. Philosophers are beginning to recognize its significance in part due to the pervasive and growing use of computer modeling and simulation by contemporary scientists. However, the new modeling practices can be interpreted as using new instruments to enrich and take to new depths and directions an existing human practice: simulative model-based reasoning. The situation parallels such developments as those of the telescope, microscope, and other instrumentation that have enabled scientists to "see" better. Such computer modeling enables scientists to "simulate" better.

A model is a conceptual system representing the physical system that is being reasoned about. As such it is an abstraction − idealized and schematic in nature − that represents a physical situation by having surrogate objects and properties, relations, behaviors, or functions of these.

The model complies with the constraints pertaining to the specific kind of phenomena being reasoned about, and simulating the model can lead to new constraints − or to recognition of previously unnoticed constraints. Examples of constraints include causal coherence (as in Maxwell's introducing the idle wheels to solve friction problem) and consistency. In the reasoning process, the model can be represented in multiple formats, including diagrams and drawings, equations, language and gesture. Inferences made in the simulation processes create new data that play a role in evaluating and adapting models to comply with new constraints. In the evaluation process, a major criterion for success is the goodness of fit of the model to the phenomena, but success can also include considerations of "fertility" or "fruitfulness" which include factors such as enabling the creation of a viable mathematical representation, as in the Maxwell case.

As noted above, a particular kind of abstraction appears to play a central role in the model-based reasoning practices in science, what I have called "generic abstraction." Although an instance of a model is specific, in the reasoning process it and the inferences made are often treated as lacking specificity along significant dimensions − or − as generic. To achieve this kind of inference, the must reasoner takes the specific model to represent features common to a class of phenomena. As David Hume long ago noted about imagining, one cannot imagine a "triangle in general" but must imagine a specific triangle. But, inferences are often not about the specific triangle, but about a class of triangles, such as right triangles or isosceles triangles. In reasoning about triangles one often draws a concrete representation or imagines one, but the context provides the interpretation of the concrete figure as generic. Thus, the same concrete representation can be interpreted and understood as generic or specific depending on the demands of the reasoning context. Further, the kind of creative reasoning employed in conceptual innovation and change involves not only applying generic abstractions but creating them during the reasoning process.

That there are different kinds of abstractive processes is not often recognized and, customarily, generic abstraction is not distinguished from notions of idealization and logical generalization. In idealization, features can be taken to the limit, but something specific still needs to be said about it as a relevant factor. Newtonian mechanics considers bodies in motion as point masses in frictionless space. A generic body would be one about which nothing would be specified about mass and friction. In Euclidean geometry the lines of a triangle, meet at a point, enclose a space, have no width, but are three in number. In contrast, a generic polygon has an unspecified number of sides. The kind of inferencing in generic abstraction is similar to what mathematicians call "generalization" (Polya, 1954). Removal of specificity is the central aspect of this kind of abstractive reasoning. The equilateral triangle is an idealization, but it has specific kinds of sides. The generic triangle represents those features that all kinds of triangles have in common, and this abstraction removes equality of the lengths of the sides and removes any specification of the degrees of the angles. The abstractive process leading to the generic polygon further removes specificity, in particular of the number of sides and the number of angles. Logical generalization, on the other hand, abstracts from one equilateral triangle what to apply to all equilateral triangles, but does not involve abstracting specificity of the kinds of sides and angles.

Generic abstraction plays a central role in creativity in conceptual change. The core idea of my account of conceptual change is that model-construction processes abstract and integrate information from multiple sources specific to the problem-solving situation. In this recombination, truly novel combinations can emerge as candidate solutions to problems. The results of these recombinations can be explored imaginatively – alone and in conjunction with external representations, through physical realizations, and through expression in other representational formats, such as mathematics and language. What is powerful about this idea is that constraints from multiple sources can interact in such a way that a model with a here-to-fore unrepresented structure or behaviors can emerge, as with the Maxwell case.

Modeling Practices Using Physical Simulation Devices

Science and engineering research laboratories are prime locations for studying cognition and culture in relation to practice. In the bio-engineering laboratories I and my co-researchers are studying, researchers construct technological artifacts in order to perform *in vitro* simulations of current models of *in vivo* biological processes. These perform as what they call "model-systems" – locales where engineered artifacts interface with living cell cultures – in specific problem-solving processes. These artifacts are what cultural studies of science refer to as the "material culture" of the community, but they also function as what cognitive studies of science refer to as "cognitive artifacts" participating in the reasoning and representational processes of a distributed cognitive system. My point is that within the research of the laboratories, they are both, and it is not possible to fathom how they produce new understanding by focusing exclusively on one or the other aspect. They are representations of current understandings and thus play a role in model-based reasoning; they are central to initiation rites and social practices related to community membership; they are sites of learning; they provide ties that bind one generation of researchers (around 5 years) to another; they perform as cultural "ratchets" that enable one generation to build upon the results of the previous, and thus move the problem solving forward. In sum, they are central in the cognitive-cultural fabric in which creative scientific understandings are produced. The model-systems need to be understood not just as "boundary objects" (Star and Griesemer 1989) existing in the "trading zones" (Galison 1997) of two or more communities and mediating communication, but as sites of interdisciplinary melding of concepts, models, methods, artifacts, and epistemologies, where genuine novelty emerges. We construe model-based reasoning with physical simulation devices as taking place within the problem spaces of the *evolving cognitive-cultural systems* of the laboratory.

Research laboratories as evolving cognitive-cultural systems. Traditional cognitive science research attempts to isolate aspects of cognition to control their investigation in experiments conducted mainly in vitro (Dunbar, 1995). Although traditional studies are still the mainstay of cognitive science, over the last twenty years significant investigations of cognition in vivo, that is, in contexts of human activity such as learning and work have become numerous. These various research thrusts can be characterized as attempts to account for the role of the environment (social, cultural, material) in shaping and participating in cognition. Many of these analyses make action the focal point for understanding human cognition. Human actors are construed as thinking in complex environments, thus they emphasize that cognition is embodied (see, e.g., Barsalou, 1999; Glenberg & Langston, 1992; Johnson, 1987; Lakoff, 1987; Lakoff & Johnson, 1998), distributed (see, e.g., Hutchins, 1995; Zhang, 1997; Zhang & Norman, 1995), encultured (see, e.g., Donald, 1991; Nisbett et al., 2001; Shore, 1997; Tomasello, 1999), and situated (see, e.g., Clancey, 1997; Greeno, 1989, 1998; Lave, 1988; Lave & Wenger, 1991; Suchman, 1987). I call accounts within this emergent research paradigm "environmental perspectives."

With the objective of interpretive integration in accounts of scientific practice, the trick is to create accounts that are neither primarily cognitive with culture tacked on nor the reverse, and this requires re-thinking current interpretive categories. One route to attaining integration is to reconceptualize 'cognition' by moving the boundaries of representation and processing beyond the individual so as to view scientific and engineering thinking as a complex system encompassing cognitive, social, cultural, and material aspects of practice. Environmental perspectives argue that the traditional symbol processing view has mistaken the properties of a complex, *cognitive system*, comprising both the individual and the environment, for the properties of an individual mind. In contrast to the standard construal of the environment as mental content on which cognitive processes operate, these perspectives maintain that cognitive processes cannot be treated separately from the contexts and activities in which cognition occurs. For example, in arguing for a distributed notion of cognition, Hutchins (Hutchins 1995) contends that rather than construing 'culture' as content and 'cognition' as processing, what is required is for 'cognition' and 'culture' to be seen as interrelated notions construed in terms of *process*. Such construal leads to a shift in analytical approach from regarding cognitive and cultural factors as independent variables to regarding cognitive and processes as integral to one another. One route to attaining analytical integration with scientific and engineering practices is to view problem solving in these domains as occurring withing complex *cognitive-cultural systems*.

Thus, in the research laboratory studies although as cognitive scientists we focus on what are traditionally thought of as 'cognitive' practices, these are analyzed as located within complex cognitive-cultural systems. The problem-solving activities are analyzed as situated in that they lie in localized interactions among humans, and among humans and technological artifacts and as distributed in that they are viewed as taking place across systems of humans and artifacts. Significantly, since innovation in technology and learning, development, and change in researchers are constant features of the laboratory environment, we characterize the laboratory as being and comprising evolving cognitive-cultural systems. This characterization places emphasis on the importance of the historical dimension both to the carrying out the research within the laboratories and to our analysis of how the research is conducted.

Laboratory learning. A significant part of the practice of scientists in experimental areas includes using research laboratories as sites of apprenticeship training. This is where graduate students, and more recently undergraduates, learn modes of inquiry in experimental practice. Further, in nearly all undergraduate and graduate education in science and engineering required instructional laboratory experiences make up a central part of the course of study, yet there is scant research on learning practices in these settings. Additionally, learning in interdisciplinary fields such as bio-engineering challenges students (and experienced researchers as well) to merge concepts, models, and methods from different disciplines, which often seem or are at odds with one another. Often, the research laboratory serves as a prime locale for developing practices that lead to successful melding through solving interdisciplinary problems. We have proposed, in accord with the cognitive constructivist stance, that studying inquiry practices in research laboratories could lead to development of effective pedagogical strategies for improving the instructional laboratory.

By various measures, such as student participation in conferences, journal publications, graduation rate, and kind of employment, the research laboratories we have been studying create effective environments for learning. We attribute much of this to the fact that in the face of cutting-edge science and engineering problems everyone is a learner – the undergraduates, the Ph.D. candidates, the post doctoral researchers, and the Laboratory Director. The sense of needing to innovate in ways of thinking and problem solving in order to carry out the research turns out to be a good thing for learners because it creates multiple locales for learning, for developing expertise, and for finding unique niches where neophytes in the laboratory can begin to develop expertise. The distributed nature of various kinds of knowledge in these settings is another feature that seems to serve learning. Given the complexity and interdisciplinary nature of the research problems, no individual has full knowledge or understanding, rather it is distributed across people, technology, texts, and various kinds of laboratory artifacts. For such reasons, we have come to characterize these settings as "*agentive learning environments.*" In so doing we mean to capture the idea that the laboratories are environments which allow the learner as *agent* to form numerous relationships with people, lab resources and technological artifacts, and which also cultivate the learner's ability to evoke the people, resources, and technological artifacts as mediators of, and participants in, their research projects. In using the word *agentive*, we foreground the person/learner who is characterized by her growing and changing relationships to other agents and by her developing ability to come to see the technological artifacts the laboratory creates in problem solving as "partners" in research during the processes of learning.

Significant features of the agentive learning environments we have observed are:
– That the requisite conceptual and methodological knowledge and skills are distributed enables everyone to make contributions, even undergraduate students
– In many respects the organization is non-hierarchical – no one person is held up as *the* expert, even neophytes find niches where they can contribute expertise (often around "building" or "designing"), which helps them quickly achieve a certain legitimacy and identity
– Interactional structures afforded by the laboratory allow for the rapid identification of membership routes into the laboratory which motivates learning
– Multiple social support systems foster resiliency in the face of the impasses and failures that are daily occurrences in this type of research

HOW SCIENCE WORKS?

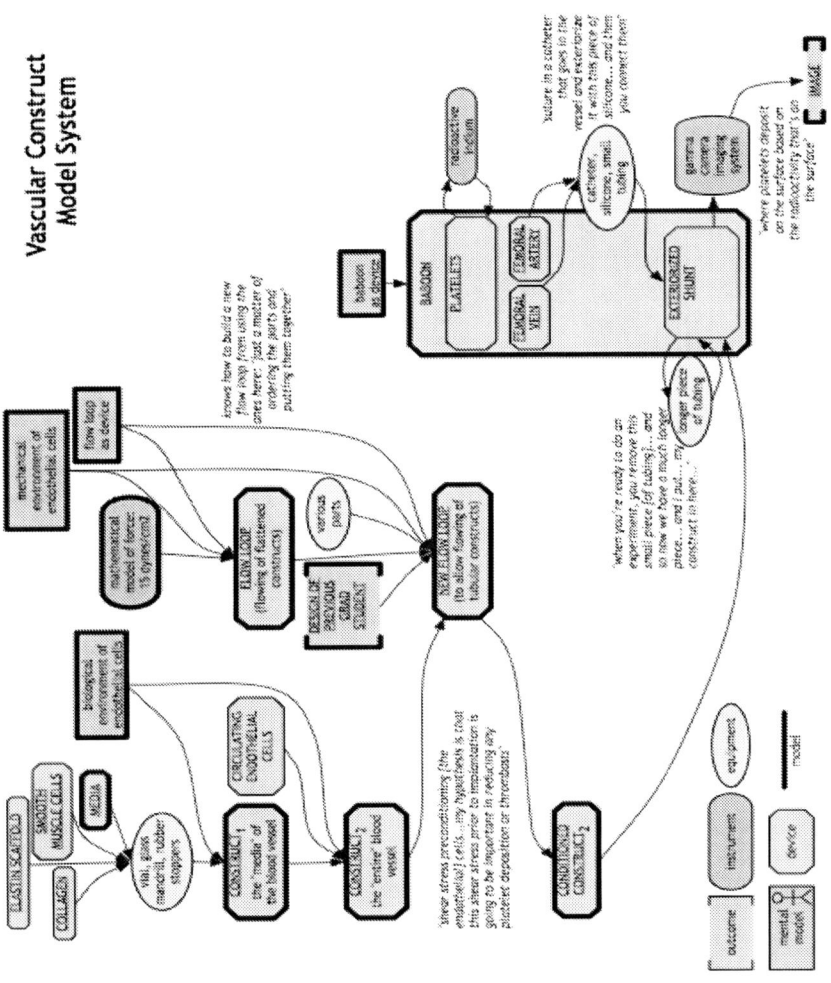

Figure 3

It is important to emphasize that although these interdisciplinary research settings focused our attention on the agentive aspect of the environment − that which encourages participants to seek out others to achieve learning, to find their own opportunities for participation, and to provide new learning opportunities for other participants − we believe it possible to design such learning environments in disciplinary and traditional instructional settings including instructional laboratories. Our current research is considering how one could transform instructional laboratories to incorporate these insights. At present we are observing current practices in two undergraduate instructional laboratories in the same areas as two of the research laboratories, tissue engineering and neuro-engineering. Additionally we are developing instruments to assess understanding of models and model-based reasoning, since our interventions will focus on these practices.

The "vascular construct model system". In this section I illustrate model-based reasoning with physical simulation devices in the tissue engineering laboratory with one example diagramed in Figure 3. Biomedical engineering is an interdiscipline in which the melding of knowledge and practices from more than one discipline is so extensive that significantly new ways of thinking and working are emerging. Significantly, innovation in technology and laboratory practices occurs continually, as does learning in laboratory researchers. Consistent with an environmental perspectives framework, we investigate the laboratory problem-solving practices as lying in localized interactions among humans and artifacts (situated) and as taking place across humans and artifacts (distributed). Thus, we characterize the laboratories as evolving cognitive-cultural systems. "Lab A" has as its ultimate objective the eventual development of artificial blood vessels (locally referred to as "constructs"). An in vivo/in vitro/ex vivo division is a significant component of the cognitive framework guiding practice in Lab A. The test bed environment for developing artificial blood vessels cannot be the human body in which they will ultimately be implanted, so the BME researchers have to design in vitro facsimiles of the in vivo environment and ex vivo implantation environments where experiments can occur. The biological mechanisms of the in vivo phenomena are known and understood both in biological and mechanical terms. The challenge is to bring together biological and engineered materials with the desired properties so as to perform properly in vivo where those mechanisms operate. The daily research is directed towards solving problems that are smaller pieces of that grand objective, such as proliferating endothelial cells within the constructs and creating constructs that can withstand the in vivo mechanical forces. The diagram in Figure 3 traces the construction, manipulation, and propagation of models within a proposed experiment aimed at reducing platelet formation and, thus, thrombosis. It is a significant experiment in that it is a first move in the direction of ex vivo research, which involves implantation in an "animal model," that is, a living creature that has been modified in some way to make experimentation possible. In this case an exteriorized shunt connecting the femoral vein and the femoral artery has been placed in a baboon (with no injury or discomfort to the animal) so that a small amount of blood flow can be diverted through a construct during an experimental process.

ONTOLOGY OF ARTIFACTS

DEVICES	INSTRUMENTS	EQUIPMENT
flow loop	confocal	pipette
bioreactor	flow cytometer	flask
bi-axial strain	mechanical tester	water bath
construct	coulter counter	refrigerator
	"beauty and beast"	sterile hood
	LSM 5 (program)	camera

computer

devices: facsimiles of the human arterial environment, engineered sites of *in vitro* simulation
instruments: extract and process information, generate measured output, enable simulative manipulation
equipment: assists with manual or mental labor

Figure 4

During a research meeting with the laboratory members, including the Director, we asked them to sort the material artifacts in the lab according to categories of their own devising and rank the importance of the various pieces to their research. Their classification is represented in Figure 4. Additional ethnographic observations have led us to formulate working definitions of the categories employed by Lab A's researchers. *Devices* are engineered facsimiles that serve as *in vitro* models and sites of simulation. *Instruments* generate measured output in visual, quantitative, or graphical form. *Equipment* assists with manual or mental labor. The devices provide locally constructed sites of experimentation where *in vitro* models are used to screen and control specific aspects of the *in vivo* phenomena they want to examine. The researchers in the laboratory call the process of constructing and manipulating these *in vitro* sites "putting a thought into the bench top and seeing whether it works or not." The "bench top", as one researcher explained, is not the flat table surface but comprises all the locales where experimentation takes place. These instantiated "thoughts" (which we interpret as mental models) allow researchers to perform controlled simulations of an *in vivo* context, for example, of the local forces at work in the artery.

Researchers refer to the experimental context where engineered devices and biological materials come together as "model-systems".

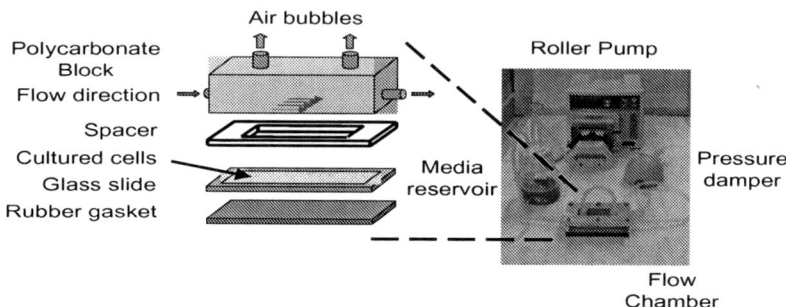

Figure 5

The simulation devices perform as models through instantiating current understanding of properties and behaviors of biological systems. For example, the *flow loop* model *(Figure 5)* is constructed so that the behavior of the fluid is such as to create the kinds of mechanical stresses in the vascular system. But these devices are systems themselves, possessing their own engineering constraints that often require simplification and idealization in instantiating the biological system they are modeling. The flow loop *represents* a first order approximation of shear stresses during blood flow in the artery. In the process of simulation, it *manipulates* constructs (Figure 6) which are tubular-shaped, bio-engineered cell-seeded vascular grafts that function as models of blood vessel walls. *Figure 3* is a schematic representation of our analysis of the *vascular construct model system* for one proposed experiment aimed at solving the problem of platelet formation on constructs, and resulting thrombosis. The models involved are highlighted by thick lines. Each physical model mimics an aspect of the cardiovascular system, for example, media and constructs mimic the biological environment of the blood vessel and the new and old flow loops mimic the shear stresses. The proposed new flow loop better approximates the *in vivo* model because constructs will be stressed in tubular shape, which is necessary for implantation. Note also that models from mathematics and physics inform the construction of the flow loop.

Finally, on our analysis the entire model system comprises interlocking models encompassing the human – artifact system, thus performing mentally as well as physically. The models in the system include
- community models of *in vivo* phenomena (biological, mathematical, mechanical)
- engineered *in vitro* and *ex vivo* physical models of aspects under investigation
- mental models of

- in vivo and in vitro phenomena
- devices as in vitro model
- devices as devices

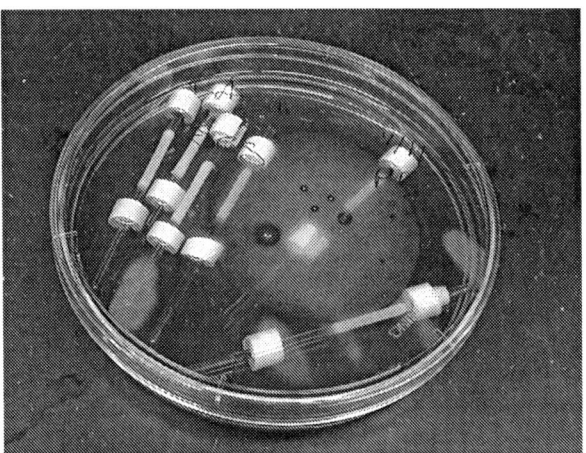

Figure 6

CONCLUSIONS: MODEL-BASED REASONING IN PRACTICE AND LEARNING

We have looked at model-based reasoning practices in quite different situations in *Section 2*. Even though the situations of practice differ considerably, there are several points that can be made about model-based reasoning across the situations. Through understanding what model-based reasoning is we can begin to determine what one needs to understand in order to be able to do to engage in it. As noted earlier, a model can be construed loosely as a representation of a system with interacting parts with representations of those interactions. For physical systems such as I have been studying, models are structural, functional, or behavioral analogs of physical objects, processes, situations, or events. Model-based reasoning is a process of problem solving via constructing models of *the same kind* with respect to salient dimensions of target phenomena (often taking several iterations to achieve this objective). Inferences are derived through *manipulation* of the model. The inferences drawn can be either specific or generic, where by 'generic' I mean that model is understood as a model-type, representing a class of models and inferences understood to apply to the members of the class. More than one instantiation/realization of a model is possible. Models are interpretations of target phenomena constructed to *satisfy constraints* drawn from the domain of the target problem and one or more source domain (e.g. Maxwell's analogical sources, the flow loop's material and engineering domains). Constraints include: spatial,

temporal, topological, causal, categorical, logical, and mathematical. Importantly, the kinds of reasoning processes include (not ordered)
- abstraction: limiting case, generic, idealization, generalization
- simulation: inferring outcomes or new states via model manipulation (mental or physical)
- evaluation: goodness of fit, explanatory power, implications (empirical, mathematical), other relevant considerations
- adaptation: constraint satisfaction, coherence, other relevant considerations

Translating this understanding of model-based reasoning in scientific practice into effective strategies for learning will require reflecting on specific questions. Among these are:

What is required to engage in model-based reasoning? What cognitive resources/capacities do children have from which to develop understanding of models and modeling? How can we assist learners in developing facility with model-based reasoning over time? How to translate the contextual nature of problem solving in science to the contexts of learning environments?

Finally, in making any translation is important to understand that model-based reasoning practices are cognitive and cultural achievements − and play roles in both. For example, learning about specific models forms part of the initiation rights into a community of practice, such as the physical simulation models in the biomedical engineering laboratory. Forms of modeling in practice are fit to the specifics of the requirements of the domain and location of the problem solving (e.g, a theoretical physics research group vs a biomedical engineering laboratory vs a psychology laboratory). The cognitive-cultural system of the classroom or instructional laboratory have their own unique constraints and affordances that need to be figured into the development of strategies for learning and using model-based reasoning. For example, in the K-8 curriculum, the kinds of models one can develop are likely to be quite different from those a scientist would use. However, the point is that the *kinds of reasoning processes* should aim to approximate those of a scientist. A good example here is Marianne Wiser's (1995) "dots-in-a-box" visual analogical models for teaching thermodynamic concepts.

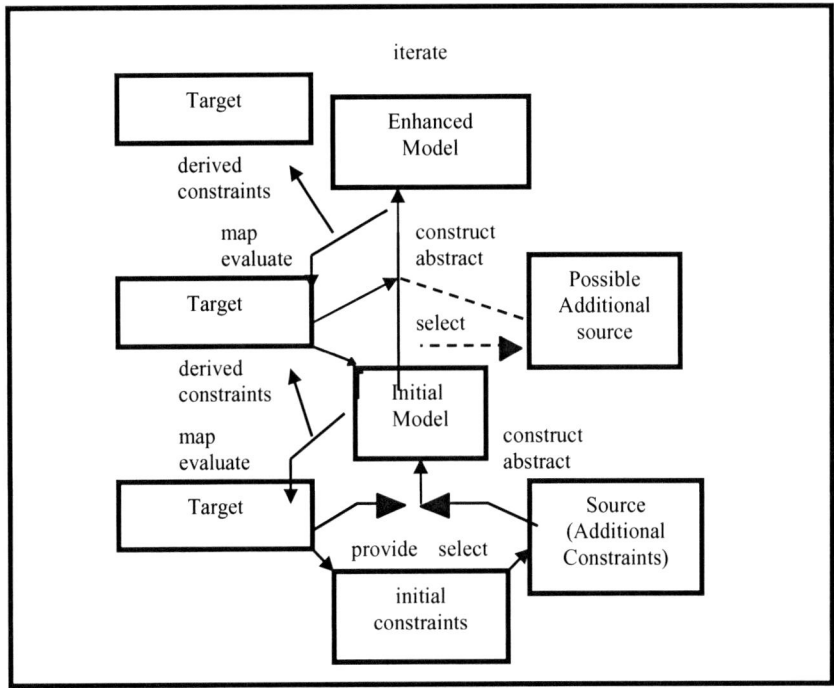

Figure 7

NOTES

[1] I will use 'cultural' as shorthand for 'social, material, and cultural' throughout as a matter of convenience. 'Social', 'cultural', and 'material' are, of course, not coextensive notions and analyses of these dimensions of scientific practice are quite diverse in the literature.

[2] One important exception is Kevin Dunbar's research on cognitive practices molecular biology laboratories. See, e.g., Dunbar (1995, 1999).

[3] This research has been conducted with Wendy Newstetter (co-PI), Elke Kurz-Milcke, Lisa Osbeck, Kareen Malone and several graduate and undergraduate students. Within our research group there is expertise in ethnography, philosophy of science, history of science, psychology, linguistics, and computer science. We thank our research subjects for allowing us into their work environment and granting us numerous interviews. We gratefully acknowledge the support of the National Science Foundation ROLE Grants REC0106773 and REC0450578.

Nancy J. Nersessian
College of Computing
Georgia Institute of Technology

FOUAD ABD-EL-KHALICK

MODELING SCIENCE CLASSROOMS AFTER SCIENTIFIC LABORATORIES

Sketching Some Affordances and Constraints Drawn from Examining Underlying Assumptions

This "Inquiry Conference on Developing a Consensus Research Agenda" is both timely and significant. Teaching pre-college science *as* inquiry and *through* inquiry is without doubt the central theme of major current reform documents in science education (e.g., American Association for the Advancement of Science [AAAS], 1990, 1998; Millar & Osborne, 1998; National Research Council [NRC], 1996). However, nothing like a consensus presently exists on what inquiry specifically means or how inquiry is to be specifically enacted in the context of pre-college science classrooms so as to achieve a variety of educational outcomes that fall under the rubric of 'scientific literacy for all students' (Abd-El-Khalick et al., 2004). Working toward a consensus research agenda on inquiry in science education is surely a worthwhile endeavor.

Despite a lack of consensus, a persistent theme cuts across discussions of inquiry in science education, namely, *authentic* scientific inquiry, which "refers to the research that scientists actually carry out" (Chinn & Malhotra, 2002, p. 177). Indeed, it is safe to claim that the underlying thesis of such discussions and probably the present volume is that: It is best that students learn science and about science through the enactment of authentic scientific inquiry in pre-college science classrooms (e.g., AAAS, 1990; Edelson, 1998; NRC, 1996). A majority of science educators shares Nerssesian's (this volume) belief that "studying inquiry practices in research laboratories can lead to development of effective pedagogical strategies for improving the instructional laboratory" (p. 16). (It should be noted that Nerssesian did not specify the level – e.g., pre-college or college science laboratories, where these pedagogical strategies can be put to effective use.) No one disputes the need to radically change current science teaching practices, which we know fail school students on multiple levels (AAAS, 1990; NRC, 1996). "Inquiry" qualifies as a promising alternative. However, a set of major assumptions underlie and probably undercut the thesis on realizing authentic inquiry in pre-college science classrooms. These assumptions include: (a) the "intrinsic motivation" assumption: Engagement with authentic scientific inquiry is inherently motivating and, thus, sufficient to actively engage children with learning science as inquiry; (b) the "child-as-little-scientist" assumption, which can be thought of in two ways. First, children can actually do authentic science, that is, can engage the same kinds of activities undertaken by practicing scientists (differences are only a matter of degree and not kind). Second, children learn in a way similar to how scientists construct scientific knowledge. (Note that this latter assumption practically renders inquiry a theory of learning.); and (c) the "knowledge-about-

Richard A. Duschl and Richard E. Grandy (eds.), Teaching Scientific Inquiry: Recommendations for Research and Implementation, 80–85.
© *2008 Sense Publishers. All rights reserved.*

science" assumption: We know how the scientific enterprise actually works, how scientific knowledge is actually created and validated, and what scientists actually do (compared to *models* of the cognitive practices of scientists and the social workings of science).

Making assumptions – a necessary component of any undertaking, is not itself an issue. Taking the above assumptions for granted might be. Scrutinizing them, I believe, could serve as a fruitful framework for discussions in the present conference. It should be noted at once that accepting or rejecting any of the aforementioned assumptions should not occupy us. This is because the answer as to whether these assumptions are valid is a resounding "yes" and "no." First, the intrinsic motivation assumption is somewhat problematic. The assumption is similar to that underlying the orthodox (Posner, Strike, Hewson & Gertzog, 1982) or 'cold' conceptual change model (Pintrich, Marx & Boyle, 1993), which assumed that cognitive dissonance was intrinsically motivating and sufficient to initiate and sustain the sort of active student engagement needed to achieve conceptual reorganization. This assumption did not necessarily hold true (Tyson, Venville, Harrison & Treagust, 1997) and cognitive dissonance was downplayed in later renditions of the conceptual change model (see Hewson, Beeth & Thorley, 1998). As Edelson's (this volume) work indicates the intrinsic motivation assumption cannot be taken for granted because students respond differently to their engagement in inquiry learning environments. From another perspective, while many of my scientist friends often express increasing enthusiasm about and satisfaction with their work, every academic year, I end up advising a number of graduate science students who decided to pursue careers in education. Any one of these students could have successfully pursued a career in science. But many indicate that research (i.e., authentic scientific inquiry) was "not for them" and provide any of a number of underlying reasons for their decision. Obviously, authentic science is not intrinsically motivating, at least, not for all graduate science students.

Second, in a sense the child-as-little-scientist assumption cannot be true because a *child* simply is not a *scientist*. Brewer (this volume) noted that if pushed to an extreme, the analogy between child and scientist breaks down. Yet, even infants and young children exhibit some behaviors that are not different from those of scientists (Gopnik, Meltzoff & Kuhl, 1999). Indeed, Brewer continued, the analogy could be salvaged if approached from a cognitive perspective: "If one takes an extreme social constructive view of science, the game is over. However, if one adopts a view that considers science to include both social and cognitive aspects ... then the question of the validity of the child scientist analogy is still open" (p. 3). But forgoing the social dimension of the construction of scientific knowledge and limiting the analogy between the child and scientist to the cognitive dimension limits the utility of scientific inquiry as a learning theory because "the social and personal elements in the construction of meaning ... are indissolubly complementary" (Solomon, 1987, p. 64): As Nersessian put it, "the cognitive practices of science should not be divorced from social, cultural, and material factors that both enable and contribute to their development" (p. 3). The lack of robust modalities that functionally link the cognitive, social, cultural, and affective aspects of conceptual change place serious limits on the utility of the child-as-scientist assumption (Abd-El-Khalick & Akerson, 2004).

Third, while it is safe to claim that cognitive, philosophical, historical, sociological, and anthropological studies of science have resulted in increasingly robust understandings of the working of science, the knowledge-about-science assumption remains, at best, partially true as evident by continued and intense scholarship within and across these fields. To cite just one example, Nersessian (this volume) noted that "although there has been abundant research on social and cultural practices in research laboratories ... there has been little that focuses on the cognitive practices employed in problem solving" (p. 16).

What is more, the above assumptions interact and generate other consequences and requirements for the realization of authentic inquiry in pre-college classrooms. First, is what I call the "composite scientist" consequence. To start with, various scholarship activities explore and characterize aspects of the work of scientists and science from different perspectives (cognitive, historical, sociological, cultural, etc.). What is more, the child is often compared with *both* the adult scientist and the cultural institution of science (Brewer, this volume). When these two facets of the approach to comparing authentic science and school science are coupled with the 'egalitarian' ideals (e.g., science for all) and practices (e.g., all students need to demonstrate mastery of the same set of educational outcomes) of school science education, the comparison leads to the creation of a composite and probably unrealistic amalgam of what the child-scientist should be able to do: The set of all activities that an "idealized scientist" practices but that no one scientist really or necessarily engages. This situation is somewhat similar to the images of the "effective" teacher that were derived from the process-product research studies of the 1970s and early 1980s. A host of independent studies identified discrete attributes of teachers and aspects of their practice that correlated with increased student achievement, which were then collated and presented as the basis for preparing effective teachers, even though teachers who embodied all these attributes and enacted all these practices never really existed. Not only does the composite-scientist consequence place unrealistic expectations on children, it also runs counter to the increasingly interdisciplinary nature of scientific research, where – unlike an often espoused homogeneity in terms of students' understandings and skills, scientists bring different expertise and skills to bear on collaborative problem solving (see Nersessian, this volume).

Second, the above three assumptions entail the need to refashion the science classroom as a microcosm of the scientific enterprise. The differences between the culture of school science and the culture of science are too well known to be reiterated here. Suffice to say is that attempts to fashion classroom or school culture after that of science will come in serious friction with a set of curricular, organizational, social, etc., practices and expectations that will necessarily place limits on the nature of inquiry that could be enacted in pre-college classrooms. Edleson's (this volume) work provides a glimpse into the sort of practical constraints (e.g., school day structure) and issues (e.g., teacher professional development) that face the enactment of inquiry learning environments in science classrooms. The issue, however, is larger than classrooms and extends to the whole school culture. For example, Norris and Phillips (this volume) presented a compelling argument to the effect that "reading as inquiry" is an essential component of authentic scientific inquiry. Fostering reading-as-inquiry in science classrooms (if and when attempted) requires some synergistic practices in language arts and social studies classrooms. However, "a simple view of reading" that usually dominates school culture detracts from, rather than support student engagement with authentic inquiry. In a sense, the enactment of authentic inquiry in pre-college classrooms entails refashioning the whole school culture.

The above discussion is illustrative, at best. The point that I want to make here is that whatever sort of inquiry learning environment we end up successfully realizing in pre-college science classrooms – though necessarily superior to current science teaching practices, will be *different* from authentic scientific inquiry. However, the *form* of inquiry that simultaneously (a) approximates authentic science, (b) is viable in pre-college science classrooms, *and* (c) has the potential to achieve a set of major educational goals still eludes us. It is this form that we should be concerned with characterizing and researching. Put differently, I believe that we should be less concerned about whether the above assumptions are true or not, and more concerned with the *specific* ways in which the assumed parallels between child and scientist and between school culture and the culture of science make or break. (This is surely a case where the devil is in the details.) Asking and answering questions within this framework will illuminate and characterize the sort of scientific inquiry that could be fruitfully enacted in pre-college science education. This conference is the right forum to do just that by forging a consensus research agenda with the above target in mind: In this volume many insightful and specific questions are being asked and significant issues are being tackled. For example, Duschl and Grandy (this volume) ask, "Where in the trajectory of science education we should introduce . . . complications of simultaneous revision of theory and data"? (p. 18), "how to contend with the transition from sense perception based science at early grade levels (e.g., K-3) to theory-driven based science at higher grade levels (e.g., 4-12)"? (p. 24), and "how do we conceptualize distinctions between perceptual and non-perceptual or hidden observational phenomenon in science lessons"? (p. 25). Brewer (this volume) scrutinizes ways in which the child and scientist are similar and ways in which they differ. Chinn and Samarapungavan (this volume) examine the several classes of difficulties associated with conceptual change and learning about and using scientific models. Of course, an additional significant dimension would be to ascertain the sort of (tacit and explicit) epistemological commitments that are prerequisite (or, at least, significant) to meaningful engagement with inquiry and ways in which students and science teachers come to internalize such commitments (Abd-El-Khalick, 2002). I believe that these and other questions and issues raised in this conference could be viewed from, and probably unified, within the context of closely examining the above assumptions.

In her paper, Nersessian (this volume) outlined an extremely important research agenda that has significant implications for the issues at hand. She argued that accounts of scientific practice "tend to line up on either side of a perceived divide between cultural [meaning social, cultural, and material] factors and cognitive factors in knowledge construction, evaluation, and transmission," and continued that "combined research on the cognitive and cultural sides shows the divide to be artificial" (p. 1). With her research group, Nersessian is investigating contemporary scientific practice within interdisciplinary university research laboratories in the biosciences and bioengineering with the aim of producing integrative accounts of scientific practice in which both cognitive and cultural factors are centerpiece and neither is simply tacked on. Such accounts, she argued, can eventually be very instructive in generating pedagogical practices aimed at improving science instruction. In this endeavor modeling and model-based reasoning practices (which she noted are not exhaustive of scientific practice) serve as a major focus because they "are the signature of much research in the sciences, both in discovery and application" (p. 7). In addition to using constructs of "situated cognition" and "distributed knowledge," she adopted an alternative approach to tackling the task of generating an integrative account: "Rather than construing culture as content and cognition as processing, what is required is for 'cognition' and 'culture' to be seen as interrelated notions construed in terms of *process*. Such construal leads to a shift in theoretical outlook from regarding cognitive and socio-cultural factors as independent variables to regarding cognitive and socio-cultural processes as integral to one another" (p. 13).

The aim of generating a practice-based modality that functionally integrates the roles of cognitive, social, and cultural factors in conceptual change in science is admirable. If successful, such a modality could help overcome some of the most crucial difficulties for the "learning as conceptual change" model (e.g., Posner et al., 1982; Hewson et al., 1998), namely, difficulties associated with solely focusing on the cognitive dimension to the neglect of social and cultural dimensions of learning (Alsop & Watts, 1997; Cobern, 1996; Costa, 1995; Solomon, 1987). Of course, even then, the implications of that much desired modality for conceptual change in pre-college students would still need some "figuring out." This latter process, again, would entail scrutinizing the parallels between some central aspects of practice in scientific laboratories and learning in pre-college science classrooms. For example, to what extent is the notion of apprenticeship, which is central to learning in scientific laboratories, comparable to the sort of best-case-scenario activities that could be undertaken in pre-college classrooms? To what extent is the distributed nature of knowledge created by the interdisciplinary collaboration of scientists and graduate students (with expertise in, at least, one field or another) in scientific laboratories applicable to the context of science classrooms with students who are relatively novices in a host of domains? (See Nersessian, this volume, p. 6 and p. 16.) Of course, this latter task does not at all distract from the significance of the research agenda articulated by Nersessian. These questions are just meant as a reminder of the major thesis outlined in this commentary, namely, that after having figured out how science and scientists work, we still need to figure out how such knowledge and understanding apply in the case of schools and children because assumptions about straightforward parallelism between the two contexts are likely to meet with some harsh reality checks.

Fouad Abd-El-Khalick
Department of Curriculum and Instruction
University of Illinois at Urbana-Champaign

MIRIAM SOLOMON

SOCIAL EPISTEMOLOGY OF SCIENCE[1]

HOW SOCIAL EMPIRICISM DEVELOPED

I first met Richard Grandy in December 1985, when he interviewed me for a position at Rice University and I was a nervous graduate student. He opened the conversation with an aggressive question: "Ten years ago, everyone was writing a dissertation on Quine. Why are you writing a dissertation on Quine now? Aren't you a little late?" Taken aback, I mumbled something about it being more important to get Quine right than to do so in a timely way. This was true, but not the whole truth. I can answer the question better now. I was preparing myself for a career in naturalized epistemology of science, indeed preparing myself to out-Quine Quine himself. Moreover, for some reason, worth speculating about, writing a dissertation on Quine has been a common starting point for those who later became feminist (or feminist-sympathetic) philosophers of science.[2] Here is the list (there may be more, but these are those I know): Sandra Harding, Alison Jagger, Lynn Hankinson Nelson, Nancy Tuana, and myself.

In "Epistemology Naturalized," Quine wrote that "epistemology ... becomes a branch of psychology" (Quine, 1969, p. 82). This paper, which is a classic statement of naturalized epistemology[3], rejects a priori and traditional analytic approaches to epistemology. As psychology developed in the 1980s on, with the growth of the cognitive sciences, philosophers were hopeful that they would get new resources for epistemology. This was especially true in philosophy of science, which was struggling to respond to the challenges of Kuhn and the Strong Programme in sociology of science. Ronald Giere was even explicit about this: he hoped that if he could explain scientific change using cognitive mechanisms, then he would not need to appeal to social mechanisms and that this would mean that concerns about rationality would be addressed (see especially Giere, 1988, Chapter 1).

I was baffled by this approach, in two ways. First, while I agreed with Quine and his followers that epistemology should be naturalized, I did not see that psychology (whatever branch) should be the favored science, rather than, say, sociology or anthropology. In my view, Quine's arguments led to the conclusion that whatever science was best for understanding human knowledge – whether it was the science of individuals (psychology), neurons (neuroscience), societies (sociology) or cultures (anthropology, history) – was relevant to epistemology. Secondly, "cognitive" – even "cold cognitive" – does not equal "rational"; and "social" or "motivational" does not equal "irrational." My work on cognitive heuristics and scientific decision making, which led to my first foray into social epistemology, was intended to show this.

In Solomon (1992), I investigate the so-called "biasing" cognitive heuristics such as salience, availability and representativeness investigated by Tversky, Kahneman, Nisbett, Ross, and others. They are for the most part "cold" cognitive

Richard A. Duschl and Richard E. Grandy (eds.), Teaching Scientific Inquiry: Recommendations for Research and Implementation, 86–94.
© 2008 *Sense Publishers. All rights reserved.*

(except for emotional salience and availability) and are pervasive in ordinary thinking and deliberation. I asked the question, are scientists affected by these heuristics, not in artificial test situations, but when they have real complex decisions to make? I took this question to the historical case of the beginning of continental drift theory, in the 1920s, and found patterns of distribution of belief and cognitive labor strongly indicative of the operation of heuristics of salience, availability and representativeness. For example, Giere cites a study in which 91% of geologists with southern hemisphere experience (who had traveled in the southern hemisphere) who wrote about drift theory supported it, while only 48% of those without southern hemisphere experience who wrote about drift supported it (Giere, 1988, p. 240). Much of the evidence for drift, at that time, was in the southern hemisphere. Those who had seen – rather than just read about – the geophysical and paleontological similarities between formerly joined land masses were far more likely to take drift seriously (salience heuristic). And, for example, those who already accepted that the earth's crust could move horizontally, as it seemed to do in forming the folds of the Alps, were far more likely to accept drift: Swiss geologists generally accepted drift (salience and representativeness heuristics). Drift was mostly rejected in the USA, where salient evidence for drift was not available.

Individual scientists, then, reasoned with what Tversky, Kahneman, Nisbett, Ross and others would call "cognitive bias." (And this is before we add pride, competition and other motivational bias to the picture, and these also influenced decision making.) Should we bring in the Rationality Police? It's hard to get steamed up about these skews in decision making, *especially if we take a social point of view* and reflect that, in this case, different scientists ended up being skewed in different directions and a division of intellectual labor resulted, with some scientists (primarily in the USA) working on stabilism and some scientists (primarily in Europe) working on drift. I prefer to use the neutral term "decision vector" rather than "bias" when describing factors affecting individual and group decision making, reserving the normative judgments for later.

This is the first step in the normative point of view I develop called "social empiricism." It is important to emphasize that it is not an "invisible hand of reason" view: that is, I do not claim as an article of faith or a priori calculation that always or even typically an epistemically good division of labor results from the skewing effects of decision vectors. In fact, in my book, *Social Empiricism*, I demonstrate this by presenting cases (such as the early history of genetics, and the theory that cancer is caused by a virus) in which division of epistemic labor could have been much better.

Social empiricism is a species of social epistemology. Social epistemologies have in common an interest in the social dimensions of knowledge (this is the definition of "social epistemology" from Goldman, 2001), and those committed to the normative mission of epistemology (as I am) need to work out what happens to epistemic advice when the epistemic subject is no longer (only) an individual. Other social epistemologies will be discussed later in this paper.

FROM DISSENT TO CONSENSUS

The second step of "social empiricism" is more difficult to understand and to explain, but it is the crux of my view, and it concerns consensus in science. In fact, it is the extension of my views for dissent in science to consensus in science – both the formation of consensus and the dissolution of consensus. I should point out that epistemologists of science do not usually treat dissent and consensus in the same way. Typically, dissent is viewed as a time when so-called "biases" are tolerated in the interests of division of intellectual labor – Kitcher, Goldman and others have put forward ideas similar to the ones I outlined above for dissent. They expect, however, that in due course, one scientific theory will become a clear winner, and all scientists (except perhaps those too "set in their ways" to change – who are therefore regarded as irrational) will recognize its superiority in terms of its superior explanatory power, greater predictive success, or whatever. For those familiar with the terminology, dissent is treated like the old "context of discovery" (where "anything goes") and consensus like the old "context of justification" (where universal logical rules must be followed). In looking at actual cases, however, I do not find that consensus forms in such a simple and general way.

The case of consensus on continental drift – or, more precisely, on its successor, plate tectonics – illustrates the distributed nature of consensus, and the need, again, to take a social perspective in order to make normative judgments about the appropriateness of consensus. Consensus on plate tectonics took place during the 1960s. Historians of the period have made remarks such as "most specialists were only convinced by observations related to their specialties" (Menard, 1986, p. 238). The order of consensus formation was paleomagnetists, oceanographers, seismologists, stratigraphers and then continental geologists concerned with paleontology and orogeny. Belief change occurred after observations confirming mobilism were produced in each specialty, and after old data were reinterpreted to be consistent with mobilism. In addition to this general pattern, there were patterns of belief change and belief perseverance due to prior beliefs, personal observations, peer pressure, the influence of those in authority, salient presentation of data and so forth.

All these patterns can be explained in terms of decision vectors (my preferred term for what others call "biasing factors"[4]). Decision vectors are so called because they influence the outcome (direction) of a decision. Their influence may or may not be conducive to scientific success, hence the epistemic neutrality of the term. Moreover, the decision vectors are similar in kind, variety and magnitude to those causing dissent in other historical cases. The result here is consensus, rather than continued dissent, because enough of the decision vectors are pulling in the same direction.

It is important to note that although there *could* have been a crucial experiment/crucial set of observations, there was, in fact, not. The paleomagnetic data, especially the data showing symmetric magnetic striations either side of deep ocean ridges found by Vine and Matthews in 1965, could have been viewed in this way. But it was viewed as crucial data only by Vine, Matthews and a few well-placed oceanographers. Paleomagnetists tended to be persuaded earlier, and seismologists, stratigraphers and continental geologists somewhat later. No single experiment was "crucial" for more than a group of geologists at a time. The consensus was gradual, and distributed across subfields of geology, countries, and informal networks of communication. Consensus was complete when plate tectonics had all the empirical successes (those that permanentism, contractionism and drift had, and more besides) – but not because any individual or individuals saw that it had all the successes. Moreover, empirical success was not sufficient to bring about consensus. Decision vectors (sometimes tied to empirical success, sometimes not) were crucial.

Consensus is not always normatively appropriate. Sometimes scientists agree on theories that do not deserve universal support. Consensus on the "central dogma" that DNA controls cellular processes, via messenger RNA and protein synthesis in the 1950s left out important work on cytoplasmic inheritance, genetic transposition, genetic regulation, and inherited supramolecular structures (work of Ephrussi, Sonneborn, Sager, McClintock and others). Sometimes scientists agree on theories that do not deserve any support at all. For example, the extracranial-intercranial bypass operation, developed in 1967 by Gazi Yasargil, became the standard treatment for stroke patients before there was any robust data for it. The operation became "the darling of the neurosurgical community" because it required special training and was very lucrative. The rationale was good enough: since the blood vessels feeding the scalp rarely develop atherosclerosis, it was reasoned that the suturing them to arteries that feed the brain would increase blood flow to the brain and thus help to prevent strokes. In fact, good supportive data was never produced and the surgery fell out of favor after 1985. But that was almost 18 years of consensus on a theory with no empirical success.

Consensus is not the *telos* of scientific research. Consensus is not a marker for truth or even for "happily ever after." So I also look at the dissolution of consensus from a social perspective, and use the same normative framework, social empiricism, to make normative judgments. Cases I discuss in my book include the dissolution of consensus that cold fusion in a test tube is impossible, and the dissolution of consensus that ulcers are caused by excess acid. It's time to state the normative framework of social empiricism – not to justify it here, because there is not time, but to state it, so that you can see that it is a normative framework:
Three conditions for a normatively appropriate dissent:
1. Theories on which there is dissent should each have associated empirical success

2. Empirical decision vectors[5] should be equitably distributed (in proportion to empirical successes)
3. Non-empirical decision vectors should be equally distributed (the same number for each theory)

Note that consensus is the special case in which the amount of dissent goes to zero.

Three conditions for a normatively appropriate consensus:

1.[1] One theory comes to have all the empirical successes available in a domain of inquiry
2.[1] This same theory comes to have all of the empirical decision vectors, since all scientists working productively (with empirical success) are working within the one theory.
3.[1] Any distribution of non-empirical decision vectors is OK, but typically more will develop, over time, on the consensus theory, as the old theories fade away. During dissent, and thus in the early stages of consensus formation, dissent's requirement of equal distribution of non-empirical decision vectors holds.

Three conditions for a normatively appropriate dissolution of consensus:

1.[11] A new theory has empirical success that is not produced by the consensus theory. (So, the new theory deserves attention.)
2.[11] Empirical decision vectors come to be equitably distributed.
3.[11] Non-empirical vectors come to be equally distributed.

The justification for these conditions is not given here, but it rests on the superior outcomes (better scientific results) when the conditions are satisfied. That is, the justification is instrumental (some would use terms such as "teleological" or "consequentialist"). Case studies are used to show this.

Social empiricism evaluates aggregate decisions on a continuum between ideally normative, in which there is a perfect distribution of decision vectors over empirically successful theories, and completely normatively incorrect, in which theories with empirical success are completely ignored and theories without empirical success are championed. Most of the time, the distribution of research effort is somewhere between these extremes. This leaves room for normative suggestions for improvement.

Social empiricism demands *less* than most epistemologies of science. In particular, it does not require individual scientists to make overall impartial assessments. Most traditional epistemologies of science, and even most social epistemologies of science, require that at some stage in scientific reasoning (typically, consensus) individual scientists make overall impartial assessment of the merits of competing scientific theories. Social empiricism does not require this. What matters is not how individual scientists reason – *it's* not *the thought that counts* – but what the aggregate community of scientists does. It is more realistic to expect normative suggestions to yield improved distributions of research effort than to expect them to correct individual scientists of their partiality.

Social empiricism also demands *more* than most epistemologists of science. It requires a social level of evaluations and recommendations. It is no longer enough for individual scientists to strive to be perfectly rational, reasonably Bayesian or properly cognitive (to give some examples of contemporary individualist epistemologies of science). It is not even enough for individual scientists to check and correct one another's work and consult with experts (a common prescription of most social epistemologists). Rather, and instead, what is required is evaluation and change at a systems level. What is required in remedying poor distribution of decision vectors is what I call *epistemic fairness*.

This is all I will say about *social empiricism* here. It's almost time to place it in the context of competing perspectives. First, a comment about the educational implications of *Social Empiricism*, especially at the grade school level. Since I say things like, "It *isn't* the thought that counts," and I tolerate all sorts of individual bias, indeed positively incorporate it with the term "decision vector," what kind of message would *Social Empiricism* send in a high school science class? Students would demand A's, on the grounds that their errors were due to natural bias ("hey, the geologist Wegener also reasoned by using a terrible analogy (as a result of the representativeness heuristic)!"). Students could argue, furthermore, that in a natural community of scientists this natural bias can have beneficial effects ("my friend is biased another way, so together we have a distribution of effort etc"). And would they be wrong? This concern is continuous with Stephen Brush's 1974 question in *Science*, "Should the history of science be rated X?" (some scientists are not good role models for recruiting future scientists) and Douglas Allchin's 2004 question in *Science Education*, "Should the Sociology of Science Be Rated X?" (can grade school students understand sociology of science without losing their respect for the objectivity of science?). So, I suppose I am asking, "Should the social epistemology of science be rated X?" (Can grade school students be taught the social epistemology of science without losing individual epistemic responsibility?"). My instinctive answer is NO! (It should not be rated X.) And YES! (Grade school students can be taught the social epistemology of science without losing individual epistemic responsibility.) This instinctive answer comes from my general commitment to telling children the truth, rather than from any specific ideas about how to teach social epistemology at the pre-college level. I hope that is an interesting challenge to educators, to teach the social epistemology of science in a meaningful way to grade school students; I look forward to hearing what Nancy Brickhouse has to say.

OTHER SOCIAL EPISTEMOLOGIES OF SCIENCE

Many who work in the area of "social epistemology of science" do not take the "social" part very far. It is common, for example, to explore "social processes" or "social factors" such as collaboration, criticism, expert consultation, and competition and their conduciveness to individual rationality and scientific success. (See e.g. Goldman, Kitcher, Kornblith, Thagard for examples of this.) This is a useful enterprise, but it should be noted that the normative standpoint is still individual. Some (Kitcher, Goldman, Hull) go a little further, and allow a social normative standpoint at times of dissent, when what is needed is an appropriate distribution of cognitive labor, and not any particular decision by any particular individual. This is why I think of social empiricism as a *more* social epistemology than most. Few social epistemologies are willing to bite the bullet and take a consistently *social* normative stance (for both dissent and consensus). Most that do are feminist social epistemologies. Helen Longino's work (Longino, 2002) is one of these, and I want to take the time to explain it, because, apart from taking a social normative stance, it is very different from *social empiricism* and is a useful contrast for our discussions (also because Helen is here and can set me right!).

Longino's social epistemology of science is currently referred to by her as "critical contextual empiricism" (henceforth CCE). There are four CCE norms, which Longino claims are obtained by reflection on the meaning of "objectivity". These norms are:
1. equality of intellectual authority (or "tempered equality" which respects differences in expertise)
2. some shared values, especially the valuing of empirical success
3. public forums for criticism (e.g. conferences, replies to papers in journals) and
4. responsiveness to criticism.

To the degree that a scientific community satisfies these norms and also produces theories (or models) that conform to the world, it satisfies conditions for scientific knowledge. Note that the norms are satisfied by communities, not individuals, although some individuals have to satisfy some conditions (e.g. they have to be responsive to particular criticisms).

The result of satisfying these four norms, according to Longino, is, typically, pluralism. Empirical success is the one universally shared value in scientific research, but it comes in many forms and is not sufficient to arbitrate scientific disputes. Thus theories are underdetermined by the available evidence and more than one theory of a single domain can be empirically successful. Different theories can be empirically successful in different ways. And the best criticism usually comes from scientists who are working on different theories. Hence pluralism is the typical and the preferable state of scientific research.

Longino's account has a certain intuitive plausibility. The requirements of greater intellectual democracy would seem to give all scientists a fairer hearing, thus increasing the chance that good ideas will be considered. It also has moral appeal. Everyone gets a chance to speak and every genuine criticism gets a response.

Compared with *Social Empiricism*, Longino demands much of scientists. In fact, she demands more than traditional logical empiricists. Her acceptance of values and ideologies in science is not (as it is in *Social Empiricism*) an acknowledgment of their inevitability so much as it is the result of an argument (in chapters 1-3) for their logical necessity. Longino expects (I do not) that "individual biases" can be, and should be, eliminated by social criticism. That is,

Longino employs a distinction between ideological factors (which she celebrates as "values") and individual factors (which she regards, together with more traditional epistemologists, as "biases" to be eliminated). Longino's requirements of "equality of intellectual authority" and "responsiveness to criticism," are more than Hempel, say, would ask for.

Longino has said that CCE has intuitive plausibility. The more important question, in my view, is whether or not it is supported by naturalistic investigation. That is, Longino needs to show that the closer that scientists are (in fact) to satisfying the norms of CCE, the better the science (in fact) is. Longino is beginning to consider case studies to show this; in conversation she has evaluated recent feminist work in primatology for how well it satisfies the CCE norms.

When I read Longino's work, my own work seems cynical in contrast. I do not expect scientists to be able to establish that much of a democratic community; I think many of them do not listen and certainly do not respond well to criticism; and I do not think it important to distinguish – for *epistemic* purposes – between (ethical/ideological) "values" and (individual) "motives." *As philosopher of science*, the primary epistemic goal is empirical success, and, for this, the norms of *social empiricism* suffice. Longino may be writing with feminist political goals as much as she is writing with epistemic goals. If the goals are also political, there are other constrains, and the CCE norms may be more appropriate. Nevertheless, we need to see that the CCE norms are a realistic goal – that humans *can* attain them, and that they are no less elusive than the standards of logical empiricism.

Longino's ideas would do well in the grade school classroom. Students could be taught "responsiveness to criticism," as more than an empty recital of "thank you for your criticisms, I respect where you are coming from." They could be encouraged to think about the appropriate level of intellectual authority to grant their teachers. The only danger would be overwork for the conscientious student: responsibility for themselves and responsibility for the epistemic democracy of the community.

CONCLUSIONS

Social epistemologies of science go beyond acceptance of social processes in science. They acknowledge that, to a greater or lesser extent, our epistemological accounts of science need to change in order to accommodate the findings of social processes in science. I have focused on the more social epistemologies of science because I am persuaded by case studies in the history of science that these are needed. If goals are purely scientific, I argue for my own *Social Empiricism*, although I think it is a creative challenge to develop the ideas for grade school classroom use (because it is aimed at the system/policy level). If goals are also political, Helen Longino's critical contextual empiricism may be more suitable, and it has the advantage of having more straightforward applicability at the individual level.

NOTES

[1] Some passages in this paper are reprinted from my book, *Social Empiricism* (MIT Press, 2001). This is done with permission of the MIT Press.
[2] Quine has also been a common point of return for feminist thinkers. Think of Louise Antony (1993) and Nelson & Nelson's (2003).
[3] It is a classic statement of naturalized epistemology, but not an especially precise one. Which branch of psychology? And does Quine intend that epistemology become a descriptive (rather than a normative) enterprise?
[4] Well, not quite. The precise definition of decision vectors in *Social Empiricism*: "anything that influences the outcome of a decision" is wider than the scope of "biasing factors." This difference is not important here. See a more precise discussion in footnote 5.
[5] Empirical decision vectors are causes of preference for theories with empirical success, either empirical success in general or one empirical success in particular (e.g. a salient empirical success, or a preference for theories with novel empirical success). Non-empirical decision vectors are other reasons or causes for success (e.g. preference for theories with hierarchical ideology, preference for simpler theories, holding onto a theory because of pride). These concepts are discussed at length in the book *Social Empiricism*.

Miriam Solomon
Department of Philosophy
Temple University

NANCY W. BRICKHOUSE

SHOULD THE SOCIAL EPISTEMOLOGY OF SCIENCE BE RATED X?

Should Social Epistemology be Rated X? Miriam Solomon raises the question and then answers with a "NO! and YES!" She's right on both counts. But perhaps I should explain. One of the major purposes of this conference is to try and reach some consensus on what inquiry is. Miriam Solomon's paper contributes to this effort by describing a social epistemology of science that uses empirical evidence of scientific practices to consider attainable norms for scientific inquiry. The cases from which she draws her evidence to test and refine her model are cases of professional science – cases on which she then makes normative judgments. Solomon's social epistemology argues that the reasoning of individual scientists is not terribly significant. What matters is the distribution of decision vectors within the community of scientists and the evaluations that can improve the distribution of research effort.

There are at least two approaches to considering social epistemology, or Solomon's social empiricism in particular, and how it might help us think about inquiry. One is as a model of inquiry that students would learn to enact. The second is as a model of inquiry that is useful for helping students understand *about* scientific inquiry. I will argue that a literal translation of social empiricism into a model of inquiry for students is not possible or desirable. However, social empiricism and, more broadly, social epistemology seem much more promising in terms of helping students learn *about* scientific inquiry. I will close by following up on some of Solomon's remarks comparing the relative educational usefulness of social empiricism with that of critical contextual empiricism.

SOCIAL EPISTEMOLOGY AS A MODEL OF INQUIRY TO BE ENACTED

If part of our jobs as science educators is to understand how students gain access to professional science, then it is critical that we understand what competencies are essential for success in those communities. For those of us who think about learning from a situated cognition theoretical stance (Brickhouse, 2001), and consider science education as a form of enculturation into scientific practices, then once again, we have to understand the target practices well enough to know how to help students acquire the competencies they need to gain access. (Of course, we also have to understand our students' incoming competencies as well, including their epistemological, linguistic, and cultural resources relevant to scientific inquiry. But, that is a discussion that should be reserved for other papers, e.g. that of David Hammer and William Brewer.)

Richard A. Duschl and Richard E. Grandy (eds.), Teaching Scientific Inquiry: Recommendations for Research and Implementation, 95–98.
© 2008 *Sense Publishers. All rights reserved.*

Miriam Solomon's social empiricism is a potential source for us in understanding the practices of scientific inquiry. However, Solomon also worries about teaching social empiricism because she is not sure she wants students acting like scientists and actually getting "A's" for it. The irony here is wonderful. I believe, however, that she need not worry too much. I think she may be concerned about a much more literal translation than would ever be either possible or desirable. School science communities are vastly different from professional science in terms of their purposes, intellectual and material resources, organization, and terms of evaluation. In the United States, as well as in many other countries, school is compulsory (to age 16); everywhere, professional science is highly selective. While some students' motivations may overlap scientists' (e.g. the desire to answer a question well; the desire to be recognized by others, i.e. getting the A), others are not motivated in the same ways and may even be in science classrooms simply because it is safer than being on the streets. Daniel Edelson writes quite well about how his desire to bring research tools used by scientists to school science students was met with a very different culture of schools that has motivated a continuous modification of his work and movement away from the goal of students in classrooms acting like mini-scientific communities. The notion of truly authentic science taking place in school classrooms is hopelessly Utopian. The best we can really hope for is that we help students gain some competence at some of the most critical skills of inquiry (perhaps model-based reasoning is a good candidate).

I do believe, however, that Solomon is at least somewhat justified in her concern regarding what might happen with a literal translation of social epistemology in which individual reasoning no longer matters. What matters is how the community reasons. Greg Kelly, in the following session, will be addressing the issue of distributed cognition within a science classroom. In his enactment of inquiry, individuals and small groups of individuals are not all thinking about the very same things. Different groups may test different lines of inquiry. The individual, however still matters. In education, individuals always have to matter.

Equity is an espoused core commitment of educators in general and science educators more specifically. This is an important difference between the purposes of scientific and educational communities. Science is about the generation of new knowledge; equity is not a core commitment. The community may be the important unit of epistemic authority, but an educator could never accept that this permits the erasure of the needs, desires, and rights of the individual to learn scientific practices that can potentially provide access to valued science-related communities outside school.

The reality is that school communities are not always idyllic spaces where children have the psychological safety to engage in inquiry in ways that generate important truths (or at least some empirical successes) and where everyone benefits from having played the game of science. Instead, they are complex social spaces where most often there are winners and losers and those who refuse to play at all.

Perhaps my concern is that school science looks a bit too much like real science in this regard – and that it works against normative goals of educational equality.

As I write this it seems odd to me that anyone would argue against my point that a literal translation of social epistemology is impossible and individual reasoning is important in educational settings. However, one can find instances of collectively enacted scientific inquiry that expects very little of individuals. For example Roth and Lee (2004) argue for a division of labor in scientific inquiry where the relevant unit of analysis is the community, not the individual. Furthermore, they argue for a model of scientific literacy where there are no distinctions between the goals and values of the community and those of school.

> This implies that science educators no longer seek to stack educational environments to coax *individuals* into certain performances, but that they set up situations that allow a variety of participatory modes, more consistent with a democratic approach in which people make decisions about their own lives and interests. (Roth & Lee, 2004, p. 267)

While I think there is much to commend aspects of this instructional approach, the results of how the division of labor plays out are deeply troubling. Although all participatory modes are valued, no distinction is made between those forms of participation that might be sufficiently generative to lead to new forms of participation in scientific communities and those that appear to me to be dead-ends. And, as one might expect, it is largely students who belong to already marginalized groups (aboriginals, girls) who engage in seemingly dead-end activity. It is for this reason that there have to be explicitly articulated and understood norms for what counts as valued engagement in science. I'm not arguing for no differentiation in contribution. There should be multiple ways of engaging in science, but none of them should be dead-ends.

SOCIAL EPISTEMOLOGY AS A MODEL FOR LEARNING ABOUT SCIENTIFIC INQUIRY

Learning *about* scientific inquiry is, of course, different from engaging in scientific inquiry. Engaging in scientific inquiry, even if that engagement is at relatively high levels, does not assure a sophisticated understanding about inquiry. However, understanding *about* inquiry is arguably the more important goal if our concern is for the large majority of students who will not become professional scientists, but who will play a role in the future of science in a democratic society.

If we want our students to understand how scientific communities actually engage in inquiry, then we need to understand how scholars in the field describe and understand scientific inquiry, including the issues on which they disagree. Allchin (2004) recommends the use of carefully crafted case studies. This kind of teaching strategy certainly has interest since if would engage students in some of the same kinds of evidence that sociologists, historians, and philosophers of science use in constructing arguments about scientific inquiry. This is, of course, still a form of inquiry, i. e. inquiry into the nature of science.

An aspect of Solomon's social epistemology that would be particularly appealing in this regard has to do with the way in which she illustrates how consensus is not always good, and in how the distribution of research effort makes a difference in the quality of scientific inquiry. Cases could be used to illustrate the ways that consensus does not guarantee empirical success (much less truth) and how diversity is essential for good scientific decision-making.

Although Allchin's idea of using a case-based approach has enormous intuitive appeal, we know very little about what students learn from such case-based approaches. I would very much like to see research that attempts to articulate how students might develop ideas about scientific inquiry and how these cases might be crafted in ways that support students' understandings. Such research would need to provide us with rich descriptions of learning and move us past the more typical research on the nature of science that relies on responses to surveys.

SOCIAL EPISTEMOLOGY AND CRITICAL CONTEXUTAL EMPIRICISM: WHAT ARE NORMS FOR?

Solomon states that the epistemic goal of social empiricism is empirical success and notes that Longino's critical contextual empiricism (CCE) also has feminist political goals. Not surprisingly, when goals are asymmetric, norms vary as well. Solomon suggests that Longino's account may be more suitable for a grade school classroom since the goals of CCE are political. I agree with her. To elaborate, let me return to my prior discussion regarding the ways in which the purposes of science differ from the purposes of education. While science is supposed to be concerned primarily with empirical success and with generating models and theories that can drive inquiry, science education has different goals. Students in science class rarely generate genuinely new ideas or models. Science education instead aims to provide students with access to scientific practices, perhaps in preparation for students to participate in professional science, but almost certainly in order for students to engage in science-related issues as citizens in a democracy that values and promotes science. We aim to provide this access to all our students; thus, equity in participation has to be a primary concern. Schools have additional moral obligations as well. We want students to be taught to respect and learn from those whose values and ideas are different from their own. Thus, Longino's requirement that scientists be responsive to criticism is a good one for our students as well. In many respects Longino's feminist goals of equity and diversity overlap those of education in ways that social empiricism alone does not.

Nancy W. Brickhouse
School of Education
University of Delaware

GREGORY J. KELLY

INQUIRY, ACTIVITY, AND EPISTEMIC PRACTICE

In "Reconsidering the character and role of inquiry in school science: Framing the debates" Duschl and Grandy (this volume) draw from the philosophy of science, the learning sciences, and educational research to discuss a number of important and converging results from these multiple disciplines that can inform the practice of science education. The argument I put forth in this paper is that this convergence results from an underlying shift in the locus of reasoning. The philosophy of science, the learning sciences, and educational studies focused on designing learning environments converge in the following manner: They shift the epistemic subject from an individual knower to a community of knowers with sociocultural practices derived from a common history of activity. This shift has potentially profound implications for research on inquiry in science education. To examine these implications, I define epistemology for the purposes of the arguments developed in this paper, review the importance of intersubjectivity for science and science education, discuss some normative considerations of the epistemic dimensions of inquiry derived from community activity, discuss ways activity theory allows for the study of inquiry in time and space, and apply these ideas to a set of studies focused on the epistemic practices of writing scientific arguments.

EPISTEMOLOGY AND EPISTEMIC PRACTICE

Before reviewing intersubjectivity in the fields of epistemology and education, I clarify the meaning of epistemology and its use in this paper. Epistemology is the study of knowledge. This field of research concerns itself with issues about the origins, scope, nature, and limitations of knowledge (Boyd, Gasper & Trout, 1991; Sosa, 1991). As a branch of philosophy, epistemology is a discipline, with a history of common problems and arguments. Applied to science, epistemology typically examines issues such as the growth of knowledge, the nature of evidence, criteria for theory choice, and the structure of disciplinary knowledge. For the purposes of this paper, epistemology should not be confused with individual subjects' views about knowledge, sometimes referred to as "personal" epistemology (for review see Kelly & Duschl, 2002). Rather, I am concerned with the ways that knowledge is constructed and justified within and for a particular community. A community justifies knowledge through social practices. A social practice is constituted by a patterned set of actions, typically performed by members of a group based on common purposes and expectations, with shared cultural values, tools, and meanings (Gee, 1999; Kelly & Green, 1998; Lave & Wegner, 1991). When such patterns in actions concern knowledge they can be labeled epistemic. Thus, I define *epistemic practices* as the specific ways members of a community propose, justify, evaluate, and legitimize knowledge claims within a disciplinary framework. My argument is that an important aspect of participating in science is learning the epistemic practices associated with producing, communicating, and evaluating knowledge.

CHANGING THE EPISTEMIC SUBJECT

A refocusing of the subject of epistemic consideration emerges out of two literatures. The first body of literature I consider is that of epistemology from three schools of thought: discourse ethics, and feminist epistemology, and certain philosophies of science. Each of these epistemological positions re-centers the knowing subject to a relevant community. The convergence of descriptive studies of science (e.g., Knorr-Cetina, 1999; Lynch, 1993), and philosophy of science grounded in these empirical studies of scientific practices (Fuller, 1988; Giere, 2001; Longino, 2002), and normative studies of moral reasoning lends support to the important epistemic role of a relevant community for considerations of knowledge claims. A second literature focused on science learning, informed by the learning sciences, similarly supports this move toward a more social view of knowledge and reasoning.

Intersubjectivity in Epistemology

A common assumption of logical positivism, discovery learning, content/process science instruction, and some forms of constructivism, is the central location of the individual consciousness as the epistemic agent. The focus on how a consciousness comes to know (subject-centered reason) defines a certain set of epistemological problems. These problems include: translation of sense experience into knowledge, rules for valid inference for an observer, and problems of meaning transfer among interlocutors, among others. The consequence of these epistemological problems is often skepticism (Strike, 1995), sometimes radical skepticism with all its drawbacks as an epistemology of science (Kelly, 1997). In contrast, philosophies of communicative action (Habermas, 1990) and feminist epistemology (Longino, 2002; Nelson, 1993) demonstrate an interest in changing the subject to a relevant community (Kelly, 2006). This changing of the locus of epistemic concern to a relevant social group, rather than a Cartesian subject, is well documented in *The Philosophical Discourse of Modernity* (Habermas, 1987). Habermas argued for a view of reason centered on communicative action: the shared norms for argument provide a basis for critique of current traditions and theories, thus forwarding the goal of rationality through dialogue. Longino (2002) applies a similar set of ideas to epistemological concerns in science. Her view calls for some levels pluralism in science and dialogical processes of adjudication.

The move to an intersubjective paradigm for epistemology is particularly relevant for science, where the social basis of disciplinary knowledge is particularly glaring. The important roles community practices and values play in empirical research have been well documented in historical, philosophical, sociological and ethnographic research (Kuhn, 1996, Knorr-Cetina, 1999; Lynch,

1993, Toulmin, 1972). Social practices of epistemic communities in science fields govern research directions, control outlets for publication, define the socialization of new members, and construct formation of knowledge through collaborative research endeavors (Boyd, 1992; Jasanoff, Markle, Petersen & Pinch, 1995; Gross, 1990). A social epistemology has relevance for education in many ways, but perhaps most centrally in ways that people are initiated into particular frames of reference through language and participation in cultural practices (Goodwin, 1994; Gumperz, 1982; Wittgenstein, 1958). Viewing science as social knowledge (Longino, 1990; Toulmin, 1972; Zuckerman, 1988) has a number of implications for views of inquiry. I consider two such implications: the value of a dialogic community supporting inquiry, and a view of learning commensurate with a social view of inquiry. Before considering these implications, I turn to a review of intersubjectivity relevant to science education.

Intersubjectivity in Science and Science Education

Three major learning outcomes identified for inquiry instruction are "conceptual understandings in science, abilities to perform scientific inquiry, and understandings about inquiry" (National Research Council, 2000, p. 75). Duschl and Grandy (this volume) formulate the goals of science instruction similarly, identifying three domains for which teachers can support learning by providing feedback to students: "promoting the communication of scientific ideas, developing scientific reasoning, and developing the ability to assess the epistemic status that can be attached to scientific claims" (p. 3). Taken from a distance these articulations are consistent with a general move in current reform in science education emphasizing the intersubjective processes of representation, communication, and evaluation of the evidentiary bases of knowledge claims (AAAS, 1993; Kelly & Duschl, 2002; National Research Council, 1996; Roth & Lee, 2002; Yore, Bisanz & Hand, 2003). The importance of intersubjectivity surfaces in at least three ways: interpretation of phenomena, public use of inscriptions, and a social view of expertise.

First, the use and interpretation of experience in science education has been advocated for as long as science education has been considered a separate endeavor (DeBoer, 1991). The importance of intersubjectivity ties to both the generating of knowledge through interactions with phenomena and through interpretations of others' interactions with phenomena (Lynch & Macbeth, 1998; Roth, McGinn, Woszczyna & Boutonnée, 1999; Roseberry, Warren & Conant, 1992). The current focus on inquiry similarly places attention on both the experiential and social worlds by setting the goals of learning conceptual structures, epistemic frameworks, and social processes of a science (following Duschl & Grandy, this volume, p. 4). The shift from content/process models of science to an evidence/explanation focus of inquiry instruction evinces the importance of the opportunities students have to engage in epistemic discourse and dialogical reasoning (Duschl & Grandy, this volume). This entails understanding the knowledge and practices of a particular epistemic community (Knorr-Cetina, 1999), which necessarily involves some socialization into the ways that the group's knowledge and practices are employed for given circumstances. Thus, the dialogical processes of examining knowledge claims through common practices (e.g., ways of representing data, weighing evidence, making inferences) presuppose a relevant community of knowers, with a common history and language for inquiry. The conceptual, linguistic, and artifactual tools are social.

A second aspect of intersubjectivity enters into learning contexts through the interpretations of inscriptions. Science studies and learning sciences both identify the multiple ways that scientific inquiry revolves around the production, interpretation, and evaluation of diagrams, graphs, photographs, data outputs, and signs and symbols more generally (Amann & Knorr-Cetina, 1990; Pea, 1993; Roth & McGinn, 1998; Latour & Woolgar, 1986). Inscriptions are tangible; they travel through space and time, and get recontextualized through talk and writing (Bazerman, 1988; Goodwin, 1994). Inscriptions have an in-the-world status, allowing for their construction and deconstruction to take place intersubjectively (Pea, 1993). The value of consideration of inscriptions for inquiry has been identified in empirical studies of student learning (for review, see Roth & McGinn, 1998).

A third aspect of intersubjectivity from a science learning point of view concerns the role of expertise, collective praxis, and local circumstances in considerations of scientific literacy. For example, Norris and Phillips (1994) argued that scientific literacy entails both knowledge of scientific concepts as well as knowledge of how scientific texts' structures and intentions can be ascertained. The argument continues that literacy should include competence in making sense of justification and evidence in the argumentative structure of science. Thus, combined with some conceptual knowledge, understandings of epistemic criteria provide a basis for understanding the value and limitations of scientific expertise (Norris, 1995). Similarly, Roth & Lee (2002) argue that overly individualistic views of scientific literacy focus attention on the ways students acquire knowledge without considering the ways those students are members of communities. In contrast, they propose a view that scientific literacy is relative to a set of particular circumstances that demand use of scientific knowledge and collective expertise in situated ways, where expertise is collective and distributed.

COMMUNITY NORMS FOR DIALOGICAL INQUIRY

The development of norms for a social epistemology in science may have implications for inquiry in science education settings. In this section I consider how social norms for a dialogical view of knowledge construction can lead to directions for research on inquiry in science education. Nevertheless, an important caveat need first be noted. Knowledge construction in science disciplines differs from knowledge construction in education in epistemologically relevant ways.

Scientific communities establish activity systems for the purposes of generating new knowledge about the natural world, as well as for a number of ancillary purposes. The criteria for examination of claims to knowledge serve this purpose. Science education communities establish activity systems with the goals of communicating and sharing ways of thinking about the natural world. The establishment of new knowledge may be local, without expectations that the "discoveries" transfer far beyond the locally relevant discourse community. Therefore, the review of a prescriptive view of epistemology, developed for establishing democratic scientific knowledge may be only partially relevant for educational purposes. I explore these possibilities of possible relevance by reviewing Longino's (2002) social norms for social knowledge.

Longino (2002) proposed a set of social norms for social knowledge. The four norms concern the venue, uptake, public standards, and tempered equality of social practices concerning knowledge production. These norms can serve as a starting place for developing critical communities of inquiry.

The venue refers to the need for publicly recognized forums for the criticism of evidence, of methods, and of assumptions and reasoning (Longino, 2002). Public dialogue about epistemic criteria is a key component of in the version of inquiry for science education advocated by Duschl and Grandy (this volume). However, much work needs to be done to develop the norms for interaction that permit close examination of evidence among students, while maintaining a respect for persons (Strike & Soltis, 1992). Venues for public discussion of evidence and explanation seem particularly important in science inquiry and have relevance for considering the value of inquiry in science education. A common problem for small group inquiry may be the lack of public spaces to examine in detail the results of groupwork (Hogan, Nastasi & Pressley, 1999; Kelly, Crawford & Green, 2001; Bianchini, 1997). This suggests a need for venues for public discussion and corrections among members.

Uptake refers to the extent to which a community tolerates dissent, and subjects its beliefs and theories to modification over time in response to critical discourse (Longino, 2002). While one can imagine science education communities, say a science class over the course of an academic year, tolerating dissent and changing views over time, the levels of disagreement may pose problems if left unresolved. Educators have a responsibility, to some extent, to communicate specified knowledge of presumed worth. A careful examination needs to be made of both inquiry that produces claims diverging from the knowledge of legitimizing institutions, and inquiry so circumscribed that students are lead to converge to a preconceived, singular answer. Neither may serve the multiple purposes of the presumed goals of conceptual understandings in science, abilities to perform scientific inquiry, and understandings about inquiry.

Publicly recognized standards are needed as a basis for criticism of the prevailing theories, hypotheses, and observational practices (Longino, 2002). These standards would contribute to framing debates regarding how criticism is made relevant to the goals of the inquiring community. Public standards would similarly change with relevant criticism and as the inquiring community changes goals and values. Such standards may emerge in educational settings relevant to the local inquiry projects, or may draw upon standards, such as protocols for investigation beyond the local community. Holding standards and research protocols subject to critique serves the purpose of teaching honesty in investigation – knowledge claims about public standards are similarly subject to dialogue, modification, and change, despite some assumed robustness.

Finally, communities must be characterized by equality of intellectual authority. This equality needs to be tempered, so differing levels of expertise and knowledge are appropriately considered (Longino, 2002). Much has been written about the uses of authority in science classrooms (Carlsen, 1991; Lemke, 1990). Some studies have tried to separate out the epistemic authority and the personal authority of teachers with respect to their students (Carlsen, 1997; Russell, 1983). Certainly, at least some of the time, in some of the circumstances, a teacher can serve as an epistemic authority – authority derived from knowledge and experience, and presumably, capable of being applied to assist others learning. The teachers' authority needs to be tempered to support open discussions of relevant issues. Students can develop confidence and responsibility and learn the cognitive goals of science through experiences with shared authority.

LEARNING AND ACTIVITY: EPISTEMIC PRACTICES IN CONTEXT

Much of the research on learning in science education has focused on how individual students learn particular concepts (e.g., Confrey, 1990; Pfundt & Duit, 1991). The alternative frameworks movement, conceptual change, various forms of constructivism have largely focused on individual learners. This has been true even if the individual learners were learning in a social situation. There is much value in this focus, particularly as related to helping students re-organize their understanding of particular concepts to align with the knowledge of the legitimizing institutions, often schools. Nevertheless, studies of this sort tend to start from a Cartesian epistemology of a consciousness trying to acquire knowledge (Kelly, 1997). There is often a view of meaning as personal and idiosyncratic (von Glasersfeld, 1991).

My argument here is that if we take a more thoroughly social view of disciplinary knowledge, research on inquiry in science education needs to examine learning situated in sociocultural practices. The focus on evidence and explanation requires some study of how particular sort of social practices are instantiated, communicated, appropriated, interpreted, applied, and change over time. This would suggest that learning focus on ways that social groups learn together, in ways where cognition is distributed among relevant people, texts, tools and technology, and social customs and practices. Thus, as the emerging views of inquiry situate learners in a social context, in space and time, working within particular social and cultural constraints, research on learning ought to consider the epistemic agency of a relevant community of knowers, along with the associated sociocultural practices that are established over time for the group. Such a view suggests that meanings are created in public spaces through collective actions (Toulmin, 1979; Wittgenstein, 1958).

A focus on collective activity shifts attention away from an individual consciousness to the interactive social processes that establish knowledge over time. Research on how students learn inquiry – conceptual understandings, abilities to engage in inquiry, and understandings about inquiry (NRC, 2000) – requires an understanding of the dynamic relationship between individual knowledge and knowledge held in common by members of the group that shapes the opportunities members have for learning (Kelly & Green, 1998; Santa Barbara Classroom Discourse Group, 1992). The sociocultural view underlying this perspective is that as members of a group affiliate over time, they create through social interaction particular ways of talking, thinking, acting, and interacting (Gee & Green, 1998; Kelly, Chen & Crawford, 1998). These ways of being come to define cultural practices, become resources for members, and evolve as members internalize the common practices but also transform them through externalization (Engestrom, 1999; Souza Lima, 1995). The cultural practices that constitute membership in a community (e.g., standardized genres) are thus created interactionally through discourse processes (e.g., multiple educational events, peer review, editorial admonishments and so forth). For further elaboration, see Kelly & Green (1998) and Kelly, Chen & Prothero (2000).

A focus on sociocultural practice suggests new units of analyses for research on inquiry learning. Activity theory provides a candidate for the unit of analysis of such a system: "object-oriented, collective, and culturally mediated human activity, or activity system" (Engestrom & Miettinen, 1999, p. 9). A focus on activity requires consideration of the multiple actors, the ways that roles are established and positioned, norms and expectations, the mediating artifacts, and local history of sociocultural practices. This unit of analysis thus requires that the study of inquiry examine these many dimensions, through systematic, careful analysis of the concerted actions of social groups. While this work typically dovetails with ethnography of practice, the agency of individual decision makers need not be neglected. For example, Giere (1999) makes a case for the study of the evolution of scientific communities, in a way that preserve the value of individual judgment among scientists as to the fruitfulness of potential research directions. An activity view would examine such judgments within a situated historical and social context.

Centering studies of inquiry on the multiple activity systems associated with the learning processes would not be complete without consideration of the relevant time scales for investigation. Time scales should include the micro-moments of interaction, the mesolevel events constructed from interactions, the ontogenetic, and the sociohistorical periods (Wortham, 2003; Lemke 2000). Each time scale is related to others. Interactional events are constructed through discourse processes and actions (Gee & Green, 1998; Gee, 1999; Kelly, Crawford & Green, 2001).

These events draw from contexts, practices, texts, and artifacts created at longer time scales. Through such interactions, meanings are negotiated. The mesolevel time scales refer to weeks and months of collective activity (Wortham, 2003). Science classrooms have been shown to develop particular cultural practices that come to define the ways of speaking, acting, and being a member over time (Crawford, Kelly & Brown, 2000; Lemke, 1990; Reveles, Cordova & Kelly, 2004). Mesolevel time scales allow for analysis of the sequencing of discourse events, identifying how interactional events are situated and contextualized by the local history of discourse processes (Gumperz, 1982). The ontogenetic scale refers to lifetime of individuals (Cole & Engestrom, 1993). This time scale is important for studies of inquiry in two ways. First, time scales in the order of lifetimes are pertinent for consideration of how sequences of events (e.g., curriculum) can potentially initiate students into the communal practices of science and for how individual students change over time. Second, the study of how such communal practices of science change over time – or minimally, the changes in how the social mediators interpret the communal practices of science. Finally, the sociohistorical time scale is important to understand the communal practices of a social group and how these practices evolve over extended time periods. For example, the changing nature of the experimental research article in science as a genre provides one such example (Atkinson, 1999; Bazerman, 1988; Gross, 1990).

EPISTEMIC PRACTICE AND LEARNING

My argument thus far has been that inquiry learning, properly conceived, centers epistemological questions in a social context. I have argued that the demise of the subject-centered epistemologies and the emergence of dialogical, communicative epistemologies require a shift in thinking about learning from a focus on individuals learning conceptions to broader studies situated in everyday life. Furthermore, I have suggested that activity theory is well suited for this epistemic shift of the subject. The first implication for this for the study of inquiry concern how and what might be investigated. The second concerns normative considerations of how knowledge claims may be evaluated.

Social Epistemology and Research Methodology

I argue that research on inquiry needs to consider at least two foci for research on learning: the study of distributed cognition in multiple contexts and the study of collective learning of sociocultural practices by social groups. Each of these subjects of investigation can be studied across different time scales including interactional, mesolevel, ontogenetic, and sociohistorical time frames for activity. As an illustrative example of these ideas, I draw from almost ten years of study of uses of evidence in scientific writing in a university oceanography course. I choose these studies from my research program as illustrative cases, more because of familiarity than for any claim that they are of exemplary quality (Kelly &

Bazerman, 2003; Kelly et al., 2000; Kelly & Takao, 2002; Takao & Kelly, 2003; Takao, Prothero & Kelly, 2002).

The studies I draw from are based in a constantly evolving university course at the University of California, Santa Barbara. A central aim of the course is for students to learn how to formulate scientific evidence in the form of a technical paper. The underlying premise is that students will learn to use, assess, and critique scientific evidence through experiences in which they use, assess, and critique evidence based on real scientific data while working on problems with ambiguous solutions. The students were provided with a CD-Rom technology that provides multiple geological data sets that students can employ to make a case characterizing the geological condition of a chosen area using the theory of plate tectonics. Multiple earth science datasets are available to be drawn into the students' arguments. They are able to choose from a variety of data forms (e.g., elevation profile plots, earthquake location and profile plots, volcano locations, hot spot locations and ages) and create an infinite array of different inscriptions from these finite datasets. Thus, to make a persuasive case the students need to make judicious choices about the type, range, and extent of the data forming the basis of their arguments. Four of the thirteen inscriptions supporting a typical students' argument are shown in Figure 1. In this case, three sorts of inscriptions constitutive of a good argument are presented: a surface area view of the chosen location showing earthquake locations relative to landforms and a oceanographic trench (a); depth profiles, both of earthquakes and elevation (b & c); and a student sketch of the argued-for geological situation (d).

The production of a scientific argument, one goal of concern, cannot be properly understood without knowledge of the sociocultural practices framing the activity. Initial ethnographic analysis examined the cultural practices of oceanography course to situate the writing within a broader set of practices (Kelly et al., 2000). In this and related studies, the focus was on the epistemological issues associated with writing science as well as the rhetorical features of scientific argument. In this paper I now consider how the social practices, discourse processes, and cultural artifacts of the university oceanography course can be interpreted from an activity theory point of view. Rather than representing the details of the studies, I draw on these studies to support the case for use of activity theory to examine learning in complex science inquiry studies.

In its simplest form the activity system includes the interaction of students (subjects) with the activity of producing a technical scientific paper (object). This activity is mediated by two artifacts, the CD-ROM technology that offers multiple data inscriptions and writing heuristics designed to support student engagement in particular genre conventions of scientific writing, see Figure 2. While this model of the activity offers a visual representation, it fails to consider the sociocultural practices inherent in the system and to demonstrate the complex ways spoken discourse and data inscriptions support particular ways of writing science. Following Engestrom (1999), I considered other dimensions of the activity system including the rules and practices, community, and the division of labor (see Figure 3). The rules and practices of writing science (in this context) were communicated through a number of media including spoken discourse in class settings, the laboratory manual that included analysis and writing heuristics, and through a series of writing activities leading to the production of the technical paper. For example, the importance of providing an evidentiary base for knowledge claims was featured across media (Kelly et al., 2000). The community members were analyzed in terms of the practices promulgated by various members through specific actions that they took and their reflections on such actions. For example, through a series of interviews around scientific arguments with various community members (instructors, oceanography students, non-oceanography students) the degree of explicitness of cultural knowledge about scientific argument strength was explored (Takao & Kelly, 2003). Finally, the division of labor among the writers and reviewers identifies a yet-to-be examined feature of the evolving cultural practices: calibrated peer review. In this case, students take on different roles as writers and reviewers of evidentiary arguments in science. Drawing from activity theory allows for the connection of the intersubjective epistemology of inquiry, based in philosophy of science, with a learning theory dedicated to understanding individual and collective learning.

To demonstrate the application of these theories across timescales, I consider in detail the dimension of "rules and practices," which I amend to name, sociocultural practices. Central to the academic task in the formulation of evidence, thus the inculcation of conventionalized ways of writing science were focused on epistemic practices – ways of producing, assessing, communicating, and evaluating knowledge in science. These epistemic practices are of a community with a division of labor, and furthermore each is related to the subject, object, and mediating artifacts – the system is indeed interrelated as expected (see Figure 3). The relevant community in this case refers to the students, instructors (professor, graduate student teaching assistants), and education researchers. Each of these community members had different and sometimes multiple roles. This division of labor is evidence in the roles taken by instructors and students. The course professor (Prothero) is the software designer, the principal lecturer, the architect of the student activities, and one primary evaluator of the students' products. The graduate student teaching assistants are responsible for instruction in the uses of technology, oceanography investigations, and writing science. Students serve as the writers of science, audience for spoken discussions and initial peer reviewers. Educational researchers have participated in creating learning heuristics, modifying the instruction and evaluation related to student writing of science, and in creating models for assessments of student writing.

To consider learning of epistemic practices, both for cognition and collective learning, across time scales, I created Figure 4, which shows plausible sites for research into inquiry learning for the given activity system of the university oceanography course in question. In this case, the ways that discourse practices, conventions, and cultural artifacts are entered into the interactional aspects of learning are made evident. I consider first the case of distributed cognition (Figure 4, top row), and then turn to collective learning (Figure 4, bottom row).

For the case of distributed cognition as related the uses of the data inscriptions, students are situated in a social situation where a context for interaction is constructed – the multiple members and various tools become mediators for the knowledge and practices communicated over time. An example of this would be the ways that scientific writing is negotiated through discourse processes with various tools and technologies (Kelly et al., 2000). The particular interactional contexts are constructed through a sequence of intertextual events, where text made available at one time can be invoked in a subsequent interaction, identifying the importance of sequencing at the mesolevel time scales. At the ontogenetic timescale, the development of the student writers across their lifespan becomes relevant. A central part of the scientific writing hinges on the epistemological orientation (Belenky, Clinchy, Goldberger & Tarule, 1986; Hofer, 2001; Perry, 1970) adopted by the student writers. Thus, the epistemological development of the writers, and in particular, as related to their previous opportunities to appropriate and make their own, conventionalized practices becomes relevant. Finally, the uses of conventionalized genres raises questions answered through analysis at the sociohistorical scale. The experimental article in science, informing the activities of the student writers and social mediators, evolved over time through changing mores of different communities (Bazerman, 1988).

For the case of collective learning, similar analyses can be conducted. Importantly for inquiry, the focus is not only on how individuals learn, but also on how collective learning occurs for social groups. In this case, I consider the learning of the relevance of the differing timescales for collective learning. At the interactional level, analysis can focus on the ways that conversational cooperation, through use of common diction, pauses, latching, and prosodic similarity (Gumperz, Cook-Gumperz & Szymanski, 1999) demonstrates appropriation of certain aspects of scientific discourse (Kelly & Brown, 2003). For example, Kelly et al. (2000) noted how a student and teaching assistant jointly constructed a "good observation" through dialogue. Over the course of weeks and months (mesolevel), the group can be seen to incorporate conventionalized discourse practices – for example, ways of employing just enough but not too many profile plots sufficient to identify a trench. At the ontogenetic timescale, analysis can examine how conventionalized practices become taken-for-granted, and thus are internalized into the ways of speaking about the natural world. Finally, at the sociohistorical timescale, an analysis of the changing mores of genre conventions, informing the instruction in technical writing can be examined.

Normative Considerations for Review of Knowledge Claims

A key component of a social epistemology is open public forums for debate and dialogue. I argued earlier that Longino's (2002) social norms for social knowledge may offer some prescriptions for inquiry in science education communities. In the activity system just described, these social norms are only partially met. For example, the oceanography course provides venues for criticism of evidence, methods, and reasoning and often tolerates some dissent – a second writing assignment is organized around a debate from multiple points of views concerning human contribution to global warming. Furthermore, there are publicly recognized standards that serve as basis for criticism, However, this criticism is more of how students apply plate tectonic theory, than criticism of any prevailing hypotheses and observational practices of geology, as suggested by Longino. Finally, there is some equality of intellectual authority; as recently, blind peer review has been added to the course practices, where previously students only read and discussed each other's papers in public meetings.

The purpose of this paper is not to levy a critique of the inquiry processes of the oceanography course from the point of view of Longino's social norms, rather I hope to identify the value of considering these norms. In this cursory review, we see that questions are raised about the viability of the norms for students who are unlikely, for the purposes of a ten week science course, change the conventionalized knowledge beyond the local community. Thus, key distinctions between the activities of science and science education need to be considered.

REFLECTIONS ON POINT OF VIEW OF INQUIRY, ACTIVITY, AND EPISTEMIC PRACTICE

As this conference is focused on inquiry, the conveners have requested that the authors discuss how we each arrived at our point of view and what we see as the main competitors. This is an inquiry into the basis of the argument made by each author. For my part, I believe I arrived at my point of view from research and experience.

Origin of my Point of View

Reading literature and empirical study constitute the research informing my point of view. My reading of research has been concentrated in a set of related fields, predominantly science studies and discourse analytic studies of interaction. Science studies provide multi-disciplinary views of the knowledge and practice of scientific communities. While I do believe that educators' understandings about science is greatly enhanced through examination of this literature, I do not believe that a consensus position regarding the nature of science emerges from this vast literature. Rather, I believe that each of the contributing disciplines offers unique perspectives on different dimensions of science, whether from history, philosophy, sociology, anthropology, rhetoric, and so forth. When it comes to the empirical studies of science classrooms, I believe that the disciplines that study scientific practices *in situ*, generally ethnographically oriented studies (Knorr-Cetina, 1999; Latour, 1987; Lynch, 1993), offer important research methodology counsel: Focus on the everyday actions of members that come to define what counts as science for the relevant social group.

For empirical studies, my research has examined epistemological issues in science education settings. I provided one example of the approach and some results in this paper. The general orientation has been on the examination of science-in-the-making (Latour, 1987) in an effort to demystify the processes of knowledge production, communication, and appropriation for students. The large role of discourse and social processes in these studies evolved from a concern about epistemological issues to a recognition of the ways epistemological issues surface in everyday talk and actions (Kelly & Bazerman, 2003; Kelly & Brown, 2003; Kelly, Crawford & Green, 2001).

For experience, I have gained some knowledge through the lived experience as a student and teacher. Concerning the central argument for intersubjectivity and the importance of the relevant community of knowers, I can site my experience teaching science in Togo, West Africa, where many of my students had a worldview different than my own. This made a strong case for me about the ways that initiation into a framework sets the observational and inferential possibilities for observers. The ways that different activity systems afford opportunities for learning has become part of my "folk" knowledge about the lived experience of a teacher of science and a science teacher educator.

Competing Points of View

There are serious competitors to the point of view I shared in this paper. I shall consider competing views of cognition, science, and epistemology.

First, a serious contender to the positions I have put forth is derived from an alternative view of cognition and activity described earlier. I have argued for a view of cognition distributed across people, technologies, texts, cultural practices and so forth. A rival view would be cognitively oriented learning theories, such as some forms of constructivism and conceptual change theory, that view reasoning as a purely individual activity (von Glasersfeld, 1991). While I readily acknowledge that individuals engage in cognition, I view reasoning as having a public, social function with intersubjective norms and consequences. I argue against views that separate out cognition from social factors, where social refers to non-epistemic considerations. Clearly, the position I advocate considers the thinking processes of individual students. However, the focus on reasoning is not solely on how an individual thinks. Rather, reasoning is considered public phenomena, where the learned practices of a group become internalized by individual students. These learned practices can become part of a research agenda for inquiry that considers the ways that opportunities to reason about evidence are made available through educational experiences.

Second, there are alternatives to the view of science I have put forth. My general orientation is based on empirical studies of scientific practice. This orientation permits definitions of scientific knowledge and practice to emerge from the everyday life of groups that define *what counts as science* through their activities. Thus, objectivity is defined by consensus, through argumentation by relevant speakers – this is similar to Habermas' (1990) view of the force of unforced argument. I remain agnostic about central ontological issues related to scientific knowledge and nature – issues of realism versus instrumentalism and so forth. Nevertheless, one strong alternative to my point of view is that of scientific realism, in the following way. A realist response to my anthropological view might argue that the results of unforced argument converge to at least a loosely defined consensus because it just so happens that the results of scientific investigations, properly conducted with reasonable checks and balances and so forth, lead to assertions about the world that are at least approximately true for features existing independent of human thinking about them (Boyd, 1992). In this case, the objectivity of science is achieved through by social consensus, but shows ways that nature enters into such considerations in a direct way.

Third, my argument is for a sustained focus on epistemic practices is derived from a philosophical view of epistemology. This perspective entails viewing epistemology as the branch of philosophy that examines the origins, scope, nature, and limitations of knowledge (Sosa, 1991). A competing point of view would be one that views studies of epistemology as referring to individuals' beliefs about knowledge (Duell & Schommer-Atkins, 2001). This contrasting view focuses on knowledge in individual minds rather than disciplinary communities. These studies tend to examine how individuals' understanding about knowledge (personal epistemologies) influences their learning (Hofer, 2001; Schraw, 2001). While these studies may have value for understanding aspects of cognition, the conception of epistemology focuses on beliefs rather than ways that knowledge is communicated, assessed, and accepted or rejected in a social setting. Presumably, these two views of epistemology can be integrated into research programs that are mutually informative and complementary. For example, there is nothing about studying how learners conceptualize knowledge that presupposes that the social bases for knowledge justification cannot be studied for such learners.

CONCLUSION

In this paper I identified the lone observer as the assumed epistemic agent in three pedagogical approaches, discovery learning, content/process learning, and constructivism that served as precursors to the current versions of inquiry. This common assumption is shared by positivism in philosophy of science (Longino, 1993). I then sketched some reasons why a social epistemology is needed to fully appreciate the knowledge and reasoning issues in inquiry learning. Drawing from philosophy of science, I reviewed how social norms for social knowledge (Longino, 2002) can support dialogic reasoning in science education. Shifting the epistemic agent to a community of knowers has implications for science education research focused on inquiry. I described how, to understand inquiry thoroughly, research needs to consider the learning processes through distributed cognition and by communities as they transform the conditions of their members and reality. These foci necessitate considerations of multiple timescales for research, so that the interactional, mesolevel, ontogenetic, and sociohistorical time frames for activity are considered. I have argued that activity theory provides a view of learning that anticipates this emerging complexity, and meshes well with the communicative epistemic agent of the social epistemology of science inquiry. I attempted to illustrate these arguments with an example from my own research program focused on the construction of scientific arguments in university oceanography.

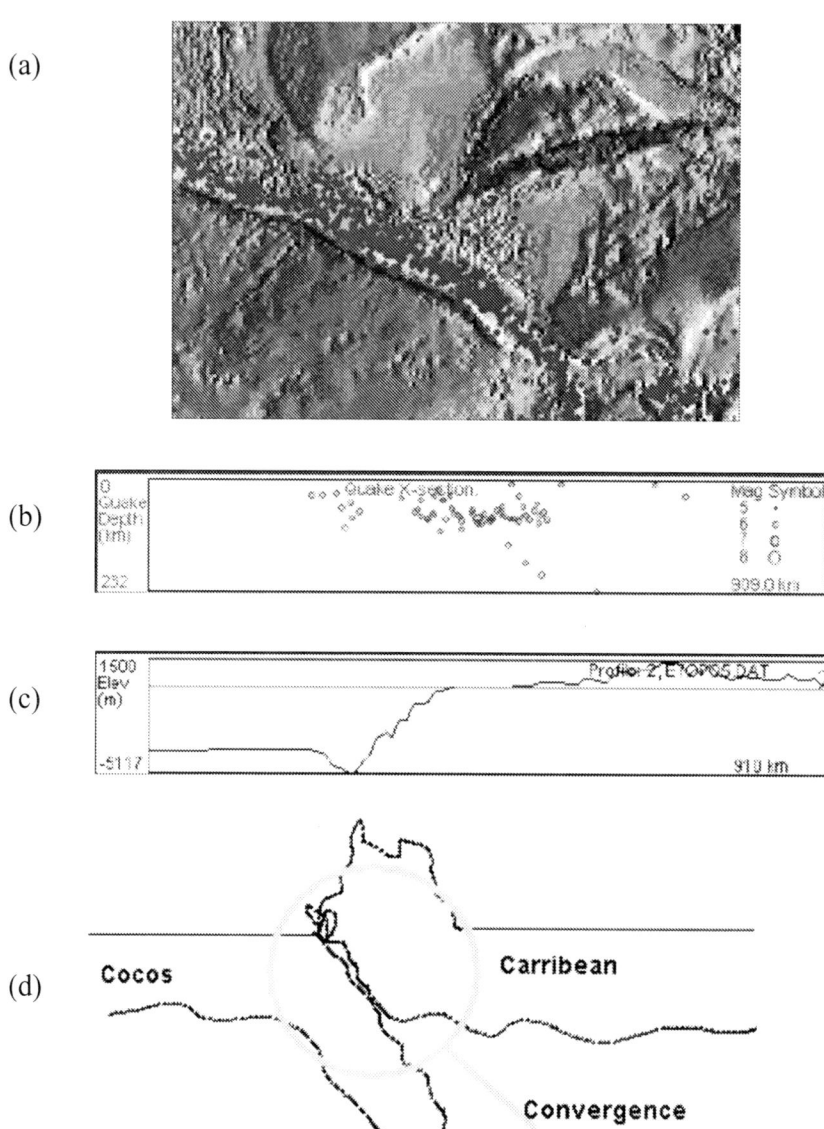

Figure 1. Sample (4 of 13) of student produced inscriptions from technical paper created with CD-ROM "Our Dynamic Planet." The four inscriptions are (a) earthquake locations, (b) earthquake depth profile, (c) elevation profile, and (d) student model of plate dynamics for El Salvador, in Central America.

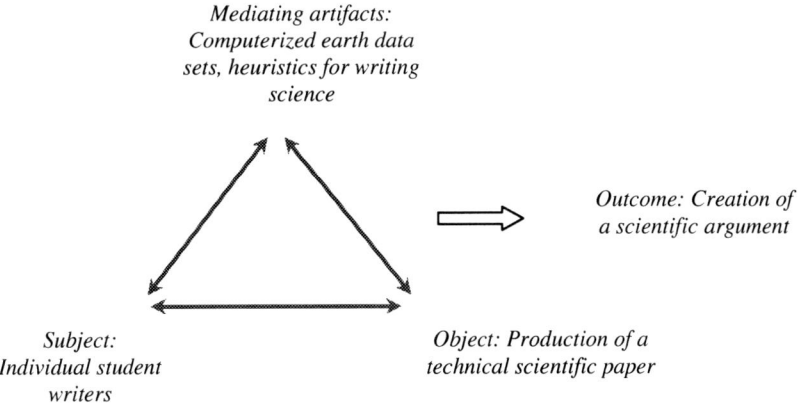

Figure 2. Triadic representation of mediated activity of scientific argument creation in oceanography.

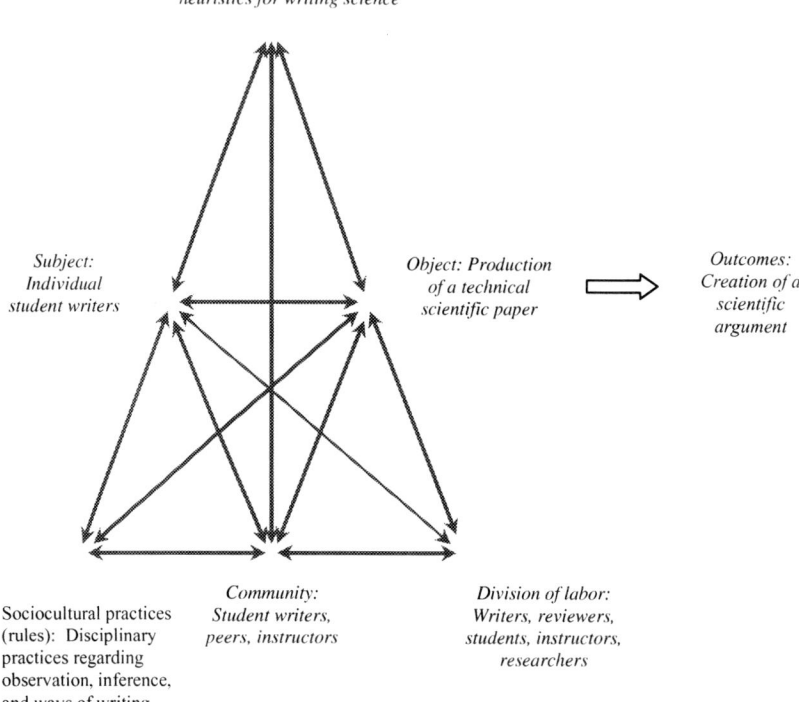

Figure 3. Complex model of activity system leading to the creation of a scientific argument in oceanography.

		← sociohistorical →			
		← ontogenetic →			
		← mesolevel →			
Time frames for activity:		← interactional →			
Focus of research on inquiry learning	Distributed cognition in multiple contexts	Examination of the ways that students change their discourse and actions, involving computerized inscriptions, conventionalized practices, instructors' actions, and each other as mediators. *Example: Discussions concerning the meaning of scientific writing, with multimedia resources, including knowledge reified in technology.*	Examination of the ways that cognition around conventionalized practices is communicated sequentially through discourse processes – situates instances of discourse in framework of ongoing meanings. *Example: Sequence of intertextual events, contributing to context for learning.*	Examination of ways students as writers evolve through changing psychological and epistemological development. *Example: Epistemological development of student writers, as related to previous opportunities.*	Examination of ways historical development of scientific genres enter into cognitive processes. *Example: Conventions of the experimental article in science.*
	Social groups' collective learning	Examination of conversation cooperation, as related to appropriation of scientific discourse. *Example: Conversational cooperation such as the joint construction of a "good observation."*	Examination of appropriation of discourse features of scientific knowledge and practices. *Example: Appropriation of conventionalized ways of presenting data – using enough but not too many elevation profiles to identify a trench.*	Examination of ways participating members' understandings of conventional genres become taken-for-granted and common practice. *Example: Internalization of practices through use as taken-for-granted ways of being.*	Examination of ways scientific genres and conventions for argument evolve over time. *Example: Changing mores of genre conventions.*

Figure 4. Sites for examination of inquiry learning of epistemic practices across timescales for university oceanography.

Gregory J. Kelly
College of Education
Pennsylvania State University

JOHN L. RUDOLPH

COMMENTARY ON "INQUIRY, ACTIVITY, AND EPISTEMIC PRACTICE"

In the paper "Inquiry, Activity, and Epistemic Practice," Kelly makes a number of interesting points and highlights key bodies of scholarship that provide productive ways of thinking about inquiry teaching in light of the goals of this conference. His primary contribution is to articulate a research agenda focused on the social and cultural practices of communities as they relate to the construction of knowledge. This focus on the interactive social processes through which knowledge is judged reliable, what he refers to as "social epistemology," is sensible given the historical shift in our understanding of science pointed to by Kelly, and Duschl and Grandy in their plenary paper, from seeing science as an individual process of rational engagement to seeing it as an interactive, socially mediated process of knowledge construction. In what follows, I outline my understanding of what Kelly is calling for specifically, offer some comments, and raise some questions related to the paper and some of the common assumptions underlying inquiry instruction in science education.

Using activity theory as the unit of analysis, Kelly argues that what we need to know more about is how groups of individuals come to appropriate the content knowledge, skills, norms, and intellectual goals that characterize a functioning inquiry community. Drawing on our understanding of epistemic communities in the sciences, Kelly suggests that we see our classrooms in similar terms and work to gain insight into how knowledge and skills are acquired in such settings. His focus, as I understand it, is on the classroom practices that are commonly employed in teaching. While he talks about the importance of argumentation, standardized discourse, epistemic norms in a given disciplinary field of inquiry for student understanding, it's not clear that student inquiry is the primary instructional approach to be used to accomplish this. He writes that "research on inquiry in science education needs to examine learning situated in sociocultural practices" (p. 10). Presumably the sociocultural practices in question are those of the classroom, or learning community. This would be the "relevant community of knowers" to which he refers shortly thereafter. But he also refers in the same sentence to the "sociocultural practices that are established over time for the group." Now these practices it seems are those characteristic of a particular expert group of researchers rather than of the students in the classroom. Kelly says little, though, about the distinction between the epistemic practices of the learning community and those of the research community. In the end, the specific role of inquiry as an instructional approach in all this – where students would themselves take on the role of researchers in a simulated expert community – is ambiguous.

In an attempt to clarify some of these issues, let me proceed with the assumption (which is mine and not Kelly's necessarily) that the aim of understanding is some

Richard A. Duschl and Richard E. Grandy (eds.), Teaching Scientific Inquiry: Recommendations for Research and Implementation, 118–122.
© 2008 *Sense Publishers. All rights reserved.*

scientific practice with its associated goals, techniques, discourse patterns, norms, etc. (such a community might be thought of as an established research group operating within some scientific discipline or sub-discipline), and that the means of achieving that understanding is through participation at some level in the practices of the community itself (what I'm referring to as an inquiry approach). In this context, educational researchers would study, among other things, the way data are deployed to support knowledge claims, how particular forms of argumentation are modeled, and the way in which group conventions are communicated. The focus of research would be on all the various aspects – material, intellectual, social, etc. – of the process of enculturation into that community.

A research program focused on understanding this process like the one proposed by Kelly is clearly desirable, especially given the inherent complexity involved in learning the ins and outs of an expert community. The simplest case to consider would be something akin to a university research lab, where novice but willing individuals apprentice themselves to an expert community of practitioners. Obviously there are various processes and mechanisms, formal and informal, through which the uninitiated come to master the skills and techniques and learn and adopt the epistemic norms and cognitive goals of the expert community. In a case such as this, the process of enculturation is relatively straightforward and research aimed at learning more about how it occurs would undoubtedly enable one to better facilitate the process. When this model of learning is moved into the school setting, though, things become more complicated.

One complicating factor is that school classrooms, as Kelly readily acknowledges, are themselves communities of practice, and, despite whatever social and intellectual environment a teacher seeks to establish in his or her classroom, educational settings (particularly at the pre-college level) have enduring cultural norms, discourse patterns, and power relationships all their own with which one must contend. As inquiry communities, classrooms are much more heterogeneous than university research groups. Students go to school for a variety of reasons, some social, some intellectual. Many, of course, are not completely willing participants in classroom activities, and their goals are often focused minimally on meeting teacher expectations. Activities in this type of situation require little in the way of coherence – students are simply seeking teacher approval (which is often manifested in the corresponding grade that comes with it) rather than being driven by an intrinsic desire to attain knowledge or to become fully assimilated within a scientific research community. There are, in other words, a variety of individual goals being sought in a classroom that can and often do operate at cross purposes.

This pre-existing classroom community creates a situation where instead of facilitating the relatively unmediated process of enculturation found in a university lab, a teacher needs to create an instructional environment in which compliant but perhaps not enthusiastic students will engage meaningfully in inquiry activities. The best that can be expected in this situation is that the students will make a good faith effort to understand and participate in the activities of the target research community. To ensure this result it's necessary, however, that the expectations, power relations, and reward structure of the existing classroom culture provide a coherent set of incentives for students to play the inquiry game. Crucial for student learning as they play this game is the ability of students to distinguish clearly between the aspects of the classroom environment that encourage and facilitate the inquiry experience and those that represent the practices and norms characteristic of the inquiry experience itself.

The complexity of the classroom situation described above should be apparent, and any research program aimed at understanding the process of enculturation into an inquiry community within such a setting needs to look at more than the idealized process through which students interactively or dialogically appropriate discourse patterns or internalize epistemic norms or particular cognitive goals. What needs to be examined as well is the way students recognize and negotiate the boundary between their existing classroom community with its culture, norms, and folkways and the simulated or target research community. Unless students are helped to see these distinctions, they may view inquiry activities as just another school task that has little relation to what scientists actually do. Kelly rightly alludes to the importance of these distinctions between contexts of inquiry as opposed to contexts of science education, but it's not clear how important this distinction is in terms of the research program he describes.

There is another complicating factor worth mentioning and one that bears more directly on the conference theme of consensus. In the simplest case outlined above (the idealized apprenticeship model of a university laboratory) all one needs to worry about is understanding how newcomers appropriate the various components of practice of the expert community in question. In this setting, the expert community is there ready and willing presumably to guide the newcomers along. In a classroom, however, no such expert community exists. What exists, rather, are teachers (who belong to a different sort of expert community) along with a variety of material and intellectual resources such as textbooks, computer software, curriculum developers, and, perhaps, educational researchers (this is the distributed cognition part of what Kelly describes in his paper). From those resources a teacher might *simulate* a research community. But any such simulation will inevitably be partial and incomplete. This raises important questions about what aspects of the simulated community's practice are the most desirable to have students appropriate, internalize, or learn about. How robustly must those community norms and practices be represented for the learning experience to be authentic? Is it even possible for any individual or group of individuals outside the target community of practice to accurately model the relevant sociocultural and epistemological aspects of that community? The most likely outcome is the production of an artificial, highly circumscribed inquiry experience for students. This isn't to say that such an experience isn't educative. What needs to be done, though, is to develop some idea of what the most desirable outcomes from such an experience should be and build the learning environment accordingly.

Kelly's piece also prompted me to think about some broader issues that have largely been overlooked so far. Much of this talk about inquiry presupposes the educational efficacy of the inquiry experience for students. At the outset of his paper, Kelly reiterates the commonly stated goals of inquiry instruction – that such an approach should lead to "conceptual understandings in science, abilities to perform scientific inquiry, and understandings about inquiry" (p. 4). Few would dispute the claim that inquiry instruction, where students themselves are engaged in the process, has the capability of producing both conceptual understanding and, naturally, abilities to perform inquiry. The potential of this approach to achieve the third goal – understandings *about* inquiry – is open to question, however. This third goal is concerned primarily with more meta-level understandings of the nature of inquiry and the scientific enterprise more generally. A number of recent studies, though, have suggested that simply participating in research, or becoming proficient in the discourse and practices of a community of researchers, in the absence of explicit attention to these meta-level issues contributes little to student understanding about inquiry (Khishfe & Abd-El-Khalick, 2002; Bell et al., 2003; Sandoval & Morrison, 2003). If the goals of science education include all three outcomes and inquiry experiences are sufficient to generate student understanding in only the first two areas, then by what means are we to achieve understanding in the third?

One could argue that inquiry experiences need only be supplemented with explicit instruction aimed at achieving the third primary goal. That certainly might be enough if those understandings have to do with general patterns of argumentation or how knowledge claims are typically articulated with empirical evidence, things in the epistemological here and now. But there are other things worth knowing as well, and this becomes apparent when looking at the time scales Kelly offers as a way of organizing research on epistemic practices (see pp. 11-12). I have in mind specifically his last category – the sociohistorical timescale, which allows one to consider how research practices change over time. Kelly rightly points out understanding at this level as a worthy educational goal. But it seems that such a goal could never be achieved using inquiry as the instructional method – if by inquiry we mean *participation in or simulation of the research practices of a given disciplinary community*. Here I'm arguing against the general idea of inquiry teaching as panacea, as a universal tool that can accomplish all of our educational objectives, which, to be clear, is not a point argued by Kelly. The adoption of skills, techniques, and science content understandings in the here and now as part of an actual or simulated research group – the process of enculturation – reduces the possibility of gaining any longer view of how that practice has changed over time.

Just as the changing nature of inquiry in time can't be captured solely via an inquiry approach, the same can be said for understanding a more topographic distribution of epistemic practices. By this I mean the range of research practices in space so to speak. In his paper, Kelly provides us with an example of student learning drawn from his research in an undergraduate oceanography course (see Kelly, 2000). He describes how one might study the way a group of students gradually appropriates conventionalized discourse practices (from the field of oceanography presumably). As valuable as this may be for future oceanographers, for the general public it provides only a very small look at the variety of epistemic practices that are currently ongoing in the world. The issue of generalizability or broad representativeness of learning outcomes is left unaddressed. This, I admit, was beyond the scope of the paper. But it's an issue that needs to be raised at some point. The challenge, if one of the goals of science education is to develop among the public understanding *about* inquiry, is to think about how one can describe and communicate to students' knowledge of the great diversity of practices that constitute the natural sciences. And here again it seems that an inquiry approach, if one takes this to be some sort of active participation or immersion experience, would be inadequate if the goal is an understanding of inquiry writ large.

Thinking about these broader learning goals regarding inquiry in both time and space (to say nothing of social, cultural, and political context) suggests that there may be a false sense among science educators, researchers, and policymakers of inquiry's efficacy as an instructional approach. Perhaps the role of inquiry in science instruction should be more limited than what's currently envisioned in our leading policy documents (e.g. Rutherford and Ahlgren, 1990; National Research Council, 1996). At the very least, we should consider carefully what inquiry as an instructional approach might *realistically* accomplish in light of the complete range of desired learning outcomes in science education.

John L. Rudolph
University of Wisconsin-Madison

STEPHEN STICH

PANEL A

I don't really know why I am here, since I don't pretend to know anything worth knowing about teaching science. Mostly what I do is work in the area where various of the cognitive sciences overlap with philosophy. Like Bill Brewer, I'm interested in how the mind works, and I've made the occasional attempt to say something about how the mind works when it's doing science. But unlike Bill, I'm an amateur in that business, not a professional.

As an amateur in thinking about how the mind works when it's doing science, I have a certain luxury. I get to do a lot of fascinating reading in a variety of disciplines, and I get to attend a lot of conferences like this one. So perhaps the most useful thing I can do is to sketch very briefly some of the work that I think may be important in gaining a deeper understanding of how the institution of science works and how the minds embedded in that institution work when they're doing science.

I think Bill and I may be at opposite ends of one kind of spectrum. I think the processes we are trying to understand are very complicated and that there's a lot of empirical and theoretical work that is potentially useful. But I don't have any *integrated* story to tell you. Rather, what I'm going to do is point to a bunch of topics which, it seems to me, would lead to a better understanding of how science works, if only we knew more about them.

Two themes are intertwined in what I've got to say in the rest of this talk. The first is cultural transmission, the second is the importance of understanding norms. I'll address them in that order.

I start with cultural transmission because I believe that if we're going to get a serious understanding of what science is and how it works, we're going to have to understand cultural transmission at a fairly deep level. Humans are the only species that has a *cumulative* process of cultural transmission. There is some recent intriguing evidence about cultural transmission in nonhuman animals, including not only primates but also creatures with much simpler minds, like fish. But as far as *cumulative* traditions are concerned, it looks like humans are unique.

Science is one of the most powerful examples of cumulative cultural transmission, but there are a lot of other examples, like kayak building, which I mention because it is Rob Boyd's favorite example. The technology that goes into kayak building and the technology that goes into the space program are both impressive examples of cumulative cultural transmission.

What do we know about this process? In the last fifteen or twenty years, we've learned an enormous amount. So let me begin by recommending to you the work of two people who have revolutionized this field: Rob Boyd and Pete Richerson. What Boyd and Richerson have been doing since their important book, *Culture and the Evolutionary Process,* appeared in 1985, is building sophisticated and increasingly detailed models of how cultural transmission evolved. The question they have begun to answer is: Why is it that having processes of cultural transmission enhances the biological fitness of a species like ours. As a result of their work, we now have at least the beginning of a deep understanding of the evolutionary background of cultural transmission. We also have, thanks to the work of Boyd and Richardson, a very rich understanding of the ways in which biological evolution and cultural evolution can influence each other. What's so impressive about Boyd and Richerson's work is the sophisticated way in which they build mathematical models and then assemble a rich array of empirical findings to test those models.

Fortunately, Rob and Pete have recently published a very accessible book explaining their theoretical work and the evidence that supports it. (Richerson & Boyd, 2005) (Their 1985 book is, I should warn you, anything but accessible. Indeed, I had to hire a mathematically gifted graduate student to learn it and then teach it to me.) Among the things that Boyd and Richerson emphasize, both in their original theoretical work and in the new book, is how cultural transmission, once it's established, can be enormously productive and can also sometimes run amuck. In the new book they use the terminology "rogue meme" to help explain the latter phenomenon. It seems to me that one of the things we want to understand when comparing science to pseudo-science is just that phenomenon of the rogue meme, since it is very plausible to suppose that various bits of dangerous pseudo-science are examples of rogue memes. There is, I submit, a lot to be learned in this area that is important for understanding both science and science teaching.

Another thing I think we can learn from Boyd and Richardson that is of importance for our current concerns falls under the heading of what they call "biases". But be warned, that's a technical term and a misleading one. I am a bit grumpy with them for using that term because what they mean isn't what you'd intuitively think of as a bias. Rather, the idea is this. In their jargon, a cultural transmission strategy in which one *randomly* chooses who to learn from is an "unbiased" strategy. By contrast any cultural transmission process in which one "cultural parent" or teacher or cultural model is preferred to another is, in their terminology, "biased".

There is an issue here that surely should be of interest to folks concerned to understand and improve science education: How do people go about determining what biases to use, whose information to take on board, whose strategies or skills or norms they should internalize? Though we know relatively little about that, Henrich and Gil-White have recently reported some interesting empirical evidence about the ways in which prestige, authority and local consensus affect who people decide to learn from and who they decide not to learn from. (Henrich & Gil-White, 2001) Getting a better understanding of these processes will, I think, give us important insights about the spread of scientific knowledge – and also about the spread of the sort of pseudo-scientific claptrap that seems to be so alarmingly widespread in our culture.

PANEL A

To a first approximation, cultural transmission is a process in which people get information from other people's heads. In this process, the information transmitted can be modified or distorted in various ways. One question that is inevitable as soon as you start thinking about this is: What are the constraints on the process of cultural transmission? There are a lot of intriguing models that are being actively explored. And it seems to me that exporting that kind of research into science studies would be an eminently valuable thing to do. What kinds of constraints are being explored? Well, one kind of model, inevitably, given its visibility and importance and power elsewhere in the world of cognitive science, is inspired by Chomsky's account of language acquisition. The idea goes something like this: In any given domain, there is a set of innately determined options. It might be a large set which can be depicted as a tree structure through which the learner can move. But whatever the details, the important point is that learners have a relatively strongly constrained set of innately specified options. Everything the learner ends up with has to be in that set.

In the domain of moral norms, Mark Hauser at Harvard is exploring this sort of model and has done some extraordinary work. In fact, if I'd known I was going to be giving this talk today, I would have brought along the URL for Mark's online study of norms. I'll make that information available after the conference, and I recommend you take a look at it (Hauser, 2005).

But the Chomskian model is not the only proposal about the constraints that affect cultural transmission, and indeed my own view is that it's likely to be mistaken in most domains apart from language. Another, very different kind of constraint is what my students and I have referred to as Sperberian constraints or Sperberian biases, after the anthropologist Dan Sperber who has been instrumental in developing the idea. In the Chomskian model, there are a few hundred or a few thousand (or perhaps a few million) different kinds of things you can acquire in a domain, and that's it. The Sperberian idea is that because of the way the mind works, certain features of theories or skills or norms are *attractors*; cultural variants which have these features are easier to notice or easier to learn or easier to remember. And therefore, gradually over significant stretches of cultural evolutionary time, the prevailing cultural variants will increasingly exhibit those features.

An excellent example of the application of this idea can be found in Pascal Boyer's work on religion (Boyer, 2002). Boyer argues that many features of religious beliefs are the result of the sort of attractors that the Sperberian approach posits: He sets out a very plausible list of features that, he argues, the mind finds attractive for a variety of reasons. And after centuries of cultural evolution, that's where religious views settle.

A very different kind of example of this phenomenon is to be found in the work of Shaun Nichols (2004) on, of all things, the evolution of etiquette norms. Disgust, Nichols argues, plays a central role in that process. If people find the transgression of an etiquette norm to be disgusting, it is more likely that that norm will survive over time. So disgust functions as a Sperberian attractor.

One thing I can't resist saying, mostly to provoke Bill Brewer since it poses problems for his Cartesianism, is that I recommend the work of Paul Harris (2002), who is one of the first developmental psychologists to look at the psychology of testimony in little kids. Harris has important things to say about questions like: how much information do kids learn information via testimony?, which testimony do they accept and which testimony don't they accept?, and what are the principles governing the way children acquire information which is not acquired through their senses in any direct way, but, rather, through the testimony of others?

Of course it's not news that we acquire beliefs and theories and norms via cultural transmission. But one body of work that I think has important things to say to those of us interested in the structure of science and how science works goes far beyond that. The work I have in mind is the research of Dick Nisbett and his colleagues (2003), who have shown that the effects of cultural transmission run much deeper than cognitive psychologists have thought. More specifically, what Nisbett has shown is that memory structuring is in part culturally transmitted; attention patterns are culturally transmitted; reasoning strategies are culturally transmitted; and – the one that blows everybody's socks off – important features of the perceptual system are culturally transmitted. Nisbett focuses on contrasts between Western culture and East Asian culture. He reports that Westerners and East Asians react significantly differently even in traditionally rod-and-frame and figure-ground perceptual tasks. So it is literally the case that Westerners and East Asians are perceiving the world differently. And it is important to stress that these differences are *not* genetic. We know this because after about three generations in the United States, East Asians behave, in all of these experiments, just the way Americans of European ancestry do. So the effects that Nisbett finds are the result of a cultural transmission process. But it's a cultural transmission process that's affecting the mind at a much deeper level than merely influencing the beliefs or norms that people accept. Does all of this have implications for science education? I'm betting that the answer is yes, though what those implications are remains to be discovered by researchers like those assembled in this room.

Helen Longino and Miriam Solomon were talking earlier about the use of *intuitions* in normative epistemology – intuitions about how science should work, about what norms we should accept and things of that sort. Those of you who have some exposure to philosophical epistemology will know that in much of the literature gathering and systematizing intuitions is a central methodology. I have never been much of a fan of this methodology, which I often describe as "intuition mongering." Some years ago, I was bemoaning the state of epistemology with some like-minded students. We had all recently learned about Nisbett's cross-cultural research, and this suggested an idea. If there are important differences in the ways the minds of East Asians and Westerners work, maybe their epistemic intuitions are different, too. So we ran some experiments of our own, with very generous help from Dick Nisbett. And it turned out that the epistemic intuitions of

East Asians do indeed differ from those of Westerners (Nichols et al., 2003; Weinberg et al., 2001). Moreover, much to our surprise and delight, there are systematic cross-cultural differences even in so-called "Gettier intuitions" which are the kinds of epistemic intuitions that dramatically changed the philosophical landscape about 50 years ago by persuading philosophers to abandon a 2300-year-old tradition in epistemology according to which knowledge is justified true belief. East Asians, it turns out, don't have Gettier intuitions, nor do Indians, Pakistanis and Bangladeshis. So approximately two thirds of the population of the planet do not have them. And it's not only that they don't have them; they actually have counter-Gettier intuitions, their intuitions go in exactly the opposite way.

What does the fact that epistemic intuitions differ systematically across cultures tell us? Could it possibly be telling us that the right epistemic norms for East Asians are different from the right epistemic norms for Westerners? That's one possible conclusion. Another conclusion that might be drawn from these facts – and this is the one that I favor – is that we should be very skeptical about any philosophical methodology that relies on intuition. Although Helen Longino's project is unique and very different from what goes on in traditional analytical epistemology, she shares one important aspect of the methodology of traditional analytic epistemology: she relies heavily on intuitions. So the work I've been sketching poses a challenge for Helen as well.

Let me end by spending a minute or two talking about norms. As has been mentioned several times today, science is, of course, a norm-governed activity, and many of the crucial norms that seem to govern science emerged during the Renaissance under circumstances we don't fully understand. But it's important to see that the norms governing science continue to evolve; and micronorms, as it were, in specific parts of science also continue to emerge and evolve.

My favorite example of this is a contemporary debate about appropriate norms in research on decision making, with many experimental psychologists on one side and many experimental economists on the other. A crucial issue in this debate is the use of deception, which as we all know, plays a central role in lots of classic demonstrations in psychology. But the use of deception has become a hot topic of debate, and most experimental economists oppose it. It is also the case that in experimental economics you can't publish your research unless your subjects have made choices that really affect how much money they get at the end of the experiment. In some of the classic Kahneman and Tversky studies, subjects were asked to make choices among imagined alternatives with payoffs that they would not really get. But you can't do that in experimental economics – even though Kahneman and Tversky's work is an important part of the foundation of experimental economics. The prevailing norms in experimental economics require that you put your subjects real experimental game situations, and you've got to pay them pay them with real money. (For more discussion of all this, see Hertwig & Ortmann, 2001, 2003.)

So science is a norm-governed activity and the norms continue to evolve. There is something deep and important to understand here, though I think we currently understand very little of it. We don't know much about the psychology of epistemic norms, or about their evolution, or about the biological mechanisms that gave rise to the fact that we have epistemic norms. Indeed, we don't even know whether epistemic norms are psychologically special or whether they are subserved by the same psychological mechanisms that account for moral norms.

On the psychology of moral norms, let me recommend the work Chandra Sripada who has done some very impressive research sketching out a framework for the psychology of moral norms, which he supports with a remarkable array of interdisciplinary evidence from anthropology, psychology, experimental economics, the history of law and other areas (see Sripada and Stich, forthcoming). If Sripada is right, then part of the moral norm system is subserved by an innate dedicated mechanism with important built-in ties to the emotion system. On the acquisition side of the norm system, we know relatively little about what the constraints are, though certainly understanding those constraints is a crucially important project.

Now here's the question I'll leave you with: From a psychological point of view, are epistemic norms, including the kind of norms that we find in science, just a species or moral norms? Are epistemic norms just one among many kinds of norms subserved by the sort of norm mechanism that Sripada describes? Or are epistemic norms psychologically and evolutionarily different from moral norms? It seems to me we don't have even the beginning of an empirically supported answer to that question. And until we do there will be something really central and really important about science that we don't understand.

Stephen Stich
Department of Philosophy
Rutgers University

HARVEY SIEGEL

PANEL A

The papers on this panel were great today. I've really enjoyed them, and I've done my best to mention them all in these comments. I might not quite get everybody, but we'll see how it goes. I'm delighted to be on a panel with Steve Stich and Helen Longino. And I want to thank Rick and Dick, or Dick and Rick, for inviting me to participate.

The aims of science education: what are they? Well, according to the NSES, they are the ones we've seen all day: Learning to do science inquiry and developing an understanding of the nature of science inquiry. But as John Rudolph alerted us in his commentary earlier today and as Helen just reiterated, the notion of inquiry is unclear. On certain understandings of it – for example, the understanding that inquiry is a matter of immersion in authentic scientific practices – it can't get us everything that we want. I think John is right that we can't get everything we want out of inquiry. We can't achieve those two aims that the NSES mentions if we think of inquiry in that narrow way.

So what do we want? What should we want? I think that what we really want is for students to understand science. We want them to understand it, we want them to be reflective about it. What do I mean by "understanding science"? At least they should understand this: They should understand both the process of scientific practice, that is, how it's done – presumably that's what they get out of the immersion experience – and something that I think needs more emphasis that it has received today, namely the epistemic status of the fruits of that practice: that is, what results from that practice, the products of that practice, and what might be called the 'epistemic upshots' of those products.

Now, I don't want to take a stand on what these products are. They could be theories, they could be hypotheses, they could be models, they could be explanations. They could be all sorts of things like that. But whatever they are, surely we want students to have some *understanding* of them. So we want them to know how these products come about, what's the process, and also to have some intelligent understanding of, or at least be able to reflect intelligently on, what's produced.

I think that a key part of this understanding consists in their understanding of the goodness of relevant reasons and the quality of arguments. And here I am stealing from Greg Kelly who talked about argumentation in an interesting and important way today. Students need to be able to evaluate scientific claims and arguments. They need to be able to tell the difference between a crappy argument and a good one, and not just by consulting their intuitions about it, because we can't, as Steve has long urged, have 'intuition mongering.' They have to have some theoretical understanding of what constitutes a good reason in science.

I think that requires that they know something about the philosophy of science, but I want to postpone discussion of that point for just a second. Because you might want to ask, and I need to answer, a prior question, which is: Why is this understanding of the goodness of reasons and the quality of arguments a fundamental aim of science education? I think the answer is: because the ultimate aim of education is the fostering of student rationality.

I think this is the ultimate aim of all education. So science education is just a special case. In history education we want to get students to be able to be rational about history, historical judgments, historical processes, same in math, same in literature. Same across the curriculum. So that's a big claim. I couldn't conceivably give any sort of adequate defense of it in the two or three more minutes I have, but it's a claim I'm making anyway, so I should at least give you a thumbnail sketch of a justification. Why should we think that the ultimate aim of education is the fostering of student rationality? I think it's because that's the only way we can educate in a way that treats students with respect as persons. (So my appeal is to Kant's moral theory.)

Why is that? Suppose we teach students in ways that don't try to foster their rationality. We might, for example, indoctrinate them or we might simply get them to memorize and accept without critical reflection whatever it is we want them to believe either about Shakespeare or about atoms or about the French Revolution or whatever. It seems to me, however, that if we do it in any way other than trying to foster their ability and their dispositions to reflect critically and intelligently on what we're teaching them, then we end up putting them in a position in which they are not, so to speak, the masters of their own epistemic fates. They're not able to decide for themselves competently and well what to believe, how to act, how to live.

So it seems to me, if we respect students as persons – I take that to mean at least in part treating them as centers of individual consciousness and centers of moral standing with as much right to determine their own fates as we take ourselves to have – then this is what we have to do. Little children, infants, newborns, are not able yet to think intelligently about what to believe and how to live. So they start out dependent upon us. We can teach them in such a way as to keep them dependent upon us, or we can teach them in such a way as to try to liberate them from us. And it seems to me that we are morally required to do the latter.

If you buy that general story, then the case for science education just falls out as a special case and there's nothing special about science for that. I think this is related to Leona Schauble's plea earlier for liberal education. I want to say something more about that in a minute. Let me turn now to a slightly different point. I'll come back to what I was just talking about in a second.

Several of the speakers today, in particular Nancy Nersessian and Fouad Abd-El-Khalick, argued that teaching students science properly requires immersion in actual science practice and, in particular, modeling and the development of model-based reasoning competence. I'm myself unclear about what to make of this. Fouad said that immersing them in that way may be necessary – I think he said it is necessary but certainly not sufficient. I agree that it's not sufficient since, after all, somebody could be immersed in that activity and yet come away without any understanding of science or with a flawed understanding of science.

I'm not sure if it's necessary, either. I think the case needs to be made in more detail that it's necessary because I can imagine, although this is typical intuition mongering now, counter-examples; that is, people who haven't been immersed in this way and yet who have perfectly legitimate understandings of scientific practice, which is what immersion is supposed to get them. So I'm not sure that it's either necessary or sufficient. Nevertheless, it's clearly very important. I'm not speaking against immersion. I'm only wondering whether or not it's rightly thought of as either necessary or sufficient.

What about Steve Norris's question about which theories, which philosophies of science we assume in the science classroom? As Dick Grandy pointed out last night, these are controversial. That was Steve's point, but Dick made it last night as well. I think there isn't any straightforward answer to this question. It's not as if we can simply assume positivism or any of the other, many candidates. I think a more sensible approach is to think that if our idea is to get them to understand science and if we think, as at least I do – but I bet most of the people in the room do – that understanding science requires having some understanding of the philosophical issues that are discussed in the philosophy of science, then our students' understanding science must involve some serious exposure to the philosophy of science. Science educators talk about NOS, the nature of science, and it's important to teach the nature of science. I agree with that. But that raises Steve's question: which 'nature of science' are we supposed to teach?

I think the answer is that we should teach them what we think we know about it. We should make it clear to students when we are unclear about aspects of the nature of science, and we should be completely explicit about those aspects of scientific theory and practice that are controversial in the philosophical community, for example, the nature of evidence, the problem of induction, the general character of justificatory status, the epistemic rationale for particular experimental designs, the nature of explanation, and so on. All those questions that bother philosophers of science, it seems to me, ought to be part of the public school science curriculum.

I realize that that's an outrageous claim, and in a way I don't expect you to take it seriously, but I'd like to argue that at least it should be an ideal we should strive for. Now, that raises Dick Grandy's question about the practical tradeoffs involved. After all, there's only a limited amount of time, and the curriculum is already jam-packed with stuff. How could I possibly be arguing for, how could I possibly justify, the inclusion of a substantial amount of the philosophy of science in the public school science curriculum? And my answer is: I have no bloody idea. I don't know and I wouldn't hazard a guess as to what should be removed in order to make room for philosophy of science.

But I think it's worth it for science educators to think about the following question. Whatever the total content of science content there is in, say, tenth grade biology class or twelfth grade physics class or sixth grade earth science class, whatever that content, are students better off having that year's worth of content or are they better off having, say, 80 percent of that content or 75 percent of that content, with the other 20 or 25 percent devoted to these philosophical questions?

My inclination is that they'd be better off with part of their syllabus being devoted to the philosophy of science, because this is what will afford them the understanding that I think we all think they should have. Now, I know Fouad is going to run out and do an empirical study on this...

I said I was going to get back to something Leona Schauble said. I couldn't write it down fast enough, so this might not be an exact quote, but she said something like: what we need is to create learning environments in which there are classroom norms involving reasons and justification. I take it that her idea was that if we want students to reason well and to think about justificatory questions, then we have to create learning environments that encourage them to do so. And Leona talked about this as a matter of development. And that seems exactly right to me, that if we really take seriously the idea that we should foster student rationality, then we need to create the kinds of learning environments that do that. So I say godspeed to all you science educators who can figure out how to do that, because I have no idea.

I just want to say one more thing. I had hoped to talk about more stuff – there were so many interesting things today. There was the social epistemology dimension of the papers, Miriam Solomon's paper and Fouad's comment on it. And several other papers, including Greg's paper. The question of the relationship between the classroom and the research lab: how they're the same, how they're different and how that matters to science education? The question of naturalism, philosophical naturalism, which didn't arise explicitly but was underneath several of the presentations today,

Instead, I'll just say one thing about Steve Stich's presentation in which he talked about intuitions and intuition mongering. I have to admit that this hits me where I live, since I'm the kind of philosopher who does intuition monger, even though I'm reluctant to admit it in Steve's presence. I think the problem is harder than he acknowledges, for the following reason. It seems to me that whenever we reason, we rely ultimately on intuitions. I mean, even if it's the intuition that Q follows from P, ultimately, I think, intuitions are ineliminable from philosophical argumentation and indeed from any argumentation. If an argument is a string of utterances or propositions of the approximate form "this follows from that," then we're going to rely ultimately on our intuitions about logical form. I think that even Quine, a major naturalist and empiricist (and so presumably an opponent of intuition mongering) granted that.

If I'm right that intuitions go all the way down and there's no removing them from the scene altogether, then that poses a really deep question for philosophy: How can we do philosophy in a way that acknowledges what's problematic about intuitions – I agree with Steve completely that there are problems with relying on intuitions – without throwing the baby out with the bath water? I don't know what the answer to that question is. I just throw it out as my way of saying that it's not quite as easy as Steve was suggesting, that we can just chuck intuition mongering and move into the lab. I think it's harder than that, because even those who work in the lab rely on intuitions.

Harvey Siegel
Department of Philosophy
University of Miami

HELEN LONGINO

PANEL A

I took my brief to be to think about the proceedings earlier in the day, to think about either common themes or common questions that seemed to be emerging or things that I had noticed. Actually, I want to talk about what I took to be an absence which surprised me; and then I'll talk about a couple of tensions I thought were recurrent through the discussion.

This morning we got two pictures of inquiry which I thought, interestingly, had on the one hand some Popperian echoes and on the other some Kuhnian echoes. I thought that our first paper about the child as scientist where the child is represented as constructing explanations and possibly not going that extra step and testing them was a more Popperian conception, the idea that what you do as a scientist is just conjecture and then test your conjecture. To turn the child into scientist was to turn the conjecturer into tester.

The other model we had was the Nersessian kind of problem solving, the construction of models approach in model-based reasoning, which struck me as Kuhnian in not taking seriously or in seriously rejecting the distinction between discovery and justification, that feature of the Popperian view, focusing instead on those aspects of scientific inquiry that are problem solving aspects. Nersessian's model was also characterized by an inseparability of the cultural and cognitive aspects of inquiry. So there I saw echoes of Kuhn, not that either of these thinkers was trying to give us either a Popperian or a Kuhnian view, but it struck me as an interesting contrast.

But on to absence: What I was most interested in here was that there wasn't much of a focus on questions, that we were thinking mostly about answers to questions. So there wasn't much of a focus on how to elicit good questions, how to ask good, fruitful questions, which did strike me as odd in an investigation of inquiry. Our first topic was about learners engaging in scientifically ⌑oriented questions. I went and looked at the journals. Although for the first item – students or children pose questions – there were a number of entries, in many cases fewer entries than there were for some of the others. But for the second topic, children select questions or refine questions, there was very little. I thought that was really interesting.

There was a question in discussion about what prompted the children's theories, what were the contexts in which they came up with these pre-Socratic conceptions of the universe. But we didn't really stay focused on that question. One commentator called into question the assumption that children are intrinsically motivated to do science once they're introduced to the strategies of scientific inquiry. It struck me that we may have a similar assumption about questions, that somehow asking the right questions comes naturally or that it's evident what they are. So if there's not a lot that's already known about asking questions then I suspect it might be a fruitful area to explore in our efforts to characterize inquiry – after all, to inquire is to ask.

Common themes were really common tensions, as far as I could tell. The biggest one seemed to be about what the aims of scientific education are. As Fouad Abd-El-Khalick pointed out, there's this "through and about" mandate. "Through" and "about" are very different. Do we teach to produce scientists and engineers who will solve social problems like our energy problems, global warming, hunger, you name it? Or do we teach to produce literate laypersons who can be intelligent users of scientific information, critical users of scientific information, whether it's with their doctors or it's in city hall or it's in the voting booth when they're deciding who to vote for? Or are we educating them to be good policymakers about science, which to some extent they're supposed to be in the voting booth. But then there are also professional policymakers. So there's a whole apparatus about science, supporting science that doesn't involve the actual practice of research.

Are these aims really incompatible? Or is the issue one of when certain topics are to be introduced? These aspects do seem to be differentiated at the college and post-college level, although possibly problematically differentiated. I think part of the paradox here is that one can start to inculcate meta-level questioning and reflection at the lower levels. There were some examples of this. But they're not necessarily part of the culture of scientific practice.

There were questions about what immersion is, what immersion entails, what's the immersion unit. Well, of course, I don't know; but some of that seems to be how much of the overall culture of science we are trying to teach when we're teaching inquiry. If we're teaching to make scientists and engineers who are going to solve our technical problems, it might be inappropriate or even self-defeating to introduce philosophical, social, and political economy perspectives. On the other hand, if we're teaching to produce literate laypersons who are going to be responsible citizens, it's irresponsible not to. I think that's the problem.

I have to say I have a great deal of sympathy for educators who are trying to navigate this paradox. But thinking about it, I guess it seems useful to ask whether there are certain core competencies that one wants to inculcate in children. Steve Stich is telling us that, of course, we don't know what those are, that we really don't know what is at the heart of science. I think that's probably right.

But there are some things we do want children to learn. We want them to know how to frame and refine a question. We want them to be able to screen possible answers. We want them to be able to design and conduct experiments, to set up observational tests. We want them to have a respect for basic empirical procedures, it seems to me. That's one of the things that we want; and we want that not just for potential scientists, but we want that also for the users of science, because we think it's critical to being an intelligent consumer of science. So it seems to me, not necessarily thinking about science with a big S, but teaching empirical procedures might be a way to focus some of these questions.

I also want to say something about the social dimensions. I was thinking, well, what's the relevance of thinking about the social character of science. Science is a social phenomenon, but it's social in different ways. In one sense, it's both collaborative and competitive. Especially contemporary science is very collaborative; people work in groups most of the time. And in contemporary science, the overall structure within which scientists work is a competitive structure. There's competition for funding, but there's also competition to be first, competition for credit, competition to be right, competition to have one's ideas be the ideas that are accepted.

So it's both collaborative and competitive, and to be successful in that collaboration and competition, one needs to have basic knowledge and basic skills as well as specialized knowledge and skills. So we've got both commonality and distribution of competences. That aspect is certainly important for problem solvers who need to know that they need to draw on the work of others. In the kind of interdisciplinary groups that Nancy Nersessian is studying, for instance, that's really evident.

Some of us go further than talking about the collaboration and competence and want to say that somehow knowledge itself is social. In that proposal we're saying a couple of things, that what matters is what the scientific community comes to agree or disagree on; that is, what the scientific community thinks is important, because that's what scientific knowledge is − that's the knowledge that's transmitted. That's the base from which future knowledge is developed.

So in making that claim, we're rejecting the view that scientific knowledge is a collection of results produced by individual researchers that all gets published and put into the great book of knowledge. The claim that science is social knowledge − that scientific knowledge is social − is also a response to the diversity of content and the diversity of norms that one can see if one looks around in the contemporary community, so it's the kind of move to solve the epistemological puzzle that's created by that diversity.

Now, Miriam Solomon and I have really different ways of representing that sociality; and if we were to include more of our colleagues in science and technology studies, we'd find ever more diversity. But the point of agreement is that there are many more factors involved in problem choice and hypothesis acceptance, than empirical considerations. Even if the appropriate consensus is the one that has all the empirical successes, groups and individuals are moved to acceptance by many factors other than purely empirical considerations.

That may not be very important for problem solvers to know, but it probably is important for the critical user of science, the critical user of scientific information. It's also important for the citizen policymaker. It's probably also an important understanding of science, in order to be critical. It's also useful in inculcating a tolerance of the plurality and diversity that we do see. I think that too often we, the culture, expect a scientific solution to be unique, and that may not be the case. The result of not getting unique solutions can be disaffection with the only process that is going to give us any solutions at all. So I think that it's important to communicate this somehow.

Another way in which science is social is that modern science and modern industrial and post-industrial science and institutions are interdependent. Society needs science. Our economies now are dependent on science-based technology and engineering. And science needs society. Science cannot proceed without social support. It takes a lot of money, a supply of appropriately educated personnel and a community that finds value in scientific knowledge. So it's not just the interactive parts, the collaboration and competition; that make science social. Science needs money, it needs clients, it needs users. It needs a whole structure of support.

Another aspect of the interdependence is that scientific research or research labeled as scientific is used to justify social arrangements and social policy, which policy and arrangements then either in the near time or the long term have an effect on research.

We were struggling with whether or not to introduce the sociality or how to introduce the sociality of science in teaching. It seems to me the question is which of these aspects of sociality should be introduced and how it should be introduced. And we should seek some clarity as to the purpose we're trying to achieve when we try to inculcate some appreciation of these areas of social dimension.

Helen Longino
Department of Philosophy
University of Minnesota

DAVID HAMMER, ROSEMARY RUSS, JAMIE MIKESKA, AND
RACHEL SCHERR

IDENTIFYING INQUIRY AND CONCEPTUALIZING STUDENTS' ABILITIES

INTRODUCTION

Duschl and Grandy (this volume) set the goal of determining areas of consensus and lack of consensus in understanding goals of promoting inquiry in K-12 science education. There are probably a number of points on which we can all agree.

First and foremost is that promoting inquiry is not simply a method for teaching but part of the substance itself our instruction should address. That is, we want students to develop abilities and inclinations for science as inquiry, and we see engagement in inquiry as essential for genuine understanding of the body of knowledge professional science has produced. Of course, "we" here means the science education research communities represented by the invited attendees, and if we are in consensus on this point we will do well to make it clear, since that view does not generally shape existing instructional practices and curricula. These typically distinguish "inquiry" from "content," the latter being the traditional canon. Sometimes that distinction has inquiry as a pedagogical choice, its effectiveness judged against other approaches through assessments of students' understanding of the canon; sometimes the distinction has inquiry as an add-on set of objectives, to pursue in various side activities – science fairs, egg drops, and so on – separate from the actual syllabus.

We may also have a consensus, as Duschl and Grandy (this volume) suggest, that scientific inquiry involves some form of dialogue among evidence, theory, and models:

> [W]e recognize that the dialogic processes by which theory evaluation and data construction occur in the community are an indispensable part of the rational structure by which science unfolds.

That view does not shape existing practices either; most science curricula remain heavily empiricist in their tacit and explicit epistemologies. It conflicts as well with some views of science evident in the NSES Standards, especially in chapter 6, which generally receives the most attention (Labov, 2004): Describing "science as inquiry" objectives at different grades, the content standards clearly focus on empirical investigations rather than on theoretical inquiry or modeling. This emphasis may be as much or more for developmental reasons as for philosophical ones, and if so it bears on Grandy's and Duschl's important question:

> One central question is where in the trajectory of science education we should introduce these complications of simultaneous revision of theory and

Richard A. Duschl and Richard E. Grandy (eds.), Teaching Scientific Inquiry: Recommendations for Research and Implementation, 138–156.
© 2008 *Sense Publishers. All rights reserved.*

data. We know that six year olds come to the classroom with concepts, theories and belief in data. Can we already introduce some elements of the dialogic processes at that level?

This is where we would like to focus our paper, on six-year olds. What do they bring to the classroom, and can we "introduce some elements of the dialogic processes?"

We organize this chapter around a first-grade class discussion about falling objects, hoping to accomplish three things. The first is to use the case to ground discussion and possible consensus with respect to what we consider inquiry. Duschl and Grandy (this volume) discuss the myriad of ways educators characterize inquiry, from asking questions to engaging in "hands-on" activities to conducting experiments. A compact definition is hard to find at all, let alone one for which there is a substantial consensus. Perhaps, however, we can identify examples and build consensus in that way, agreeing on instances a definition should include. In this we use the class conversation as a case-in-point, not as data but as reference; we could generate hypothetical examples to the same purpose.

The second thing we hope to accomplish is to use the conversation as data to support claims about these children's abilities, as a modest contribution to the extensive body of evidence that children have abilities for dialogic processes of inquiry (Chinn & Malhotra, 2002; Duschl, 2004; Gopnik & Sobel, 2000; Karmiloff-Smith, 1992; Metz, 2004; O'Neill & Polman, 2004; Samarapungavan, 1992; Sodian, Zaitchik & Carey, 1991; Tytler & Peterson, 2004).

The third is to use the case to support claims that these abilities are not stable features of the children's reasoning (Siegler, 1996) but variable with context, even within a single conversation (Hammer, 2004). Over the course of the conversation we see the students shifting from one sort of activity to another, each with its own local coherence. Some of what they do looks like the beginnings of scientific inquiry; some of what they do does not, and they make the transitions from one to the other with apparent ease.

These claims about the first-graders' abilities bear on the question of what children bring to the classroom. It is an ontological question, asking what we as researchers or teachers should attribute as existing in children's minds. In the closing section of this paper we discuss the implications for curriculum and instruction of attributing *manifold resources* that may or may not be active in any moment, in contrast to attributing stable features of mind in the form of concepts, theories, and beliefs. On this view, stabilities and coherences may be local, and we understand them in terms of the students' being able to *frame* an activity in different ways.

FIRST GRADERS DISCUSS FALLING OBJECTS

This account is based on videotapes and transcripts of the discussion, the teacher's case study (Mikeska, 2006), and an analysis of the students' reasoning (Russ, Mikeska, Scherr & Hammer, 2005).

Mikeska prepared her case study of this class as part of a project, *Case Studies of K-8 Student Inquiry in Physical Science* (SIPS), a collaboration among K-8 teachers and university faculty, post-doctoral research assistants, and graduate students. The project was funded by the National Science Foundation[1] to produce case studies of children's inquiry for use in pre-service and in-service teacher education. Over three school years, the teachers and staff met every other week in two-hour sessions, working in groups to discuss "snippets" of data from the teachers' classes, typically videotapes of classroom conversations. The teachers then chose their favorite snippets to revise and expand into case studies. Our purpose with these cases is to provide materials to help prospective and current teachers gain experience interpreting student reasoning and thinking about possible ways to respond.

Mikeska had several objectives for the particular class we use to organize our comments about scientific inquiry. She describes those goals in her case study:

> *Making observations* and *drawing conclusions* are two essential components of the county's science curriculum, but my students were already doing both. I also knew they would not be challenged by the county objectives that they be able to "give examples to demonstrate that things fall to the ground unless something holds them up" and to "describe the different ways that objects move (e.g. straight, round and round, fast and slow)" (Curriculum Framework, MCPS, 2001).
>
> I wanted them to go further than these basic objectives, to use their abilities to make observations and draw conclusions to delve deeper by asking "why" questions. I consulted the science specialist for ideas, and he suggested a classic experiment: Drop a flat piece of paper and a book at the same time from the same height, and then drop a crumpled piece of paper and a book at the same time from the same height, and compare the difference in results. That seemed perfect. (Mikeska, 2006)

In this section we present the case study in three segments: (1) Predictions for the book and the flat paper, (2) Discussing the experimental results of the book and the flat paper, and (3) Discussing the book and the crumpled paper. Following each segment we make comments regarding our first two purposes in this paper, using the segment to illustrate possible points of consensus with respect to what we consider inquiry, and using it to make claims with respect to these children's abilities.

Predictions for the Book and the Flat Paper

On the first day, Mikeska posed the question to the students, "What will happen if we drop a book and a [flat] piece of paper at the same time?" She did not have them try it, just talk about it, and the students excitedly predicted the book would fall first. Autumn said "The paper will go down after the book," and Allison agreed, saying it was because the book "had more force." Ebony explained that the book would fall first because "the book is going to fall down first because the book is more stronger than the paper." Other students agreed; with many children speaking at once: "Yeah!" "The paper is so light." "The book is heavier than the paper."

Mikeska settled them down to have an orderly conversation, and asked Ebony to begin:

> Ebony: Um. The book is going to fall down first because, um, they have more strength, um, than the paper because the paper, it has no strength. Because if you put it on, um, the things that you weigh yourself on, the paper is going to weigh nothing but the book is gonna weigh like one or three pounds.

Mikeska decided to have the students focus for a moment on understanding Ebony's idea, asked them to paraphrase what he'd said, and then asked whether they agreed that "the book would hit the ground first because the book has more strength than the paper." Autumn agreed because she'd seen this happen at her house, and Allison offered that they "could test it at our house this afternoon and then tomorrow morning we could tell you what happened."

Mikeska kept them on the topic of explaining their reasoning, instead of shifting to the topic of collecting empirical evidence. Ebony elaborated that "the paper will fly down like this," gesturing back and forth with his hand to show how the paper will move. Several other students mimicked his gesture and agreed with him. When Mikeska asked, "What's making the paper go like that?" several students answered immediately, "The air!" Mikeska asked what the air was doing, and Allison answered, "It's blowing it." For a few minutes the students reiterated and reaffirmed this prediction, Allison emphasizing that the book would land first even if "you dropped the paper faster," or released the book from higher than the paper.

Toward the end of this prediction discussion, Ebony added another idea.

> Ebony: The paper won't, when it hits the ground, the paper, it won't make a noise, but the book will.

Several students agreed and took the opportunity to demonstrate, with their hands, the sound the book would make hitting the floor. When Mikeska turned them back again to talk about what the air does to the paper, several students said that it "blows it." Some began talking about differences between the paper and the book, in particular saying that "the paper is thicker than the book."

Examples of inquiry? There are two aspects of the students' thinking in this first segment of activity that, we expect, would be consensus examples of inquiry.

The first is the students' generating causal explanations. When Mikeska first posed the question, several students quickly responded not only by predicting the book will hit the floor first but also with causal explanations for why: The book has more "strength" or "force" or "weight" than the paper. The students were not using these words in the same ways a physicist would, but they were expressing a sense their peers (and we) could understand. When they explained further that the paper moves back and forth while the book falls straight down, and the teacher asked what makes the paper move that way, the children identified "air" as the causal agent. Asked what the air does, they added that the air or "wind" is "blowing" the paper.

The second respect in which this might be a consensus example is the children's explicit reference to evidence to support their answer. When they demonstrated how piece of paper moves, they were clearly making use of what they had seen. Autumn reported having seen this happen at home, and Allison followed this with a suggestion that they all try it at home and report what happened. Ebony made a rhetorical suggestion for empirical evidence in saying what a scale would read measuring the weight of a book or a piece of paper, to support his contention that a book weighs more. Later, he cited a different kind of evidence based on things he already knew to support his prediction, that when a book hits the ground it makes a sound, but paper does not.[2]

Children's abilities. Note that Allison and Ebony, with others following, responded to Mikeska's first question ("What will happen if we drop a book and a piece of paper at the same time?") not only with phenomenological predictions but with causal explanations for why: The book has more "force" or "strength" or "weight," which moves it more quickly to the ground. That is, the children were spontaneously and explicitly mechanistic. Later in the conversation, describing how paper moves as it falls, they did not move immediately to explain the cause, but they understood the question when Mikeska asked, "What makes the paper move that way?" They responded immediately, "The air!" Clearly the children were capable of mechanistic reasoning, entering into it on their own, and of understanding a request for mechanistic explanation.

Discussing Experimental Results

The next day, Mikeska reminded the children of their conversation from the previous day. In fact, as she had it recorded on video, she showed them the tape and asked them, as they watched, to keep track of the ideas they had raised. With her help they produced a list on chart paper, as shown in Figure 1.

Later in the day, she had the children break up into small groups to try the experiment, having demonstrated what they should do: Hold the book and the sheet of paper at the same height, and let them go at the same time.

After several minutes of this activity, she called them back into one group to discuss what they had found. Starting the discussion, she called first on Ebony.

> **What will happen when you drop a book and a flat piece of paper at the same time?**
> The book will fall first.
>
> **Why?**
> 1. The book has more force.
> 2. The book is stronger than the paper.
> 3. That's what happened when she did it at her Dad's house.
> 4. The paper will float

Figure 1. Ideas on chart paper (from Mikeska, 2006).

Teacher: What happened when you dropped the book and piece of paper at the same time, at the same height? What happened? OK, Ebony why don't you go ahead and begin.

Ebony: To me, first, the paper fell first.

Several students: No way, No! Whoa!! The book fell first.

Ebony: No, to me the paper fell first.

Student: It fell at the same time.

Ebony: No, the book, um… the paper fell… the paper fell first to me!

Student: Yeah, but not to me!

Jorge: Yeah, but did the book fall first, just like the paper?

Ebony: No, the papers fell first. No, the paper – to *me*.

Alison: To Ebony – to Ebony the paper fell first.

Jorge: And to all of us the book might of fell first to us.

Students: Yeah.

Jorge: Our paper – our paper goes slowly. It's, it's a little bit, ah, ["out of practice"?].

Rachel: With me and Julio twice the book and the paper tied – twice.

Diamond: Went at the same time.

Rachel: They both tied twice.

Allison: What do you mean 'tied'?

Student: They went both at the same time.

At the time, Mikeska naturally expected that the students would come back from trying their experiment having confirmed their unanimous prediction. To be sure, that was what she had seen happening, as she circulated in the room watching the students conduct their experiments. Of course, she could not be everywhere at once, and the first students to speak were several she had not seen during the activity. To have Ebony saying that *to him* the paper fell first, and then Rachel reporting that *with her and Julio* the book and paper tied, *twice*, was surprising.

After the fact, we have the advantage that we can go find the video of Ebony dropping the paper and the book. He held the paper under the book and dropped them so that the book pushed the paper down ahead of it as they fell. (Rachel and Julio were not on camera.)

Mikeska asked others to say what happened when they did it, and several students reported that the book hit first. Diamond remarked that Henry was "putting the paper so that it could fall first." When Mikeska asked her to explain, she said he was dropping the paper first, but Henry denied it. With this as a possible direction for the discussion – here was a plausible answer for different outcomes – Mikeska pressed for an explanation for how it could be that "we got all these different results." Allison reviewed how her group had gone around two times, with different outcomes, and Mikeska asked, "Why did that happen?" Ebony answered, "Cause we did it twice," and then Rachel had a new idea.

Rachel: Forces of gravity?

Teacher: Rachel has an idea.

Rachel: Forces of gravity?

Alison: Yeah!

Diamond: What are forces of gravity?

Rachel: Gravity –

Alison: Gravity – you know how when we jump we always land back on the ground.

Rachel: Exactly. It's what keeps us down on the ground.

Student: And like ground magnets.

Ebony: No gravity. No gravity is when you're like in space and you can never fall down.

Student: You know, you just float in the air. [Ebony nods]

Alison: Gravity – see how when I jump [jumps] I'm just landing at the same place on the ground that... because gravity, gravity is just pulling me down.

Other children began to jump up and down, giving gravity a try, and Mikeska asked them to stop.

> Teacher: Oh, we're not all going to – They already, they already showed us. (Students jumping.) You can sit down. (7 second pause)
>
> Students: [Laughter.]
>
> Teacher: [to Rachel] Okay, so, so what you're saying is that a for – what is a – you're saying the force of gravity –
>
> Rachel: Is pulling it down at different times.
>
> Teacher: So, you're saying the force of gravity is pulling the book down at a different time than the paper.
>
> Student: Yeah.
>
> Rachel: Yeah, probably. And, sometimes it's pulling it down at the same time, or pulling the paper down
>
> Student: Before the book.

Mikeska asked why gravity would pull down differently at different times, but no one had an answer.

Examples of inquiry? It may be more difficult to find consensus examples of children's scientific inquiry in this segment than it was in the first. One could be Rachel's emphasizing "twice" as an early version of valuing replication. Another might be the way Allison and Diamond asked other children for clarification: "What do you mean, 'tied'?" and "What are 'forces of gravity'?"

But the segment is more interesting for its contrast from the first. We would like to try for a consensus that the children slipped out of the sort of inquiry educators would like to promote as the beginnings of science. Ebony, to begin, seemed to have shifted in his purpose and approach from the previous conversation, when he had been so articulate in justifying his prediction that the book would hit first. Coming back to the group, he simply announced his result to the class, and did not explain the reasons for its difference from their prediction. From the first, he emphasized that the paper fell first *to him*, probably anticipating his result would differ from others', and that was how he responded to their disbelief, by reasserting his claim and emphasizing "to *me*." Allison supported his stance, affirming that "To Ebony, the paper fell first," and Jorge agreed, adding "and to all of us the book might of fell first."[3]

This is not to say the children were being irrational or unintelligent – and we need to be clear about that or we may be uncomfortable excluding anything children do from our objectives. On this interpretation, Ebony was being *creative;* he was telling the story of *his* experiment much in the same way he might tell a story about what happened to him on the playground the day before. He may have chosen to tell that story because he felt the answer to the original question was obvious and uninteresting. Allison and Jorge were providing a socially harmonious resolution to the difference of opinion, allowing Ebony to have his opinion and others to have theirs, much as they might over a matter of preference: Different people have different experiences, and they are entitled to different views.

So the children's approach here had its own rationality; we might characterize it as a version of show-and-tell, where there would be no reason to require different experiences to align. But it was a different approach from earlier when Ebony and other children had explained their prediction that the book would fall first in terms of causal mechanism. Here, neither Ebony nor Rachel nor Allison attempted to explain why their results (the paper falling first or the paper and the book tying) differed from the rest of the class' results, or from their predictions on the previous day. For the rest of the class, except for their initial disbelief at Ebony's result, most seemed content to accept disparate results; they accepted that everyone could have her or his own answer.

Mikeska was concerned by this resolution (or lack thereof) and tried to make the inconsistency problematic, asking how they could come up with different answers. Finally, Rachel tried "Forces of gravity?" as a possible answer, looking at Mikeska with her question. In this moment, it seems, she was trying a different kind of answer, one based in "science words," on the possibility that was what the teacher wanted. By that interpretation, this was another example of something we should not consider as scientific inquiry. Again, this is not to say it isn't rational: The students have heard this term, and it is perfectly reasonable to wonder if it is relevant here, 'Is this related to these words I've heard people use?' However, this behavior cannot be considered valuable inquiry because when students use vocabulary without being able to articulate a meaning or relevance, they will not be able to make progress in reasoning about the mechanism at work.

A more liberal interpretation would be that Rachel was thinking "forces of gravity" are variable like the wind, and their variation leads to the variation in results. By that interpretation, we might see her reasoning as more scientific. Like the students' previous use of "strength" and "weight," this would connect to some other aspect of their experience. Note, though, we need to read that in to the thinking: In the earlier discussion, children explicitly referred to measuring weight on a scale as well as to personal observations of how paper falls and the sound a book makes hitting the floor; we had a clear basis for seeing their reasoning as connected to their kinesthetic experience of weight. Here, we cannot be sure what meaning Rachel attached to the term "forces of gravity" so we cannot directly assess whether she could use it in a way that would be helpful for making sense of the situation.

For a moment, the children digressed to a discussion about gravity, and we might see some of this as explanatory: Gravity pulls objects to the ground, "like ground magnets"; without gravity things do not fall. In this conversation we find some evidence that Rachel knows the role that gravity plays in other situations, although we cannot judge how she relates that to the problem of the book and the paper. Only when Mikeska explicitly asks Rachel to apply her understanding to the problem does Rachel complete Mikeska's sentence by explaining that the force of gravity "is pulling it [the paper and the book] down at different times."

Children's abilities. If there may be consensus that, for the most part, the children were not engaging in productive science in this segment of the conversation, it would be hard to understand that in simple terms of ability. It would be difficult to make a case, in particular, that Ebony lacked the ability to explain why his results were different from other students'. He had been quite articulate the previous day, in justifying his prediction that the book would hit first, based on mechanism (the book weighs more) and on consistency with other observations (the paper moves back and forth, the book hits with a sound). And there is no reason to suppose he was not able to follow Mikeska's directions for how to drop the book. Instead, it seems more reasonable to say that although these students were capable of reasoning mechanistically (as they had done on a previous day), they did not always do so.

The Book and the Crumpled Paper

Mystified for the moment by how the class was going, and hoping to get back to a productive discussion, Mikeska demonstrated dropping the book and paper together three times. Several children shouted "I knew it!" and "The book fell first"; someone said, "See Ebony?" Mikeska asked Ebony again – "you found out something different" – and he simply repeated "the paper fell first," shrugging his shoulders. Allison brought them back to the idea of gravity, suggesting that "gravity probably just pulled the paper a little more than it pulled the book." Ebony joked "Maybe it's tired," and others laughed and repeated the line: "Yeah, maybe the gravity's tired!"

With the discussion beginning to deteriorate, Mikeska moved the students on to the next question: What if they were to crumple the paper? Then which would win? Jorge began by saying that he knew they would tie because he had tried this at home. Diamond and several other students agreed, and Allison explained that the crumpled paper "has a little more weight," which the students had decided on the previous day was a reasonable explanation for why the book falls before the flat paper. After this brief discussion, Mikeska asked the children go back to their small groups and try the experiment.

When the class reassembled, the students quickly agreed that the book and crumpled paper hit at the same time, and they again focused on why, without any questioning by the teacher. Brianna opened the conversation saying, "They will fall at the same time. 'Cause they both got the same strength together." Rachel explained that crumpling the paper gave it more strength, and Julio said that crumpled paper is "kind of heavy." This explanation became a point of contention, and the students discussed it for several minutes without intervention from the teacher. In response to Julio's idea, Brianna said, "If it's balled up it's still not heavy, it's the same size" and reached to grab the ball of paper. She un-crumpled it, held it in her hand and bobbed it up and down, feeling its heft, and said "It's still the same size!" Allison tried to insist that the balled up paper was no longer light, arguing that her "dad could probably throw it up to the ceiling and he wouldn't, and he wouldn't say it's light." But more children, after Brianna's demonstration, agreed with her that the crumpled paper was "still light."

Having ruled out weight as a possible explanation for why the crumpled paper now falls at the same time as the book, the students continued to converse for a few more minutes. Diamond had two more ideas, one that the crumpled sheet "had more pages [?] when it's balled up... it looks like the book have a lot of pieces of paper." A bit later, she suggested that with the paper crumpled it could "roll." Diamond explained the importance of the paper's shape by saying that the crumpled paper falls more like the book, straight to the floor.

> Diamond: 'Cause the piece of paper was balled up, it don't go like this no more (Shows a rocking motion with her right arm.)
>
> Brianna: No, yeah. It don't, yeah. It just drops, kind of like the booklet.

Mikeska decided to let the exploration continue to another day, challenging the children to do the same experiment with things they find at home.

Examples of inquiry? With the new topic of crumpled paper, the students moved back into a form of inquiry we contend science educators should support. The terms "forces" and "gravity" disappeared, and the children have returned to ideas of causal mechanism, discussing whether and how the crumpled paper has more strength or weight. Brianna's move in particular, to open the ball of paper and show that it is the same size, should be something we all recognize as productive inquiry. She drew on their experience and intuitions about an everyday activity (crumpling things up), to refute an idea that initially seemed a plausible explanation. Diamond's comment about the shape of the paper also seems like valuable scientific inquiry, identifying a specific factor other than weight that might have contributed to the change in outcome. As well, she reminded the class of their reasoning from the previous day, that the flat paper took longer because it moved back and forth on the way down.

We suggest that Ebony's joke is another example of inquiry, although here we do not expect consensus. It may seem odd to consider a joke as an "element of the dialogic processes" of science, but we believe Ebony's that "maybe [gravity]'s tired" was a productive contribution. In particular, by making fun of the idea, Ebony was not only ruling out an explanation but also calling attention to it as ridiculous. In this way, the joke helped express a sense of what could and could not constitute an acceptable explanation. Other children seemed to get it, although it is possible some of them were only picking up on the laughter rather than understanding and accepting the tacit premise.

Finally, Allison's comment that her father could throw the crumpled ball up to the ceiling but "he wouldn't say it's light" seemed an attempt to draw a distinction in meaning. There is the question of whether something is easy to lift or throw in the air, and there another question, which she does not articulate, in which something might be light or heavy. If she was trying to distinguish shades of meaning, she was doing something else educators should support as inquiry.

Children's abilities. Several children expressed some version of the idea that crumpling the paper makes it heavier (or gives it "more strength"), and from a developmental perspective the most obvious interpretation would be in terms of classic Piagetian conservation tasks. On this view, the children's reasoning is a sign they are still pre-operational.

Another interpretation would be that some of the children were thinking of "strength" and "weight" in a different sense they could not clearly articulate. That several children explained that the book and the crumpled paper have the "same strength" suggests they may have meant something different, in that moment, than what adults call "weight": It is unlikely that they suddenly came to believe the crumpled paper would have the same heft in their hands as a book. So, for example, a child picking up a small piece of lead might describe it as "heavy," although she could lift it easily or even "toss it to the ceiling."

Brianna's flattening the paper, however, is an especially clear indication that she was able to reason about reversibility and conservation. Moreover, the class's general acceptance of the idea that crumpling the paper does not change its weight suggests other children were able reason in this way as well.

A WORKING DEFINITION OF INQUIRY

To review, using this discussion as a case-in-point we are suggesting some examples of what science educators might agree are examples of what we mean by inquiry. Note that for this purpose it is not essential that everyone agrees that these interpretations are correct for these students at this time; everyone need only agree that if these interpretations are correct, they describe the sorts of ideas and reasoning educators should support.

The examples we suggested were of children:
– arguing from a sense of mechanism to explain a prediction, that the book will fall faster than the paper because it is heavier, apparently thinking of the greater agency downwards causing the greater effect (speed) or, at another moment, thinking of the air or wind blowing on the paper as slowing its fall;
– arguing from a sense of mechanism to explain a result, that the book and crumpled paper must have the same downward agency acting since they fell at the same rate;

- drawing connections to other observations, of the paper falling in a flutter back and forth rather than in a straight line like the book, and of the book hitting the floor with a noise rather than quietly like the paper;
- asking for precision and clarity in their explanations;
- valuing replication, in emphasizing repeated results;
- drawing distinctions between two similar ideas – one (difficult to articulate) sense of how "light" the paper is versus how hard or easy it is to lift;
- checking for consistency among lines of reasoning: There is no more paper, so how could it be heavier?

We also suggested some examples we would all agree are not the sorts of ideas and reasoning educators should support, namely of a child or children
- altering an "experiment" to produce a desired result, and presenting the result as comparable to others;
- accepting inconsistencies in results as matters of personal choice or perception;
- using terminology ("forces of gravity") without tangible meaning.

onsidering these and other examples of what we see as inquiry, we suggest that all of the positive examples fit the description of children *trying to form coherent, mechanistic accounts of natural phenomena.* "Mechanistic" here means bearing on the children's sense of causes and effects, from what they have experienced or inferred.[4] "Coherent" here means the account is internally-consistent and stable: different parts of the account do not contradict each other, and the account "holds together" over time.

In these examples, the students' reasoning about "strength" and "weight" is mechanistic: The stronger downward cause on the book produces the effect of faster motion. It is also coherent: The account fits with what they remember seeing when paper falls, with what they remember hearing when a book hits the floor, and with what they remember feeling, holding a book compared to holding a piece of paper. It fits as well with their more general experience of pushing and pulling on things to make them move, although no one actually makes this explicit. Other examples in the list fit the description as well: Asking for clarity, valuing replication, distinguishing shades of meaning and checking for consistency among lines of reasoning are all aspects of trying to develop a coherent account.

And, we suggest, the description excludes the examples of what is not inquiry. Making a change to an experiment to produce a different outcome – and not reporting the change! – does not contribute to forming a coherent account; neither does accepting inconsistencies without explanation. Using technical vocabulary without tangible meaning is not attending to mechanism.

These examples illustrate what we believe we could find more generally, that moments we assess as inquiry fit this general description. So perhaps we could generate a consensus for this as a working definition: *Inquiry in science is the pursuit of coherent, mechanistic accounts of natural phenomena.* It is a fairly broad definition, as it could include a wide range of epistemic activities – from consulting an authority to conducting an experiment to imagining an outcome – all of which can be part of a student's or scientist's attempts to form a coherent, mechanistic understanding. We contend that this breadth is appropriate.

The definition will clearly need refinement, at least to understand professional science. The idea of "causal" becomes problematic, for example, in describing various ideas and approaches to physics (e.g. Lagrange's formulation of classical mechanics, Heisenberg's formulation of quantum mechanics, etc.). The idea of "coherent" may be problematic for reasons Duschl and Grandy (this volume) discuss, citing philosophers of science such as Cartwright (1983), who argues against a view of physics as the pursuit of principled coherence. We believe these refinements are possible, however, with perhaps a more generalized notion than *causal* and a more qualified version of *coherent*.

WHAT CHILDREN BRING TO CLASS

Our first purpose in presenting the case study was to provide some examples to ground discussion about what science educators think of as inquiry. Our second purpose was to provide a bit of evidence regarding children's abilities. In particular, we have argued that these children showed abilities for inquiry, including reasoning from their sense of mechanism, drawing connections to aspects of their experience, and attending to consistency among ideas and observations.

In this we contribute to an extensive body of evidence that children bring abilities to begin inquiry in science, such as pre-school children's abilities for reflecting on knowledge (Gopnik & Meltzoff, 1996; Karmiloff-Smith, 1992); kindergarten-second graders' abilities to distinguish hypotheses and evidence (Sodian et al., 1991; Tytler & Peterson, 2004); second graders' abilities to analyze uncertainties in experiments they design (Metz, 2004) and for attending to empirical and logical consistencies in theories (Samarapungavan, 1992); third-sixth graders' abilities for generating, assessing, and working with analogies (Atkins, 2004; Louca, Hammer & Bell, 2002; May, Hammer & Roy, under review); and fourth-sixth graders' for distinguishing observation from theory and seeking consistency in reasoning (Chinn & Malhotra, 2002).

There seems to be a broad consensus that children bring a range of abilities for scientific inquiry. These include abilities educators have at times classified as "abstract" or "theoretical," such as for reasoning about the consistencies and inconsistencies among ideas, for drawing connections to different parts of their understanding, for thinking in terms of causes and effects.

Variability

There is not yet consensus with respect to how educators should conceptualize the form in which students bring those abilities. We use the term "ontology" to refer to this conceptualization: What forms of cognitive structure do educators attribute as existing in students' minds? A number of authors who ascribe more sophistication to children than depicted by traditional accounts continue to model children's reasoning abilities as stable and consistent, as stages of development and/or tacit theoretical frameworks (Burr & Hofer, 2002; Gopnik & Meltzoff, 1996; Kuhn, Black, Keselman & Kaplan, 2000). We describe these as *unitary* ontologies in that the forms they ascribe to minds are intact units, seen as either present and functioning or not present. Other authors have argued that reasoning is more variable than unitary models depict (Feldman, 1994; Karmiloff-Smith, 1992; Metz, 1995; Siegler, 1996).

Unitary accounts compete in the literature with *manifold* ontologies that also ascribe cognitive units but at scales that need not correspond with ideas or behavior. Minsky (1986) coined the phrase "society of mind" to capture this view of cognitive phenomena arising from the activation of many "agents," seen as existing in the mind but not necessarily functioning in any given moment. DiSessa's (1993) "phenomenological primitives" ("p-prims") is a manifold model of intuitive physics, specifically intuitive sense of mechanism. We have suggested a similar approach to modeling student epistemologies (Hammer & Elby, 2002). In that work and here we use the term "resources" to refer, generically, to the elements in these models of mind, including agents, p-prims, and schemata (Rumelhart, 1980) as particular models for these elements.

Our analysis of this case study argues for views of children's reasoning as variable, in several respects. First with respect to the substance of the children's thinking, there was certainly evidence of children believing that heavier things fall faster than light things. But the children's first prediction for the crumpled paper had it hit the floor at the same time as a heavier book. No doubt they were influenced in that prediction by contextual cues – the crumpled paper question followed the flat-paper question, and the children likely expected contrast. Still, the instance puts pressure on the attribution of a robust misconception that heavy things fall faster than light things: Why would the hint of contrast with an earlier question be sufficient to suppress it?

There was also evidence of variability in the children's use of their abilities for inquiry. During the first segment, the children quickly agreed that the book would fall first and began to give various reasons to support this answer, including mechanistic arguments for why the book falls first and connections to their experience: the book is heavier; the air blows on the paper; the paper moves back and forth; the book makes a loud noise when it hits. During the second segment, after they had tried dropping the book and paper, the conversation shifted dramatically, to an unproblematic acceptance of different people having different answers and then to the obscure and unclear notion of "forces of gravity." During the third segment, discussing the book and crumpled paper, the students returned to a scientifically more productive mode of sense-making, discussing reasons the paper might have fallen more quickly this time, debating whether or not the crumpled paper could be heavier and considering how shape affects an object's descent.[5]

From our perspective, this should not be surprising: Children are able to understand and engage in many different kinds of epistemic activities, from fact-finding ("go see if it's raining") to tangible explanation ("if clouds are all full of water how do they stay up?") to playful explanation ("the clouds are crying because the wind punched them") to story-telling ("once upon a time..."). Different activities call on different abilities; why would we expect children to stick to one kind of activity? In this example they seemed to move from an activity of tangible explanation, for why they expected a book to fall faster than paper, to an activity of show-and-tell, of what happened when they tried dropping books and paper. In the first segment, their explanations needed to make sense, in some common way, with respect to what they knew about causes and effects. In the second segment, after a moment of puzzlement — some children possibly still thinking about what makes physical sense — most seemed to shift into a tolerant stance. Soon the conversation would revolve around "forces of gravity," and one child's jumping prompted others to try that activity, too.

Newton and Newton (1999) found similar phenomena in 10 year-olds sometimes being unable to recognize the "kind of understanding" that was appropriate to a particular context. Diakidoy and Ionnides (2004) found that second graders in their study confused hypotheses with statements of preference. Rather than understand these confusions in terms of levels of development in the children's abilities, however, we understand them as arising from shifts in the students' understanding of what they are doing, from the first activity of tangible explanation to the next activity of show-and-tell. Metz (to appear) discussed the social context of elementary school, valuing harmony over argumentation. She gave examples of teachers having children vote to resolve disagreements and of children using "rock, paper, scissors." In the case above, the children resolved the conflict with a shift in the activity, one that allowed each student her or his own perspective.

On this account, the children have resources for understanding different kinds of knowledge, including intuitive versions of hypotheses and statements of preference, as well as many others — questions, observations, fictional stories and "real" stories, and so on (Hammer & Elby, 2002). Different contexts, including fine-grained differences within a given conversation, activate different epistemological resources: Children are capable of understanding knowledge in a variety of ways and of engaging in a variety of epistemic activities.

Coherence

It may be important to emphasize that we are not describing children as randomly incoherent in their epistemic activities. In this case, the first graders moved from one relatively coherent activity to another. Within the first segment, when they seemed to frame the conversation as one of tangible explanation, most student comments make sense within that framing. During the second segment, a series of student comments make sense within a framing of show-and-tell, and so on. That is, we are not describing the children as varying randomly in the ways they think about the question; we are describing them as varying among different local coherences (Russ et al., 2005).

Work in two research projects at the University of Maryland has described similar phenomena of students shifting among different epistemic activities from young grades elementary school (Louca, 2004; Louca, Elby, Hammer & Kagey, 2004; May et al., under review) to introductory college physics (Lippmann, 2003; Lising & Elby, 2005; Tuminaro, 2004). Redish (2004) has suggested these phenomena connect to accounts in linguistics and anthropology of *framing* (MacLachlan & Reid, 1994), meaning an individual's forming a sense of the activity. In this description, for example, the first graders were at first framing the activity of discussing the book and paper as one of tangible explanation; later they framed the activity of reporting their results as one of show-and-tell.

These local coherences are consistent with a resource-based ontology of mind. In fact, Minsky's (1986) model included "frames" as cognitive structures comprised of agents in a particular organization. We take "framing" to refer to the activation of a set of resources that is self-consistent and locally stable (Hammer, Elby, Scherr & Redish, 2005; Redish, 2004).

For example, Rosenberg, Hammer, and Phelan (2006) analyzed an eighth-grade conversation of the rock-cycle (Phelan, in preparation) to show different stabilities in terms of mutual activations among epistemological resources. At the outset of their work, the students were in a "cut and paste" mode of reasoning, gathering information from their worksheets and placing it in an ordered list. Rosenberg et al. (2006) offer a simplified model of local coherence to this activity in terms of a set of resources for understanding epistemic forms and activities: *knowledge as propagated stuff* (transmitted from a source to a recipient), *information, accumulation, ordering,* and *ordered lists*. These resources formed a closed system in which one activity (e.g., retrieving information from the worksheets) led to the next (placing it in the list), which called the next (retrieving more information), and so on.

The teacher, Jessica Phelan, admonished the students to "start with what you know, not with what the paper says," and the students shifted into a different mode of thinking. A simplified model of the new mode includes *knowledge as fabricated stuff* (made from other knowledge), *imagining, causes* and *causal story* (in which the chain of events depends on causation). Again, these resources call and re-call each other, forming another locally stable system. This system allows a natural explanation for how the students' difficulty identifying a causal link for their story (how does sedimentary rock get exposed to heat?) triggers a different mode focused on problem-solving and argumentation: *Causal story* requires *causes*, and if there is not one available for the next event, the story-telling must pause, to locate or construct a cause.

In the case of the eighth-graders, it was evident they were deliberate and aware of their shift in mode of reasoning. In the case of the first-graders here, we do not have this evidence, and we suspect that for some of the transitions they were neither deliberate nor aware. When Mikeska first posed the question of the book and the paper, the students entered into a mode of mechanistic sense-making about why the book would fall faster. The next day they fell out of that mode and then back into it, over the course of the conversation; there is little reason to suspect they were aware of these transitions. One aspect of progress, for these students, would be their coming to recognize mechanistic sense-making as an epistemic activity of science.

IMPLICATIONS

We have noted two points we believe deserve and are likely to achieve consensus in the communities represented in this volume.

The first is over examples of what educators would and would not consider examples of inquiry, drawing from this first-grade discussion. Citing Abd-El-Khalick (2004), Duschl and Grandy (in preparation) listed (some of) the range of ways researchers have characterized inquiry. By giving examples and, in our attempt, defining inquiry as a pursuit, we hope to clarify that inquiry is something to promote and assess in students' actions and reasoning. It is part of the substance that should pervade science instruction, rather than a pedagogical approach or an ancillary objective pursued through added, side activities.

The second point of likely consensus is a broad-brush estimation of children's abilities, of what children bring to class. Children are capable of much more than traditional accounts credit them. Certainly this has implications for early objectives in science education, including increased emphases, even for six year olds, on the dialogic processes Duschl and Grandy (in preparation) describe. Young children can and do enter into mechanistic sense making as well as comparing and assessing ideas and evidence; they can be both theoretical and empirical in pursuit of coherent accounts of natural phenomena, connecting to and expanding on their sense of mechanism.

Perhaps there is a third point of consensus in the phenomenology of variability: What we see children doing at any one moment is of only limited value in determining what we will see at another moment. We advocate a resources-based account of children's abilities, but the phenomenology may be enough for useful implications for instruction. It would mean that to the extent children are not behaving as we might like, in any moment – using anthropomorphic reasoning, e.g., or ignoring inconsistencies, or conflating hypotheses and evidence – it does not necessarily reflect a developmental or structural limitation. It also means that to the extent we discover children behaving in ways we see as valuable – mechanistic reasoning, reconciling inconsistencies, coordinating theory and evidence – it does not mean they have "arrived" and the objective is covered.

In sum, we hope to help motivate explicit instructional emphasis on scientific inquiry, as its own agenda. Educators need to learn to assess student reasoning by means focused directly on inquiry and not simply on progress toward canonical ideas. A rich collection of case studies, with examples of what inquiry looks like, should be the starting point. Especially in early grades, if children are capable of a variety of kinds of activities, which they move into and out of with ease, educators should help guide them to those that represent nascent scientific inquiry.

Still, we find it easiest to speak of and think about this variability in terms of a resources-based ontology: What might help activate the children's sense of tangible, causal explanations? In this sense we might see progress in education in terms of helping children learn to activate productive aspects of their abilities with respect to inquiry in science, to move into these productive modes of thinking not only by accident or for context sensitive reasons but because these are forms of reasoning they understand as intrinsic to science.

NOTES

[1] Grant #ESI-9986846, D. Hammer and E. H. van Zee.
[2] Ebony may have been making an argument for consistency among observations: that the book makes a noise but the paper does not agrees with the idea that the book moves more quickly. He may also have simply been listing something else he knew would happen.
[3] A number of students in the class spoke English as a second language, which added to the challenges of interpretation, both for us and for them. Jorge's question a moment earlier ("Did the book fall first, just like the paper?") was an example. He seemed to be trying to resolve the conflict as a matter of language, checking on precisely what Ebony meant, but it is difficult to know what alternative he was presenting. One guess is that he was thinking "first" might mean "right away."
[4] Machamer, Darden, and Craver (2000) provide a more precise definition of *mechanisms* in science: "entities and activities organized such that they are productive of regular changes from start or set-up to finish or termination conditions." See Russ et al. (2005).
[5] Of course, not all the students were participating all the time, but the shift in the conversations cannot be attributed simply to different students speaking. Ebony, Allison, and Rachel all spoke in more than one segment of the conversation, and in varying ways.

David Hammer, Rosemary Russ, Jamie Mikeska, and Rachel Scherr
Departments of Physics and Curriculum & Instruction
University of Maryland, College Park

WILLIAM A. SANDOVAL

EXPLORING CHILDREN'S UNDERSTANDING OF THE PURPOSE AND VALUE OF INQUIRY

What counts as inquiry in a science classroom? How can we best characterize children's abilities to engage in inquiry? These are the main questions considered by Hammer, Russ, Mikeska, and Scherr (this volume) in their paper. The first question is a central issue to the theme of this conference: identifying areas of consensus in science education. Obviously, it is hard to study or promote inquiry if we cannot agree on what it is. In this commentary on their paper, I will focus most of my attention on their answer to the second question, as it is a question whose answer has far reaching implications for how we theorize about and study learning and how we might organize science instruction. I see the *manifold resources* view advanced by Hammer and his colleagues as an important step forward in theorizing about personal epistemology, i.e., the beliefs people have about the nature of knowledge and its production. Yet, the account put forth by Hammer et al. raises methodological concerns about how epistemic resources can be detected and described, and a theoretical question about how best to explain individual cognition in the context of complex social interaction. Reflecting on their account and subsequent discussions at the Rutgers conference also leads me to wonder how best to understand the difference between what students believe is the purpose of inquiry and the value that such inquiry holds for them.

IDENTIFYING INQUIRY

In their example of a first grade class' conversation about falling objects, Hammer et al. note a number of forms of reasoning that these young children engage in that they deem productively scientific. These include proposing mechanisms to explain a prediction, arguing the relative merits of competing mechanisms to explain empirical results, drawing connections among various observations, checking for consistency, and so forth. From their examples they propose a broad "working definition" of scientific inquiry as "the pursuit of coherent, mechanistic accounts of natural phenomena" (p. 12). This may lack enough precision to be a good philosophical definition. As a definition that might be applied to education, however, especially as something that might be understood and enacted by teachers, this definition has an important advantage. As Hammer et al. point out, their definition of inquiry differs from typical definitions of inquiry (e.g., the NSES) as a combination of specific sorts of activities. Instead, this definition focuses on the *epistemic goal* behind any activity that should be counted as inquiry: the pursuit of a particular kind of "account" of the world.

Defining scientific inquiry in terms of its epistemic goal, that is, the kind of knowledge it aims to produce, is an important re-framing of inquiry from a teaching and learning perspective. We know that the goals that learners pursue during scientific investigation affect the strategies they use and their consequent learning (Zimmerman, 2000). We also know that simply doing inquiry does not seem to change students' beliefs about the nature of science (Khishfe & Abd-El-Khalick, 2002; Meichtry, 1993). To my mind, then, Hammer et al.'s definition of inquiry provides a touchstone against which to evaluate any instructional activity, like a conversation about falling objects, in terms of whether or not it has as its goal the pursuit of a causal, mechanistic account of natural phenomena. There are corollary questions involved, of course, such as who is leading the pursuit, but I believe that implicit in Hammer et al.'s use of the word "pursuit" is the idea that students are the ones pursuing. I find it useful, however, to not get bogged down in too much detail at the start about what the pursuit might look like. The value in Hammer et al.'s definition is that their focus on the purpose of inquiry, in terms of the "what" it should produce, gives context to considerations of the processes of inquiry, the "how" of producing scientific knowledge. That is, this focus on product provides a means for evaluating processes of inquiry in terms of their ability to satisfy the epistemic goals behind the effort.

WHAT'S IN A RESOURCE?

It is in this light that I want to examine the examples provided in Hammer et al.'s paper. I am very sympathetic to Hammer's and colleagues' manifold resources view. I agree with their underlying assumption that epistemological beliefs are highly contextualized, and perhaps locally coherent. I think that students may have "practical" epistemological beliefs that guide their efforts at doing inquiry, but are quite different from their expressed beliefs about what scientists do (Sandoval, 2005). I mention this because the problems that I see in Hammer et al.'s analysis stem in part from the difficulty of the enterprise they have undertaken, and stand as general problems for anyone interested in understanding how students' perceive inquiry in school, and how school science inquiry can be organized to help students understand the nature of professional science. In this section I raise some methodological and theoretical issues with Hammer et al.'s manifold resources view as they explain it in their paper. These issues will then provide some context for the question of value that I raised above.

Hammer et al. raise the issue of how to identify inquiry, in part, to get at how children's abilities to engage in inquiry can be characterized. They also argue that any characterization of students' abilities should be able to account for how such abilities develop with experience. On this point, Hammer et al.'s manifold resources perspective, as currently formulated, falls short. Part of the shortcoming arises from what I perceive to be some methodological issues in how Hammer and his colleagues attempt to document possible resources. Moreover, there is, I believe, a deeper theoretical issue about how to account for individual cognition within complex social systems, such as first grade classrooms.

Hammer et al. propose that children develop a number of variable, locally coherent "resources" that they bring to bear when reasoning about the natural world. The argument here builds upon previous work (e.g., Hammer & Elby, 2002) in which they have suggested that specifically epistemological "resources" are organized, accessed, and used in people's minds. As outlined in the present paper, the manifold view that they propose accounts for the variability seen in children's reasoning about epistemological issues and the apparent incoherence in such reasoning (see Sandoval, 2005). From the manifold resources view, variability in children's reasoning across various contexts, or even during the course of the same conversation, is to be expected. Rather than signifying developmental limits, such variability, Hammer et al. argue, is a sign that children attend to change and respond to it, that they are capable of reasoning in a number of ways. The general argument is a "knowledge-in-pieces" approach (diSessa, 1988), and emphasizes that children's scientific thinking is coherent when viewed in its own terms (Smith *et al.*, 1993/1994). That is, students are coherently applying their "resources" to reasoning through particular problems. The manifold resources view outlined here by Hammer et al. has much to recommend it. It potentially explains recent empirical results that are not explained well by so-called unitary ontologies. As a knowledge-in-pieces kind of theory it aligns well with constructivist accounts of learning because it assumes that children's current knowledge can be productively built upon.

This manifold resources notion, however, currently suffers from at least two problems. The first problem is that the "resource" construct remains vague. This is a conceptual problem, in that without a clearer articulation of the kinds of cognitive structures that are resources it is hard to see how to theorize about how resources are organized or how they could be re-organized through experience. Hammer et al. mention that the term "resource" could include "agents, p-prims, and schemata" (p. 13). The disadvantage that I see in their current formulation is that none of the particular resources that they suggest they have seen in their own work have been categorized in this way. For example, Hammer et al. describe eighth graders shifting from a "knowledge as propagated stuff" resource to a "knowledge as fabricated stuff" resource. Yet, what exactly are these resources? What specific "epistemic forms and activities" (p. 15) do they entail? By looking at their example, it is possible to infer that certain nouns are epistemic forms and verbs are epistemic activities. For instance, "information" and "ordered lists" seem to be epistemic forms that are related to what I believe Hammer et al. would describe as a more general epistemic frame of "knowledge as propagated stuff." This is just speculation on my part, however. I have no idea whether or not Hammer et al. would agree with my formulation. I also cannot tell how they think these various epistemic forms are organized in relation to each other. For instance, if knowledge as propagated stuff is a frame that competes with knowledge as fabricated stuff, is it the case that the epistemic form "ordered list" is exclusive to the "propagated stuff" frame? How would we know?

The questions of how we would know what epistemological resources particular children might have, how they are (presumably variably) organized, and how they develop into (potentially) more coherent frames over time and through experience seem crucial to answer in order to evaluate the viability of the manifold resources theory. It reflects the second problem I see in Hammer et al.'s current articulation of their theory. I am concerned that the approach to studying these resources that Hammer and colleagues seem to be pursuing is unable to answer them. Specifically, their analyses of the activities and talk that children engage in during what we might consider scientific inquiry, by themselves, cannot enable us to confidently ascribe particular epistemic resources to particular children or groups of children. It seems a simple matter to identify, as an observer, the particular epistemic activity (or epistemic game, Collins & Ferguson, 1993) that children are engaged in and the epistemic form it produces. It is a very risky inference from such observations, however, to what the children so engaged believe they are trying to do, including what knowledge or kind of knowledge they see themselves as trying to make. It is not at all clear that the first graders they studied saw their activity as the pursuit of a mechanistic account, or which children did and which did not.

This issue of attribution, which children think what, raises a third issue in the analysis presented by Hammer et al. Their analytic approach, which entails close analysis of the talk of this group of first graders and their teacher, is not particularly well-suited to the individualistic, cognitive terms in which they articulate their manifold resources theory. Rather, their approach is much more suited to a sociocultural perspective that would frame learning as changes in participation over time. What Hammer and his colleagues present in their paper are shifting patterns of participation, shifts that occur over time and as the activity itself changes. From their manifold resources view, it is not at all clear how particular resources could be said to underlie differences in children's participation, or changes in one child's participation.

Consider the example of one student, Ebony, featured in Hammer et al.'s analysis. Ebony is one of the first children to propose a mechanistic explanation for why objects might fall at different rates. After experimenting with a piece of paper and a book, Ebony surprised his teacher and the researchers by insisting that a flat piece of paper "fell first," instead of the book. Hammer et al. describe how Ebony appeared to "shift out of" causal, mechanistic inquiry to engineer a particular outcome. They noticed that he held a piece of paper under a book, together, and dropped them, making the paper "fall" first. Does this action mean, however, that Ebony has various ideas about the purpose of inquiry that are activated by these different contexts? Hammer et al. suggest that Ebony may have thought the answer to the question of which object, book or piece of paper, would fall first was obvious. In other words, he was trying not to be bored, or he wanted to stir up a little controversy, but he was no longer really engaging in inquiry. For Hammer et al., and for readers, what Ebony may have been thinking can only be speculated about. We certainly cannot say that he was using different "resources" in any sort of consequential way.

What is clear from this example is that Ebony's surprise announcement shifted the participation of the group. Similarly, the teacher's response, or lack of it, also shaped the group's participation in the activity. The teacher could have, for example, asked Ebony to explain how he had done his experiments, and the class may quickly have concluded that his experimental procedure was invalid for answering their original question. Such an interjection on the teacher's part would have drastically changed the children's participation, by changing the subject of the conversation. As such, any potential resources that might get triggered would also change. As researchers, I think we could legitimately ask whose resources might be getting activated? Could they be said to belong to individual students? Would it be more accurate to ascribe them to the situation itself? I do not think that the methodological approach Hammer et al. have pursued here can answer these questions. I do not see how attributions about the purposes that individuals are pursuing in such conversations can be made accurately without more data about those individuals. Moreover, in this particular case I think that more attention needs to be paid to how the teacher shapes the activities themselves, and thus constrains students' participation in particular ways.

PURPOSE AND VALUE

One of the things missing from Hammer et al.'s manifold resources theory, and other accounts of epistemological beliefs (Abd-El-Khalick & Akerson, 2004; Sandoval, 2005), is that it does not distinguish purpose, the goal being pursued, from value, the worth that meeting that goal holds for the pursuer. As suggested by Ebony, the goals one pursues and the vigor which they are pursued are mediated by the value they hold for the person pursuing them. Wenger's (1998) conception of a community of practice describes how a person's changing participation within a community of practice stems from and leads to changes in the meaning that people make out of their participation and consequent changes in their identity. From this sociocultural perspective, meaning is more than a cognitive outcome of comprehension or understanding, but includes the personal meaning, i.e., the value, that people find in their participation in certain practices. Understanding children like Ebony and their participation in their classroom communities requires understanding the meaning they make from their participation and their forming identities, both as members of the classroom community and of other communities of practice.

This issue seems especially salient to efforts to study epistemological beliefs (or conceptions, or resources). As Hammer et al.'s examples make clear, observations of children's participation in activities is insufficient to distinguish children's comprehension of the purpose of an activity like inquiry from the value or meaning they ascribe to satisfying that purpose. It is an issue that has vexed studies of personal epistemology through an almost routine conflation of beliefs about knowledge with strategies for learning (Hofer & Pintrich, 1997). From my view, it would be wrong to conclude from Hammer et al.'s analysis of Ebony's participation that he did not understand the purpose of his experimentation, either for himself or for his classroom community. Rather, it seems just as likely that he simply did not care about that purpose, and so pursued another purpose more meaningful to him.

I raise these issues with Hammer et al.'s analysis as a learning researcher interested in answering the same questions they are: How do students understand inquiry? How do their ideas about what inquiry is for influence how they do it? How can inquiry experiences be organized to help students learn science, and learn about scientific inquiry as a way of making knowledge about the world? These issues are academic. They concern how we construct a theoretical account of the development of epistemological beliefs. They can also potentially relate to pragmatic issues of how to support such development through schooling. Hammer et al.'s manifold resources theory is more likely than unitary accounts to explain the variability seen in students across a range of learning activities. Their theory will become more viable and useful as the "resources" construct becomes more clear, and thus more easily observed and described. An outstanding issue for this, and any knowledge-in-pieces theory in fact, is how the locally coherent resources can be re-organized into more broadly coherent conceptual (or epistemological) frameworks. The pursuit of such an account would not only go a long way toward settling the theoretical dispute between unitary and manifold accounts, it would provide clearer guidance to science educators striving to help their students understand the epistemological pursuit of scientific inquiry.

The very fact that Hammer et al.'s definition of inquiry seems so novel indicates how rarely mainstream science education seems to ask questions of purpose. What is inquiry for? For scientists, the definition Hammer et al. suggest seems pretty workable (notwithstanding philosophical objections of narrowness or imprecision) – inquiry is for producing mechanistic accounts of the natural world. I am less sanguine that this definition of the purpose of inquiry speaks to many outside the scientific community. I suspect that questions of purpose and value may be some around which we have very little consensus in science education because we do not ever raise them. Hammer et al. do not raise questions of value explicitly in their paper. Yet, clearly children's perceptions of what they are trying to do relate to why they might want to do it. Further, children's variable participation in classroom inquiry may say more about their perceptions of the value of the activity than their conceptions of the purposes of the activity.

Another question that I worry that those of us in science education do not think enough about is: Why should children care to learn scientific inquiry? What value does it hold for them? At the Rutgers conference there was a lot of discussion about what inquiry in science classrooms should be like, what kinds of things people can learn through inquiry, and even whether or not learning through inquiry was necessary or efficient (efficiency is a kind of value, of course, but not the sort of thing I have in mind here). Such discussions raise general questions about what students should learn, including what "core" scientific ideas all children should know by the time they leave high school. I find these conversations frustratingly solipsistic.

I wonder what value students find in pursuing mechanistic accounts of the natural world. I am certain that science educators can enumerate what value might accrue to students from studying science, but I wonder how hard we try to do that in terms that children and adolescents care about. My sense is that it is more common to project a future value to understanding science: that scientific knowledge helps students to be more productive adults and better informed citizens. It seems also that the scientific ideas that we deem important are deemed so because of their relevance to particular disciplines, rather than their relevance to the people we expect to learn them. It seems rare that we consider the current value of studying science to children in their lives today. We do not really pose questions about the value of science education from children's perspectives. What is in it for the first graders to figure out how objects fall? I wonder how answers to that question would change the way science educators thought about the purposes and values of inquiry in school, and studied how students understand and appropriate those purposes and values.

William A. Sandoval
Graduate School of Education & Information Studies
University of California, Los Angeles

DANIEL C. EDELSON

ENGINEERING PEDAGOGICAL REFORM
A Case Study of Technology-Supported Inquiry

BACKGROUND

It is not uncommon for researchers to suggest that the model for educational research should be engineering and the applied sciences, rather than the social and natural sciences (e.g., Collins, 1992; Zaritsky, Kelly, Flowers, Rogers & O'Neill, 2003). However, it is less common for an individual with an engineer's training to become an educational researcher. I am such a person, and in this paper, I will describe a research program in which I have been attempting to "engineer" learning experiences for middle and high school science students that incorporate inquiry. In this historical account, I will tell the stories of three different engineering problems that arose, a software design problem, a curriculum design problem, and a teacher professional development. For each, I will describe the problem, as it presented itself and evolved over time, the products that we designed to address the problem, the design framework we articulated to enable us and others to address similar problems, and the open research questions.

This line of research began in 1992 with the Learning through Collaborative Visualization (CoVis) initiated by Roy Pea, Louis Gomez, and Elliot Soloway, which had the goal of placing scientific visualization technologies that could support inquiry-based learning into the hands of high school students (Pea, 1993, May). In the 1980s, as a result of a substantial investment by the federal government in high-performance computing, scientific visualization technologies had revolutionized the practices of many scientific disciplines, including the geosciences (McCormick, DeFanti & Brown, 1987, November November; Gordin & Pea, 1995). One of the premises of the CoVis Project was that the same properties that made visualization tools so valuable for scientists – that they harness the power of the human visual system in order to allow scientists to find patterns in complex data – would apply to students as well. The idea was that the same tools that allow scientists to expand the horizons of human understanding could allow students to expand the horizons of their personal understanding. So, in 1992 we initiated a program of research and development exploring the opportunities for scientific visualization in geosciences education that continues today.

The original goal of this work was to provide students with tools that would enable them to engage in open-ended geoscience investigations with data. The goal was for students to initiate investigations with self-generated questions and pursue them through a largely self-directed and self-regulated inquiry process. While this goal of supporting open-ended inquiry remains largely unrealized, it is

Richard A. Duschl and Richard E. Grandy (eds.), Teaching Scientific Inquiry: Recommendations for Research and Implementation, 164–181.
© 2008 *Sense Publishers. All rights reserved.*

responsible for initiating an immensely productive research and development program.

One of the hallmarks of this research program is that the problem definition has expanded with every step of the research. The problem was originally conceived as a technology challenge. We recognized almost from the start that the tools used by scientists were not appropriate for high school students. Therefore, we initially defined the problem as one in designing scientific visualization tools that would be appropriate for high school students. However, as soon as we had made enough progress on that problem to begin classroom studies, we discovered that teachers were not able to take advantage of the software unless it was embedded in curriculum. In other words, getting the technology right was just the price of admission to work on the curriculum problem. Therefore, we reconceptualized the problem as being one of both software and curriculum design. However, once we had created curricula and attempted to expand to larger number of teachers, we discovered that many teachers were not prepared to use the technology-integrated and inquiry-based curriculum units we had created. Therefore, our ever-increasing research problem grew to include teacher learning and professional development, as well. It is my nightmare that I will eventually be forced to take on the problem of organizational context and school policy, but for now, I am restricting myself to the three interrelated issues of software design, curriculum design, and design to support teacher learning.

THE ENGINEERING PERSPECTIVE ON EDUCATIONAL REFORM

To understand the perspective in this paper, it is important to understand the practice of engineering research. The goal of engineering research is to develop robust methods for designing solutions to a class of problems (in contrast to engineering itself, where the goal is the solution of specific problems). Some of the most important aspects of engineering research are:

- It rests on existing science. Engineering solutions start with what is already known reliably and reproducibly about the relevant phenomena.
- Engineering research starts where the science ends. Engineering research is necessary when the science is not sufficient to base design decisions upon.
- Engineering research is driven primarily by the design goals and secondarily by the goal of understanding the relevant phenomena. In other words design heuristics based on descriptive empirical results rather than a causal model is a satisfactory outcome for engineering research, where it would be considered unsatisfactory in scientific research.
- Engineering research proceeds through iterative, theory-guided experimentation. At each iteration, a design or partial design is developed and tested. The evaluation at each step extends or revises the theory and guides the next iteration in design.

- The outcome of engineering research is a set of guidelines for designing solutions to a class of problems, accompanied by a rationale for those guidelines grounded in science, theory, or empirical results.

This describes my own research, which I characterize as educational design research (Edelson, 2002) and is similar to what others have called design experiments (Brown, 1992; Collins, 1992) and design-based research (Design-Based Research Collective, 2003). By working on a range of specific design challenges in educational reform, my goal is to develop general guidelines for the design of a class of learning experiences. In particular, I am interested in the design of learning experiences in which students engage in scientific inquiry that incorporates authentic scientific practices in the service of improving students' understanding of both science content and science practice. I have chosen the context of Earth and environmental science because of the importance of these subjects to modern society and the insufficiency of current educational practices in these subjects to prepare citizens to make decisions they will face in their lifetimes that have implications for the environment. I include a focus on computational tools in my research because of their ability to represent scientific phenomena dynamically and because they are essential to contemporary scientific practices.

THE SOFTWARE DESIGN STORY

As stated in the introduction, this program of research started with the goal of engaging secondary school students in open-ended inquiry in atmospheric sciences supported by visualization and data analysis technologies. The first challenge that we recognized was the challenge of using tools developed for scientists (Gordin & Pea, 1995; Edelson & Gordin, 1998). Therefore, the first design and evaluation iterations were focused on adapting the tools used by scientists to be appropriate for learners. In this case, we developed tools for visualization and analysis of gridded (raster) data of the sort used by climatologists and oceanographers. The final tangible product of this research is a software environment called WorldWatcher (Edelson, Gordin, Clark, Brown & Griffin, 1997) that is still in use by schools. This design cycle, described in Edelson, Gordin & Pea (1999), resulted in guidelines and strategies for the adaptation of scientific investigation tools for use by learners. These guidelines and strategies respond to the following challenges we identified through formative evaluation cycles in the design process (Edelson et al., 1999, pp. 399-400):

1. Accessibility of investigation techniques. For students to engage in inquiry, they must know how to perform the tasks that their investigation requires, they must understand the goals of these practices, and they must be able to interpret their results. Scientific investigation techniques such as data collection and analysis can be complicated and typically require a level of precision and care that are not required of students in their everyday experiences. If students are not able to master these techniques, then they cannot conduct investigations that yield meaningful results. The need for tools to be accessible to learners across

the full diversity of abilities and prior experiences is another challenge of Learner-Centered Design raised by Soloway et al. (1994).
2. Management of extended activities. To achieve the ultimate goal of open-ended inquiry, students must be able to organize and manage complex, extended activities. A scientific investigation requires planning and coordination of activity and the management of resources and work products. Students are not typically asked to manage extended complex processes as part of traditional educational activities. If they are unable to organize their work and manage an extended process, students cannot engage in open-ended inquiry or achieve the potential of inquiry-based learning.
3. The practical constraints of the learning context. The technologies and activities of inquiry-based learning must fit within the practical constraints of the learning environment, such as the restrictions imposed by available resources and fixed schedules. While this challenge may not have the same theoretical importance as the other two for advancing our understanding of the learning sciences, it has enormous practical implications for design. A failure to work within the available technology or fit within the existing schedule in a school will doom a design to failure. Therefore, meeting the constraints of the environment is a critical consideration in design that must be considered along side learning needs in the design of curriculum and technology.

In response to these challenges, we identified a number of software adaptation strategies, for which we were able to give guidelines.

To address the challenge of accessibility, we developed the design strategy of *embedding the tacit knowledge of experts in the novice's user interface.* For example, we discovered that scientists bring knowledge of geography to their interpretation of geographic visualizations that students lacked. By incorporating geographic references into the visualizations, we were able to provide students with access to information that was tacit knowledge for scientists Figure 1. Similarly, we developed interfaces for data libraries that allow students to access data through diagrams indicating the relationships among variables instead of the lists of scientific terms that scientists use as interfaces to data libraries.

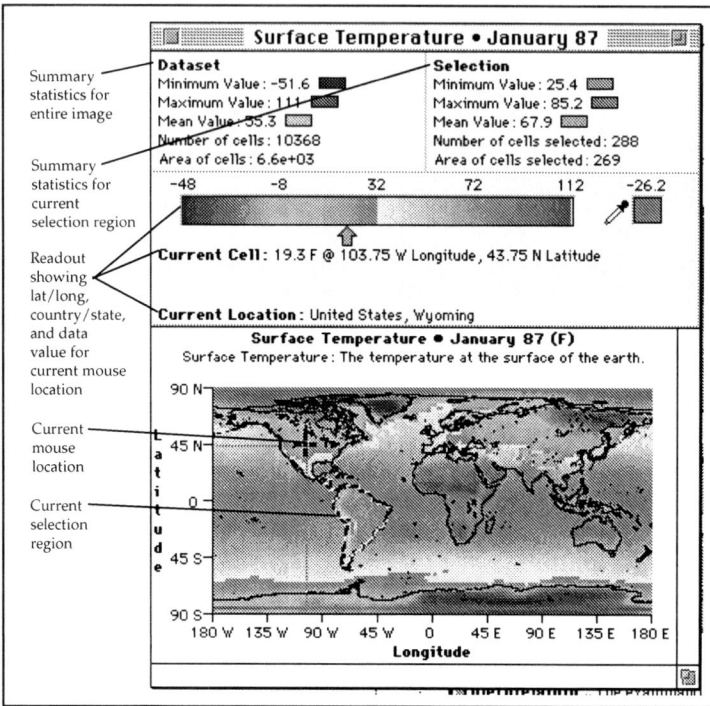

Figure 1. A Visualization window from the WorldWatcher software displaying surface temperature for January 1987.

Another design strategy for the challenge of accessibility was the *strategic selection of functionality*. The goal of this strategy is to select functionality that provides learners with the functions of an expert's tool that will provide them with the most investigative power without overwhelming them with complexity in the form of too many functions or functions that are too difficult to implement. For example, we incorporated functionality for conducting simple arithmetic operations (addition, subtraction, multiplication, division, minimum, maximum, average) between two data sets (Figure 2) representing the same quantity (e.g., subtract January temperature from July). However, we decided to only support these simple two-place arithmetic operations and did not incorporate a facility for specifying operations through arbitrarily complex algebraic expressions, as was found in the experts' tools.

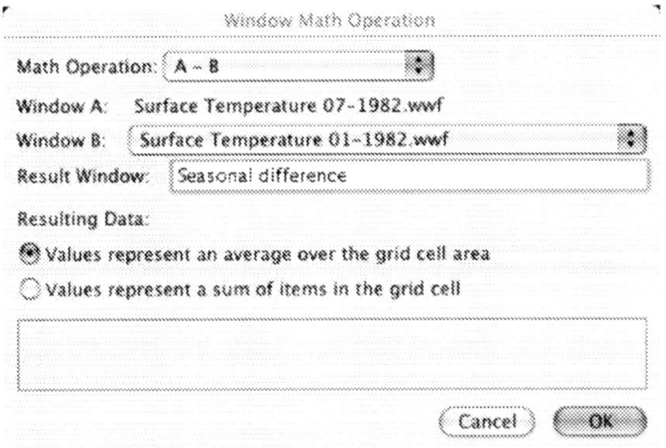

Figure 2. A Dialog for specifying a math operation between two datasets.

A third strategy for addressing the challenge of accessibility was *automating operations with minimal pedagogical value*. In an early phase of this research, we observed climate researchers engaged in data analysis with scientific visualization tools. They spent a large percentage of their time reformatting data and setting display parameters prior to actually analyzing data. The pedagogical value of students' engaging in these activities would be very low, and it would be very difficult to justify the time that these processes would take. Therefore, we made it unnecessary for students to participate in these tasks by either automating the tasks or pre-processing the data manually in advance of distributing it to students. Thus, when students select any data to visualize in WorldWatcher the display parameters have been set in advance to make it possible for them to start working with the data immediately. Students still have full control over all the display parameters (color scheme, spatial resolution, quantization, etc.), but it is not necessary for them to set them before they can begin their data interpretation and analysis.

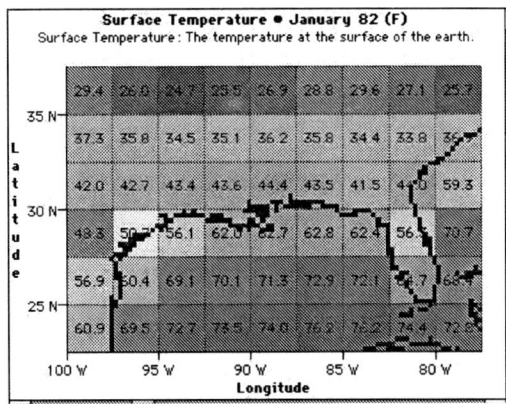

Figure 3. A visualization magnified to show the individual data values in each cell.

A fourth strategy for the challenge of accessibility was *adding bridging functions*. A bridging function helps students understand the functionality of an expert tool in terms of concepts they are already familiar with. The result is adding functionality that would not be useful to an expert but has significant pedagogical value. One example in WorldWatcher is a feature that display numerical values for colored cells when a user zooms in on a visualization far enough to display them (Figure 3). This feature helps students to bridge between their familiarity with tables of numbers (e.g., spreadsheets) and the unfamiliar colors as numbers representation of color maps. Often students approach these visualizations with the metaphor of a printed map, where the colors are ink, rather than a dynamic representation of numerical values. Once a student understands a color map visualization as an array of numbers, an arithmetic operation between two visualizations makes sense, whereas a subtraction doesn't make sense to someone with a model of the visualizations as colors without numbers behind them. Other examples of bridging functions are a feature that enables learners to modify or create data sets by drawing on visualizations using a paint program metaphor (Figure 4) and another that enables them to print visualizations as cut-and-fold images that they can assemble into three-dimensional globes (Figure 4).

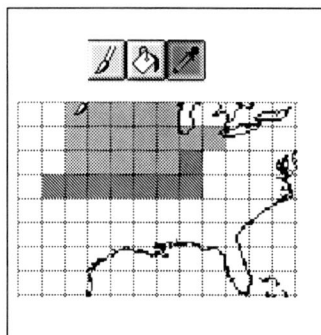 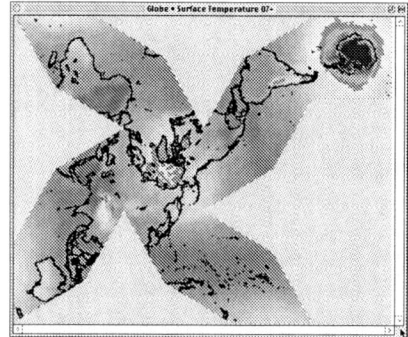

Figure 4. Images from WorldWatcher showing (a) the paint interface for drawing "data" and (b) cut-and-fold globe generated from a temperature visualization.

A strategy for addressing the challenges of open-ended tasks was to develop inquiry-support software tools that allow teachers or curriculum designers to structure students' activities and provide students with tools for planning, recording, and monitoring their progress in an investigation.

Our strategies for responding to the practical constraints of school technology infrastructures consisted of reducing or eliminating our reliance on high-end computers or reliable network connections, developing interfaces that protect students from needing to interact directly with computer file systems, and simplifying installation and administration processes.

Based on classroom use of the most recent version of WorldWatcher, we believe that we have been relatively successful in identifying challenges to the use of investigation tools to support inquiry and in developing strategies to address them. Based on that success, we have been engaged in a similar effort to adapt geographic information systems (GIS) technologies for learners for the last several years, drawing on these same strategies in the context of a technology that offers a much larger and more sophisticated set of investigation tools than the technologies that served as the model for WorldWatcher.

However, our classroom experiences also revealed that there were significant challenges to the implementation of technology-supported inquiry learning that could only be partially addressed by software design, if at all. In Edelson et al. (1999), we reported the following challenges in that category (p. 399-400):

1. Motivation. For students to engage in inquiry in a way that can contribute to meaningful learning they must be sufficiently motivated. The challenging and extended nature of inquiry requires a higher level of motivation on the part of learners than is demanded by most traditional educational activities. To foster learning, that motivation must be the result of interest in the investigation, its results, and their implications. When students are not sufficiently motivated or they are not motivated by legitimate interest, they either fail to participate in inquiry activities, or they participate in them in a disengaged manner that does not support learning. Motivation is recognized by Soloway et al. (1994) as one of three primary challenges for Learner-Centered Design.
2. Background knowledge. The formulation of research questions, the development of a research plan, and the collection, analysis, and interpretation of data, all require science content knowledge. In designing inquiry-based learning, the challenge is providing opportunities for learners to both develop and apply that scientific understanding. If students lack this knowledge and the opportunity to develop it, then they will be unable to complete meaningful investigations.

In this first phase of research, we were able to address motivation by creating data libraries as part of our software that supported investigations on topics that middle and high school students found meaningful. We were able to address background knowledge by creating a facility that enabled teachers and curriculum developers to link documents with content information to data sets and to activity guides. However, we found that we needed to turn our focus to the design of learning experiences (curriculum) in order to address these challenges.

THE CURRICULUM DESIGN STORY

Since 1998, the primary focus of my research on engaging students in inquiry has been curriculum and activities design. The delayed onset of this research relative to the software research reflects, in part, a naïve view that students would be prepared to engage in open-ended inquiry without requiring a structured curriculum that was followed by an equally naïve view that teachers would be able to develop whatever curriculum was required for their own classrooms. My own work on curriculum development began in earnest in 1998 with the initiation of two new efforts. The first was an NSF Instructional Materials Development grant to create a one-year environmental science curriculum for the high school around inquiry supported by WorldWatcher. The second was the NSF Center for Collaborative Research on Learning and Technology called the Center for Learning Technologies in Urban Schools (LeTUS), which took on as a central focus of its work the creation of inquiry-based, technology-integrated science curriculum units for use in middle schools in Chicago and Detroit.

Approaching this work from the perspective of engineering research, I set about trying to understand what the science of learning had to contribute to the design of inquiry-based curricula, and I drew on contemporary research in the cognitive sciences to develop a set of guidelines and strategies for curriculum design that I call the Learning-for-Use (LfU) design framework (Edelson, 2001). The Learning-for-Use Framework is concerned with the design of learning activities that develop what I call *useful knowledge* Its goal is to address two significant challenges to teaching that are often overlooked in the design of learning activities: fostering engagement and ensuring that learners develop knowledge that they can access and apply when it is relevant in the future. Regardless of the nature of the learning activities that students participate in, if they are not sufficiently and appropriately engaged, they will not attend to those activities in ways that will foster learning. Likewise, if students do not construct knowledge in a manner that supports subsequent re-use of that knowledge, it remains inert (Whitehead, 1929). The Learning-for-Use design framework is based on a model of how people acquire useful knowledge. The four principles are[1]:
– Learning takes place through the construction and modification of knowledge structures.
– Knowledge construction is a goal-directed process that is guided by a combination of conscious and unconscious understanding goals.
– The circumstances in which knowledge is constructed and subsequently used determine its accessibility for future use.
– Knowledge must be constructed in a form that supports use before it can be applied.

The model incorporates these four principles and their implications into a description of learning. It characterizes the development of useful understanding as a three-step process consisting of (1) motivation, (2) knowledge construction, and (3) knowledge refinement.

Motivate: Establish perceived need for new understanding[2]. The first step in learning for use is recognizing the need for new knowledge. The *motivate* step in the learning-for-use model creates a need for specific content understanding. In this context, *motivate* is being used in a very specific sense. It describes the motive to learn specific content or skills, not a general attitude or disposition to learn in the particular context. Understanding the usefulness of what they are learning provides a motivation for students to engage in learning activities and to construct understanding in a useful form.

Knowledge Construction: Building new knowledge structures. The second step in learning for use is the development of new understanding. This step results in the construction of new knowledge structures in memory that can be linked to existing knowledge. An individual constructs new knowledge as the result of experiences that enable him or her to add new concepts to memory, subdivide existing concepts, or make new connections between concepts. The "raw material" from which a learner constructs new knowledge can be firsthand experience, communication from others, or a combination of the two. This step in the Learning-for-Use model recognizes incremental knowledge construction as the fundamental process of learning.

Knowledge Organization: Organizing and connecting knowledge structures to support use. The third step in learning –for use is refinement, which responds to the need for accessibility and applicability of knowledge. In the refinement step, knowledge is re-organized, connected to other knowledge, and reinforced in order to support its future retrieval and use. To be useful, declarative knowledge must be reorganized into a procedural form that supports the application of that knowledge (Anderson, 1983). Useful knowledge must also have connections to other knowledge structures that describe situations in which that knowledge applies (Simon, 1980; Chi, Peltovich & Glaser, 1981; Roger C. Schank, 1982; Glaser, 1992; Kolodner, 1993). Refinement of knowledge can also take the form of reinforcement, which increases the strength of connections to other knowledge structures through the traversal of memory structures and increases the likelihood that those connections between knowledge structures will be found in the future.

While there is an inherent ordering among these three steps, the ordering does not preclude overlaps or cycles. For example, knowledge construction and revision may be interleaved, and knowledge construction or revision can create new motivation. Because of the incremental nature of knowledge construction, it can require several cycles through various combinations of the steps to develop an understanding of complex content. Even with this cyclical nature, the order of steps is important. To create the appropriate context for learning, motivation must precede construction, and to insure accessibility and applicability, refinement must follow construction.

Based on this model of learning, I developed the *Learning-for-Use Design Framework*. This framework provides guidelines for the design of activities that will contribute to the development of robust, useful understanding. The design framework articulates the requirements that a set of learning activities must meet to achieve particular learning objectives. The Learning-for-Use model poses the hypothesis that for each learning objective a designer must create activities that effectively achieve all three steps in the learning for use model.

The Learning-for-Use design framework describes different design strategies that meet the requirements of each step (Table 1). The different design strategies for each step can be treated as alternative or complementary ways to complete the steps. In the case of rich content, however, several learning activities at each step involving both of the processes for that step may be necessary.

Although it was designed to describe learning in general, when applied to inquiry-based science learning, the Learning-for-Use design framework represents a variant of the Learning Cycle (Karplus & Thier, 1967; Renner & Stafford, 1972; Lawson, 1995; Abraham, 1998). While they are similar in many ways, the two frameworks were developed with different goals. The Learning Cycle was developed as way to bring the process of learning from inquiry that scientists engage in to students. The Learning-for-Use design framework has been developed to highlight the need for motivation based on usefulness and the need to develop knowledge that is organized to support access and application (the *motivate* and *refine* phases in the framework), because they are too often overlooked.

The LfU model does not make any commitment to particular approaches to motivation, construction, or apply, but in my own work, I have been pursuing a specific approach that I call *Scenario-Based Inquiry Learning* (SBIL). Scenario-based inquiry learning draws from project-based science (Krajcik, Czerniak & Berger, 1999), anchored instruction (Cognition & Technology Group at Vanderbilt, 1997), and goal-based scenarios (Roger C. Schank, Fano, Bell & Jona, 1993/1994), which share a focus on authentic tasks as a way of motivating and contextualizing learning. In scenario-based inquiry learning, the motivate phase of LfU is accomplished by introducing students to a scenario in which they must achieve a goal that necessitates the achievement of the learning goals associated with the unit. This goal might be answering a driving question as in project-based science, solving a problem as in anchored instruction, or achieving a goal in a role-play as in goal-based scenarios. In addition to providing the motivation, this scenario also provides the opportunity to organize their new understanding for use by applying it. As with, project-based science inquiry plays an important role in scenario-based inquiry learning. A substantial part of the knowledge construction is supported by activities in which learners engage in inquiry.

In one example drawn from the *Looking at the Environment* high school environmental science curriculum, the scenario focuses on water resources management in California's Central Valley. In this scenario, students are asked to make a policy decision about how to manage water resources in the Fresno area, where one of the most productive agricultural regions in the U.S. and a fast-growing urban population are using surface and underground fresh water supplies at a faster pace than they are being restored. Students are given several policy options that include drilling more wells, changing irrigation practices, removing land from agricultural cultivation, restricting housing development, and restricting domestic water usage. This scenario is designed to motivate students' learning

Table 1. Overview of the Learning-for-Use Design Framework.

Step	Name	Description	Desired Effect
Motivate	Create task demand	Students are presented with a task that requires new understanding.	Creates a perceived need for new knowledge or skills.
	Elicit curiosity	Students are placed in a situation that *elicit curiosity* by revealing an unexpected gap in their understanding.	Student becomes aware of limits of knowledge and need for new knowledge to address those limits.
Construct	Direct experience	Students are provided with *direct physical experience or observation* of phenomena.	Students construct knowledge structures encoding the attributes and relationships that describe the phenomena.
	Indirect experience	Students hear about, view, or read about phenomena.	
	Modeling	Students observe another person performing a task.	Students construct knowledge structures that encode the elements of a practice.
	Instruction	Student are told or read about how to perform a task.	
	Explanation	Students are provided with explanations of phenomena or processes.	Students construct knowledge structures that encode causal information behind the relationships among phenomena or elements of a process.
	Sense Making	Students engage in explanation or synthesis activities.	
Organize for Use	Practice	Students use components of new understanding outside of motivating context.	Students construct procedural representations from declarative representations, reinforce understanding, expose limitations and need for further knowledge construction
	Apply	Students *apply* understanding in context.	Students develop indices for retrieval, construct procedural representations from declarative, reinforce understanding, expose limitations and need for further knowledge construction
	Reflect	Students articulate what they have learned and what the boundaries of that understanding are.	Knowledge is re-indexed for retrieval, expose limitations and need for further knowledge construction

about agricultural practices, irrigation, surface and underground freshwater reserves, dams and wells, ecological impacts of water engineering, and domestic water use, among others. In this 12-week unit, students conduct a range of investigations and other learning activities to develop an understanding of the science necessary to make the policy decision in the scenario. For example, in one sequence they grow Wisconsin Fast Plants in soils of different water retention capabilities, then use a spreadsheet model of a farm water budget to try to create efficient irrigation plans for different crops in different soils, and finally try to develop a plan for improving efficiency of a real farm based on data from precision agriculture that they visualize and analyze in WorldWatcher. At the conclusion of the unit, they investigate the policy options using a variety of GIS data, then select and defend a specific option using a framework for environmental decision-making that they learned in a previous unit.

In other scenario-based inquiry learning units we have developed, students develop climate models for a fictitious planet, participate in a mock international conference on global climate change, create a design for a new school building to minimize impact on the site's natural ecosystem, and select an energy source for an electrical power plant for a specific community.

These units have demonstrated a measure of success across settings that have included urban, suburban, and rural schools, and across diverse student populations. While none of the evaluations conducted on these units would qualify for the What Works Clearinghouse[3], we have collected pre/post-test data on tests involving a combination of multiple choice, short answer, and extended response questions that assess a mix of factual knowledge, conceptual understanding, and scientific inquiry practices. For example, in an evaluation of the LeTUS unit on Global Warming conducted in the 2001-2 school year, Herman and colleagues measured a pre to post effect size of .66 (n=623 7th and 8th graders; P<.001) (Herman, MacKenzie, Sherin & Reiser, 2002). In an external evaluation of one unit of the Looking at the Environment high school curriculum, researchers from SRI International measured an effect size of .68 (n=280; p<.000) (Crawford & Toyama, 2002). We hope to be able to do evaluations in a randomized field trial in the future to understand the differences in effectiveness of these curricula in comparison to more traditional approaches. In addition to evaluations of learning outcomes, we have also begun to conduct research on the motivational effectiveness of these curriculum units. The theory behind them posits that students will adopt the role and/or goal associated with the scenario, and that that will motivate them to participate in the learning activities and contextualize their learning in a way that improves learning outcomes. In the first stage of this research, we are investigating the primary assumption that students do "buy in" to the scenario by looking at individual differences among students through data collection that includes interviews, daily questionnaires, and observation (Pitts & Edelson, 2004). In our first round of small scale studies, we have found that most students do buy in, but that the degree of buy in differs significantly and the nature of that buy does as well. For example, some students described themselves as "pretending" to be a scientist in one scenario, while others reported that they were "interested" in working on the scenario because the problem seemed interesting or important. Others, reported that the scenario did not engage them, and that their motivation was based on a concern for doing well in school.

While we are able to report success for these curriculum units in classrooms, the story would not be complete without a discussion of teacher preparedness. The teachers that were included in the studies summarized above all received substantial professional development that was offered as a result of our earlier implementation efforts.

THE TEACHER LEARNING STORY

The teacher learning story is the newest and least fully developed of the three. The best way to motivate is through the early implementation efforts for the curricula developed by LeTUS in Chicago. These curricula had been developed by teams of researchers from Northwestern and teachers from Chicago Public Schools around the district and state science standards that had recently been released. Before enacting each 6-8 week curriculum unit, each teacher participated in a 3-hour session at a summer workshop and a one-day in-service workshop on that curriculum. They also had access to a professional development specialist who could answer calls by phone, visit schools for planning meetings, and co-teach lessons that teachers were not comfortable with. Despite that level of professional development and support, our experience in the first year of implementation was that the vast majority of teachers either never started teaching the unit they had signed up to teach or never completed it. This was a pretty dramatic lesson that this representative group of teachers either were not adequately prepared or did not feel adequately prepared to teach these technology-supported inquiry units.

This led us to initiate a program of research on both the content and structure of teacher professional development on technology-supported inquiry-based science teaching that is still underway. In the case of this research, we do not yet have well-developed design frameworks, but we have strategies that we are investigating.

The primary challenges that we have identified are consistent with the literature on teacher knowledge and teacher learning. Often teachers lack knowledge of specific content, particularly in the case of a curriculum unit that has a context or emphasizes depth of knowledge in a particular area that goes beyond the typical textbook. For example, the unit on water resources management described earlier goes beyond the content of a standard environmental science curriculum on agricultural practices or water engineering. Often teachers' pedagogical content knowledge, e.g., how to specify learning objectives, how to respond to students' ideas, and how to assess student understanding, is weak. For many teachers, the pedagogical approach is new. They have difficulty implementing both the scenario-based and inquiry-based elements because they don't fully understand the rationale, they lack the knowledge and skills necessary to implement them, and they don't have prior experiences to draw on. Finally, the specific scientific practices and uses of technologies are often unfamiliar and intimidating to teachers.

One observation that surprised us is that many of the LeTUS teachers who failed to implement or to complete the units had participated in substantial amounts of professional development on inquiry approaches to science teaching. In fact, they could often describe the goals and techniques of inquiry-based science teaching as well as or better than we could. This disconnect between the talk and the walk was a puzzle. Our conjecture is that the decontextualized way in which that professional development was implemented (inquiry in the abstract) prevented teachers from being able to apply it in the specific context of LeTUS units. Therefore, an important strategy we have been pursuing is that *professional development ought to be conducted in the context of specific curriculum*. We call this curriculum-linked professional development. Therefore, in LeTUS we began offering extended professional development experiences (10-week graduate credit courses) focusing on a specific LeTUS curriculum unit. While this approach carries the risk of devolving into training for a particular unit, we attempt to design our professional development experiences so that teachers develop understanding in a particular context but are encouraged to recognize how that understanding would transfer to other curriculum contexts. An important rationale from this approach comes from the fact that offering professional development in the context of a curriculum that a teacher intends to implement provides a motivation for the teacher to learn. In this way, the curriculum-linked approach to professional development fulfills the requirements of the "create demand" and "apply" strategies in the Learning-for-Use framework.

A second strategy that works in concert with the curriculum-linked approach is to *extend the professional development activities over the time that the teachers are implementing the curriculum*. In the case of the graduate courses offered to LeTUS teachers, the teachers were required to implement the curriculum in their classes over the same period that they were enrolled in the class. This gives the facilitators of the professional development to link the activities in the PD to classroom activities and experiences. While this approach has proven successful for a community that is in close proximity, as in LeTUS, it poses problems for a community that is geographically dispersed. For that reason, we are exploring the use of online tools to support this approach to professional development.

Our experience with this approach to professional development is that a much larger percentage of teachers who are taking the courses complete the units associated with the courses than those who don't. More important, they are more likely to teach the unit again and to try other LeTUS units as well. On the other hand, we only have very limited information about the quality of their classroom implementations, other than the test scores of their students. In future research, we hope to identify scalable techniques for characterizing teacher implementation and to use these techniques to study the impact of different forms of professional development.

REFLECTIONS

Reflecting back upon the research programs summarized here, I am struck by one clear conclusion and two important open questions. The conclusion is that this is demanding research requiring large teams representing diverse expertise. It falls into the paradigm of "big science". It cannot be performed by a solitary individual alone in his or her garret. It requires individuals representing expertise in science, technology, cognitive science, classroom teaching methods, and teacher professional development. On our research and development teams, we have had faculty, graduate students, and postdocs in learning sciences, geoscience, and computer science, programmers, teachers, and professional development specialists. It has involved long-term partnerships among universities, school districts, and schools. Finally, it has required individuals to get out of their usual places of work and into each others for extended periods of time. Researchers have spent extensive periods of time both observing and participating in classroom, and teachers have spent extensive periods of time in the university setting, participating in design meetings and developing materials as part of design teams.

The first question is a research methods question. It is, *How can lessons about design be shared?* I typically characterize the products of my research as being design frameworks. But, what is a design framework? In what form can and should design theories be shared? For example, a design framework can consist simply of a set of questions that should be answered, issues that should be considered, or tradeoffs that should be weighed in a design process. Or, a design framework can consist of strategies, guidelines, and recommendations, that do not simply enumerate issues, but provide guidance on how to respond to those issues. At what level of detail can and should issues and recommendations be expressed? The nature of content and pedagogy mean that educational design can not be reduced to a mechanized process. The differences in disciplines and content areas within disciplines mean that each area is a special case. The misguided efforts to teach *the* scientific method show the pitfalls of attempting to treat science or any other content area as a uniform whole. Therefore, the challenge of creating design frameworks is identifying the "sweet spot" that captures essential commonalities across content areas and disciplines, while leaving room for designers to improvise and adapt to the specifics of content area, audience, and context. Finally, what is the relative value of abstractions, like issues and strategies, compared to examples. People are naturally case-based reasoners (Roger C. Schank, 1982; Kolodner, 1993). In complex domains like design, where there can be no complete theory, it is often more natural to design by finding and adapting relevant examples then by designing by rules and principles. I have been struck that people are more likely to tell me how much they appreciate the examples in my papers than the abstract principles. It might be that the educational design community can best be served by creating large libraries of cases to consider than handbooks full of issues to attend to and guidelines to follow. I believe that understanding how designers can best be served by research is a question that educational design researchers must begin to address directly if we are to have impact on practice.

The second question is about inquiry itself. The question is, *How do we enable students to make the transition from guided inquiry to open-ended self-directed inquiry?* To me, our lack of insight on this question represents the major shortcoming of this research program to date. In the research described here, we made the conscious decision, based on our initial experiences, to develop curricula that would engage students in guided, or structured, inquiry. In these curricula, the driving questions or problems are presented to students. They engage in structured investigations of portions of the problem and are challenged to synthesize their own research results together with information that is provided to them to construct a solution to the overall problem. Their experience has important elements of inquiry, but it is not self-guided, and it is open-ended within a provided structure and context. We justified this approach based on the observation that the middle school and high school populations that we have worked with were not adequately prepared to engage in open-ended inquiry, nor were their teachers prepared to supervise them in truly open-ended inquiry. As an exception that proves the rule, one CoVis teacher did create a curriculum to engage his students in open-ended inquiry, with fascinating results described by Joe Polman in a book based on his Ph.D. dissertation (Polman, 2000). This book describes the extraordinary effort required by an exceptional teacher and is inspiring in the possibilities that it demonstrates while discouraging in the challenges that it exposes. I believe that we have made a sound decision in pursuing a developmental approach that builds inquiry skills in a scaffolded fashion with the goal of fading those scaffolds at some time in the future. The question, however, is when and how to fade those scaffolds. In my own research area, I characterize one of the skills necessary to conduct open-ended inquiry as *analytical fluency.* Analytical fluency is the ability to select and apply a particular set of analytical tools to answer a specific question. Most students only appear to gain the ability to apply analytical tools and interpret their results from the curricula they have developed. They do not appear to acquire the ability select the appropriate analytical operations for a particular question or context. I see understanding how to help students make the transition to analytical fluency as being the next major challenge to developing true inquiry skills in students. However, making this transition will require both creative educational designs and time. It is not clear that the current curricular emphasis on broad content will offer students and teachers the opportunity to spend the required amount of time on inquiry to develop those skills.

CONCLUSION

The design of learning experiences that will enable students to develop the skills and knowledge required to engage in inquiry is in many ways an engineering problem. We currently have sufficient science and theory to begin to engineer solutions to the "inquiry" problem. However, it is not a simple problem that can be solved through narrow approaches. It requires systemic approaches that attend to the challenges of both student and teacher learning and of the physical and organizational context.

NOTES

1. From (Edelson, 2001)
2. While the first step in the Learning-for-Use model is called motivate, this phase is only concerned with a small portion of what is normally thought of as motivation in education. In this context, I am using motivate to refer to a specific type of motivation—the motivation to acquire specific skills or knowledge within a setting in which the student is already reasonably engaged. Addressing the broader motivational challenges of engaging students in schooling are critical to, but beyond the scope of, the Learning-for-Use model.
3. http://www.whatworks.ed.gov

Daniel C. Edelson
School of Education & Social Policy and Computer Science Department
Northwestern University

JANICE BORDEAUX

A COMMENTARY ON "ENGINEERING PEDAGOGICAL REFORM"

The body of work that Edelson's design research paper summarizes is rich and complicated. It will help in discussing it to distinguish two kinds of design goals, which operate at different levels of abstraction. I would characterize the time-consuming work that takes place during daily operations with his collaborators and at his schools as developing a design for a learning environment that teaches science *through* inquiry activities. (Following Nersessian (this volume), let's call this a domain specific goal.) The design goal is challenging and certainly not a small one. It is to create a learning environment with a range of inquiry-based activities that increases the knowledge of middle- and high schoolers about the environmental sciences, along with their motivation to study them. In engineering this would be the equivalent of designing a system, perhaps even a lab or a factory, for the production of a specific product.

Edelson's second and more powerful body of work is educational design research (a domain general goal, according to Nerssesian). This work takes place when he and his collaborators set about determining the meaningful commonalities that occur within a class of learning environments. Its design goal is to create a framework for engineering learning environments that effectively guides the development of varied kinds of teaching and learning. In engineering, this would be the equivalent of designing a model for building different kinds of systems, laboratories or factories that generate a product group. The focus of his work, his design "testbed," has been inquiry-based instruction situated in a range of urban, suburban and rural middle and high schools.

In this commentary, I will explain my understanding of his design approach, offer some comments on the two kinds of design, and raise some questions about educational design research for science education.

CREATING A LEARNING ENVIRONMENT THAT TEACHES THROUGH INQUIRY

I have drawn a diagram (Figure 1) to illustrate some of the main features of the educational system in which Edelson's inquiry-based learning environment is situated. One thing this diagram demonstrates is the scope of the challenge to achieve student scientific knowledge through inquiry. Students arrive at class interested in some identifiable scientific topics, but equipped with prior knowledge that is incomplete (and also sometimes erroneous). In classes they are engaged in inquiry based on the 3 steps of the Learning-for Use Design framework – motivate scientific learning, construct knowledge, and refine that knowledge for use. But teachers didn't have the time and were not prepared to create inquiry-based curriculum, so curriculum development teams were needed to develop units on specific topics in the environmental sciences. However, it then turned out that teachers could not implement the new units, both because they generally lacked (or believed they lacked) the pedagogical knowledge and skills needed to teach inquiry, and the deeper content knowledge they needed to teach specific scientific topics. Moreover, teachers (like students) needed ongoing support and assistance during implementation to complete the units. I would be surprised if these challenges are not common in most school systems, particularly the systems that serve large metropolitan districts. However, it appears as though the Learning-for Use framework applies, not just to student learning, but to the professional development educators need to implement any new curriculum.

THE SPECIFICITY OF KNOWLEDGE

One of the most striking strategies that Edelson employs is to teach the *specific* knowledge and skills needed for useful knowledge. In the case of students, he claims that the desire for useful knowledge motivates the learning of specific content or skills that they can apply. He also argues that by applying them, students make knowledge and skills more accessible when they need them later. Similarly, professional development for teachers "ought to be conducted in the context of specific curriculum "that teachers intend to teach, as much to motivate them as to help them apply their new knowledge and skills.

This approach differs sharply from the assumption that general knowledge should precede more specific knowledge and skill acquisition, which is one rationale for requiring broad introductory courses before advanced topics in many curricula. At many universities it is a major reason that engineering majors are required to take general math and science courses designed for math and science majors. However, students are concerned about the relevance of the course work, and faculty are disturbed that students cannot access or apply their general knowledge in later coursework. It may be that introductory knowledge should be much more domain-specific. At my university the faculty have created a new course on optics and waves for electrical engineers, which selects content that is important *now* in Electrical and Computer Engineering, and provides many more examples of applications in engineering. In addition, the desire to teach all students about the impact of science and engineering on society, and the need to help majors understand what different kinds of scientist and engineers do before choosing careers has some schools rethinking their introductory curriculum. As a result there is considerable more interest now in creating project-based and topic-oriented courses and problem-based learning assignments.

An emphasis on specific knowledge and skills also appears to contrast with the assumption that some kinds of general knowledge and skills can be taught which then transfer widely applied across the curriculum, as in some older writing-across-the-curriculum programs. In practice, many of these programs develop units that are highly context specific; they embed writing units within existing curriculum, require unit-specific knowledge and skills, and teach the forms and styles preferred by practitioners in the field.

It appears that educators are introducing more domain-specific learning in higher education, but much more must be understood about what the optimal level of specificity should be for: 1) various domains or topics and at different levels of instruction, 2) different levels of cognitive and social development, 3) individual differences in cognitive and social ability, and perhaps 4) cultural and gender-related differences.

As Edelson notes in discussing teacher training, this approach runs the risk of devolving into training for a particular unit. Possibly, he sees generalization as an achievement that occurs incrementally as a result of his third step – knowledge organization. He cites evidence that useful knowledge is connected to other knowledge structures that describe situations in which that knowledge applies. Teachers who offer explicit explanations for these connections or guide students in developing their own explanations may be supporting the leap to generalization. I have diagramed this hypothetical process as the iteration of specific learning experiences within a domain, but with the addition of teaching that supports reasoning about the connections between concepts (see Figure 2). I hope that Edelson will share his thoughts with the group on the problem of how and when domain knowledge is generalized with this approach.

Much more knowledge is needed about the relevant age-related differences that affect intellectual inquiry at all ages in order to prepare successful curricula. There is ample evidence that younger students, even very young students, are capable of learning through inquiry-based activities. However, as a developmental psychologist, I must also ask whether students must be older and more physically mature before acquiring the abstract reasoning abilities they need to fully understand scientific inquiry and apply it to novel situations. Moreover, scientific inquiry is a social as well as an intellectual activity. When inquiry-based instruction is sustained throughout the curriculum, even elementary students are capable of understanding the central role of reasoned argumentation and appreciate that scientific controversies are characteristic of scientific debate (Smith et al., 2000). However, recent evidence indicates that the maturation of the brain involved in social judgment, reasoning, decision-making undergoes dramatic changes beginning around age 16 and is incomplete well into early adulthood (Giedd, 2004). It may well be that preparing older students to evaluate scientific issues as adults may require more knowledge of how scientific judgment develops during intermediate and secondary schools. In particular, educators will need to know more about how and when students develop social judgment as individuals and in groups and how this affects discussions about evidence, professional standards, ethical issues, and the impact of science on society.

I might add that the difficulty of developing and giving teachers access to libraries of topic-specific content can be assisted by free public access to individually and collaboratively authored content modules. The *Connexions Project (CNX)* offers an open-source web-based technology dedicated to supporting the creation and sharing of educational content free of charge worldwide. Anyone may author content, access the published modules in the repository, or revise them to suit their own goals and audience. A beta version of module editing features which includes automatic import of Microsoft Word files is available at http://beta.cnx.rice.edu/

MOTIVATING SCIENTIFIC INQUIRY

The first step in the Learning-for-Use (LfU) framework is to motivate students to learn about specific scientific topics. According to the LfU framework, either they need new knowledge to do an assigned task or discovering that their current understanding is somehow inadequate piques their curiosity. I agree that motivating and engaging students is critical, but I think there is more involved in engaging students in scientific inquiry than the framework suggests. Individual interests, beliefs, and goals are influenced by cultural and social experiences. In fact, the research teams made an effort to uncover scientific topics that appeal to students in these age groups, probably from diverse cultural groups and both girls and boys. Acknowledging more explicitly the importance of including topics that appeal to diverse cultural and social groups at these ages should help us interest more students from minority cultures and women in science and engineering careers than we have thus far.

ENGINEERING LEARNING ENVIRONMENTS

At a more abstract level Edelson's current version of the LfU framework illustrates engineering design practices (see Figure 3). For example, in the design process, problem-definition entails not only clarifying inquiry learning objectives, but also three other criteria (Dym & Little, 2000). First, designers must establish the desired characteristics and costs of inquiry-based learning, what makes inquiry worthwhile to the users, and how much time and effort they can afford. Then, they must also identify which educational standards should be met and any important practical considerations. This last design practice is particularly difficult, because it requires an understanding of the major constraints to assembling the expertise, tools, time and space to create a learning environment in the *existing* schools. For inquiry-based education, the administrative constraints of the existing school agenda, calendar and daily schedule are often major challenges. Finally, problem definition also establishes what inquiry–based learning must be able to do. That is, all the critical functions of the environment must be identified. As we have seen in Edelson's research, some critical functions have been neglected, for example, motivating scientific inquiry, developing specific curriculum, and training educators. These functions may be essential for bringing about systemic educational reform, and the failure to include them may explain why promising initiatives fail to replicate in other school systems. Engineered educational design explicitly and systematically examines all of these issues during the design process.

Finally, as Edelson emphasizes, the end-goal of engineered design, indicated by the expanded definition of the problem, generating and testing design alternatives and optimizing the efficiency of the design, is to produce a model that yields descriptive or empirical results, and does not necessarily contain a causal explanation of inquiry-based learning. This design strategy may be better suited for solving complex, systemic education problems than the traditional scientific strategy of focusing on only few dimensions of a complex problem or hunting for causal explanations.

PRINCIPLES OF LEARNING

Edelson's design framework is based upon a modest set of 4 learning principles drawn from the contemporary literature in the cognitive sciences. Three of Glaser's (1995) and Bransford et al.'s 7 principles (1999) are prominently featured: the constructed nature of knowledge, that prior knowledge structures support and scaffold new knowledge, and the idea that procedural knowledge develops in context. Another principle, the characterization of knowledge construction as goal-driven, appears in the set of 14 Learner-Centered Psychological Principles recently developed by the American Psychological Association (2003). (See Table 1 for a list of these principles. The similarities and differences among these discipline-centered lists and their implications for education are worth exploring.)

I have a murkier idea of what Edelson means by his fourth learning principle, that "the circumstances in which knowledge is constructed and used determine its accessibility for future use." Does he mean that the same environmental and developmental factors that influence learning also shape and limit accessibility? Also, what types of circumstances is he referring to? Certainly he means instructional practices and educational technology. Possibly, he means that the social or cultural circumstances affect accessibility. If so, then educators need to understand what kinds of specific circumstances are most likely to influence the acquisition of scientific and engineering knowledge, and at what ages.

In general, Edelson's educational framework is not presently focused on the social and developmental processes that affect inquiry practices within communities. A more effective model might consider the social and developmental factors that affect classroom performance. It might also add bridging functions that help students understand the social and intellectual practices prevalent within professional disciplinary communities in terms of the social practices they are already know.

REFLECTIONS

One of the two most challenging objectives of education is to transform young people into citizens capable of understanding science and engineering, making critical decisions about their impact on society, and acting together to solve the major problems facing humanity. The other is to transform some of those students into professional investigators. In many ways, the increasing pressure and discipline required for academic achievement at the highest-level leaves scientists and engineers little time for developing the habits of investigation. To determine what kind, how much and when inquiry should be embedded in instruction may well require the application of engineering design principles as well as scientific research on learning.

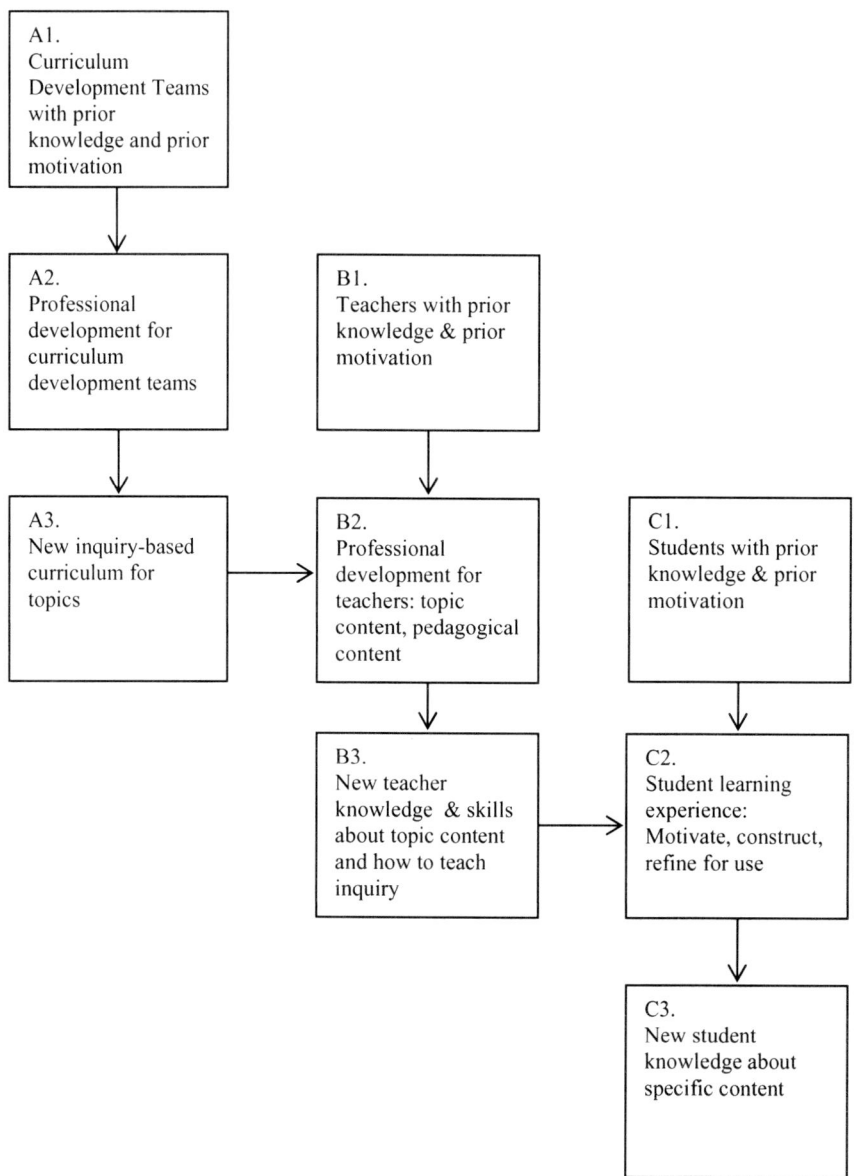

Figure 1. A Design for an Education System Containing a Learning Environment That Teaches Inquiry

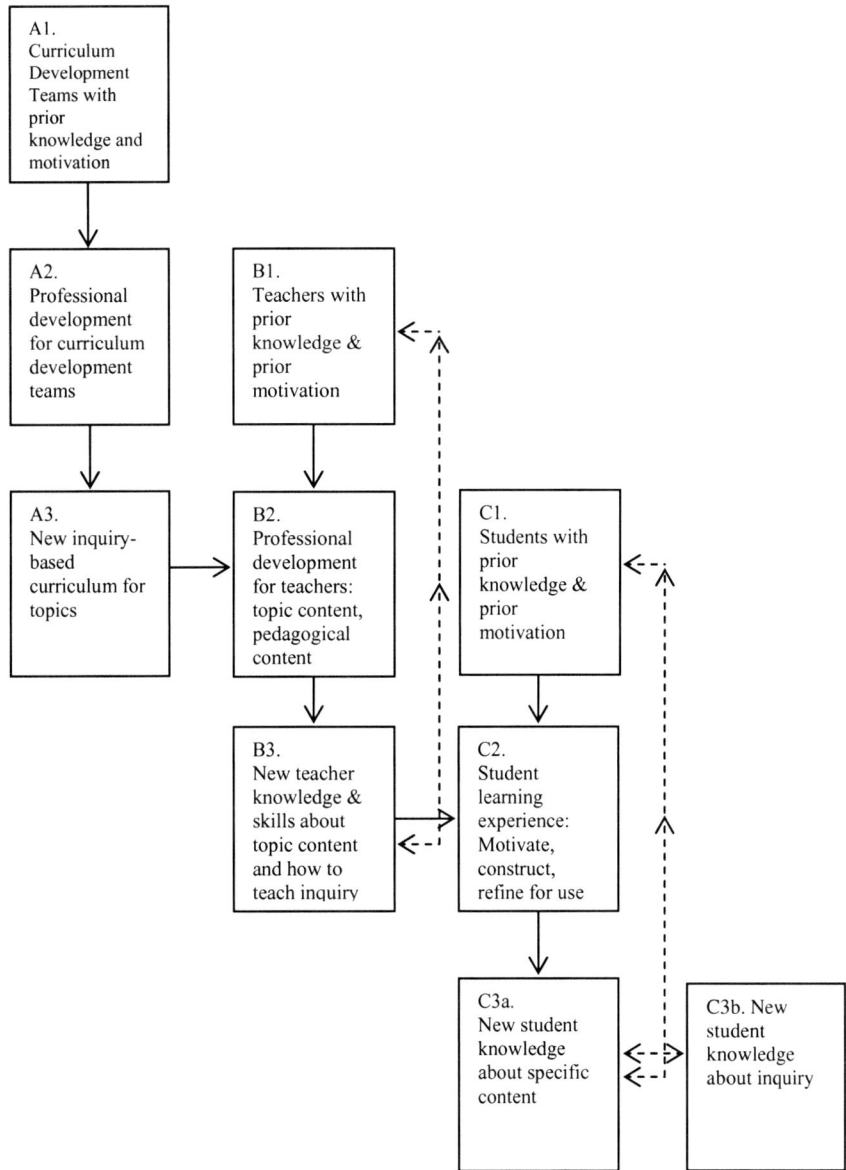

Figure 2. An Idea for an Educational System Containing a Learning Environment That Teaches Independent Inquiry

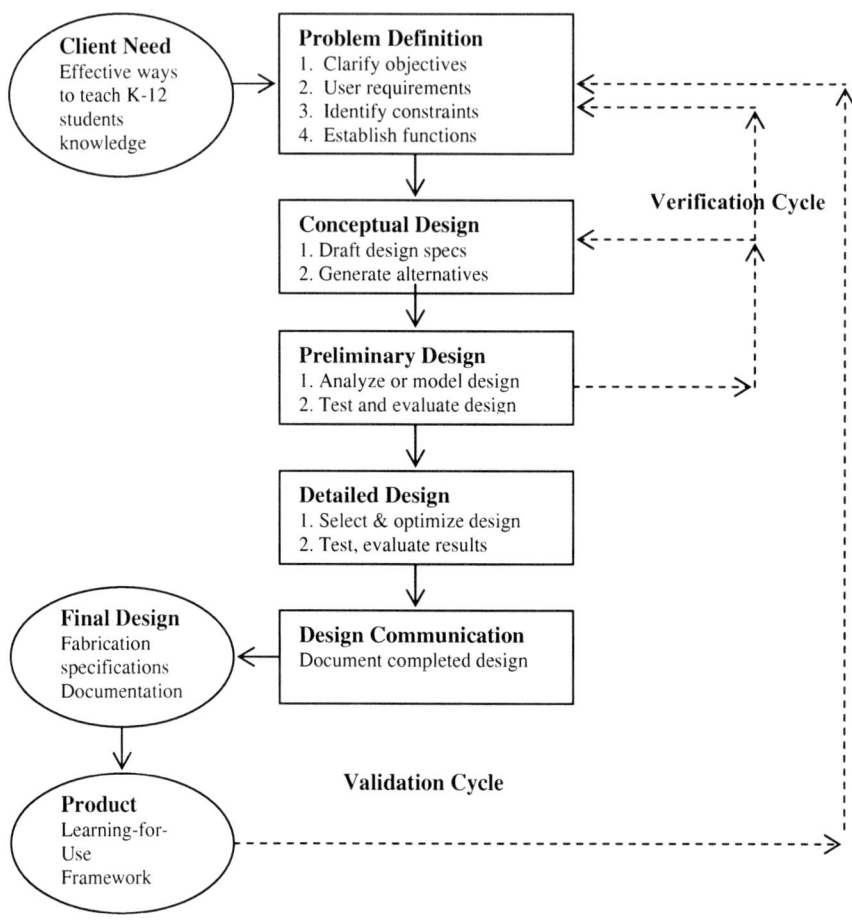

Figure 3. Engineering Learning Environments: Developing the Learning-for-Use Framework

Janice Bordeaux
George R. Brown School of Engineering and Psychology Department
Rice University

CLARK A. CHINN AND ALA SAMARAPUNGAVAN

LEARNING TO USE SCIENTIFIC MODELS: MULTIPLE DIMENSIONS OF CONCEPTUAL CHANGE

Learning to use scientific models is extremely challenging for students because learning to use models is thoroughly intertwined with the problem of conceptual change. Without conceptual change in how new models are understood, students will misunderstand and misapply the models they are learning. The central goal of this paper is to outline a theory of conceptual change that provides insights into how students can learn to use scientific models.

There are many theories of why conceptual change in science is difficult. We outline several prominent theories below. What most theories have in common is an assumption that the difficulty of achieving conceptual change can be explained by positing one or several key factors that impede change. Our alternative theory takes the opposite view: Conceptual change is extremely multifaceted. There are *multiple dimensions* of conceptual change, and hence *multiple roots* of difficulty. That is, there are many different reasons why conceptual change is difficult, and the sources of difficulty may vary from topic to topic. Attempting to characterize difficulties in conceptual change as due primarily to one cluster of factors or another is a serious oversimplification of conceptual change.

Our *multiple-dimensions* theory is predicated on the assumption that conceptual change involves both changing *beliefs* and changing *ideas* or *understandings* (Chinn & Samarapungavan, 2001). For a student to undergo conceptual change regarding Newtonian mechanics, it is necessary both to understand the central conceptions of Newtonian mechanics and also to believe that they are valid. The seminal work of Posner, Strike, Hewson, and Gertzog (1982) can be viewed as indirectly laying out the importance of this distinction. When they posited that *dissatisfaction* is a condition for conceptual change to occur, they were noting the importance of willingness to change beliefs. When the asserted that conceptual change occurs when *intelligible* new theory is available, they were implicitly taking note of the difficulty in understanding new theories. Conceptual change often requires both changes in belief and changes in understanding. Later theories of conceptual change have tended to focus predominantly on the belief dimension or the understanding dimension. Researchers have not developed integrated theories explaining both changing beliefs and adopting new understandings.

We divide theories of conceptual change into three broad groups: belief-focused theories, theories of restructuring, and theories of knowledge refinement.

1. *Belief-focused theories*. These theories focus on why it is hard to change scientific beliefs but they have less to say about why it is hard to understand new scientific conceptions. Theories that fall into this category include Chinn and Brewer's (1993) analysis of how to promote theory change in response to anomalous data and Dole and Sinatra's (1998) Cognitive Reconstruction of Knowledge Model. These models each provide rich explanations for why students resist changing beliefs in response to experiments, demonstrations, and other

Richard A. Duschl and Richard E. Grandy (eds.), Teaching Scientific Inquiry: Recommendations for Research and Implementation, 191–225.
© 2008 *Sense Publishers. All rights reserved.*

evidence, and they provide a broad range of theory-based suggestions for promoting belief change. However, they do not address how to promote understanding of difficult new conceptions, beyond asserting the need for an "intelligible" new theory, which begs the question of how to make a theory intelligible.

2. *Theories of Restructuring.* Theories of restructuring posit that science learners must undergo fundamental changes in conceptions when they learn scientific theories that involve conceptual change. Current conceptions must be restructured in some way when students learn challenging science concepts. In some theories that fall into this category, learners' initial ideas are held to be incommensurable with the target scientific ideas (e.g., Carey, 1991), in much the same sense that Kuhn (1962) proposed that rival scientific theories are often incommensurable in times of scientific revolutions.

Chi's (2005) ontological theory of conceptual change posits that conceptual change is difficult primarily in situations in which students apply concepts from the wrong ontological categories to new scientific conceptions that they are learning. For example, students' naïve theories of heat treat heat as a substance that travels from one place to another. When they learn about the kinetic theory of heat, they are learning a theory in which macroscopic behaviors are emergent from the independent, simultaneous, random, and simple behaviors of particles at a microscopic level. Because learners often try to interpret these ideas from their own ontology of substances that engage in straightforward, direct processes, they fail to understand the new ideas, which belong to a different ontology of emergent processes.

Grotzer and Perkins (2000) have presented a theory of causal reasoning that provides a powerful account of why students have difficulty learning many different aspects of scientific models. Their framework specifies models of causation that learners may have and shows how these differ from scientific models. Learners' understanding of causality may vary along four dimensions: mechanism (e.g., from surface generalization to giving an underlying mechanism); interaction pattern (e.g., simple linear causality, interactive causality, constraint-based causality); probability (e.g., from belief in deterministic systems to understanding a variety of probabilistic influences on causal systems); and agency (e.g., from central agents being wholly responsible to understanding emergent entities and processes). For Grotzer and Perkins, difficulty in conceptual change arises when students' understanding of causality differs in fundamental ways from the form of causality that is present in scientific models that students are attempting to learn. Grotzer and Perkins's framework includes as one cluster of categories many of the ideas present in Chi's theory of emergent processes; their framework is broader in that it implicitly proposes a broader range of model types that may be difficult for students to learn.

Wiser and Amin (2001) present a set of ideas about conceptual change that are similar in some ways to Chi's but also different at key points. Focusing their attention on the domain of thermodynamics, Wiser and Amin hold that students have difficulty differentiating between heat and temperature. In contrast to Chi, they argue that students have little difficult understanding the kinetic model of energy transfer and how equilibriums emerge from molecular interactions; instead, students' difficulty lies in understanding how these ideas relate to their undifferentiated everyday ideas about heat and temperature. The instructional solution is to help students develop an understanding of the differences between everyday usages and scientific usages of heat and temperature and to show that the kinetic model provides an explanation of their everyday ideas. C. Smith and her colleagues (C. Smith, Snir & Grosslight, 1992; C. Smith & Unger, 1997) have similarly focused on difficulties inherent in differentiating weight from density.

Vosniadou, Ioannides, Dimitracopoulou, and Papademetriou (2001) have argued that a key to conceptual change is to focus instruction on entrenched presuppositions that constrain students' models in a domain. For instance, when learning about the shape of the earth, elementary school children's ideas are constrained by their presuppositions that the earth appears flat and that things fall downward. If one tries to teach the scientific model without addressing these presuppositions, students will simply develop a new but still mistaken model. Vosniadou et al. (2001) agree with Wiser and Amin (2001) about the importance of teaching students about the differences between everyday language and scientific language.

All of these theories view learning science as a process of helping students construct new conceptions that are very different from their prior conceptions. The theorists differ in how they characterize the conceptual differences between the old and new conceptions and in how they propose to address the prior conceptions. Chi's approach is largely to sidestep prior knowledge and build new theories from scratch by teaching students the new ontological categories and encouraging them to refrain from using their prior ontological categories inappropriately when learning new theories. Wiser and Amin, in contrast, argued that the new conceptions provide an explanation for students' old conceptions, so that students' old conceptions should be actively engaged. Vosniadou et al. proposed that it is essential to explicitly work with students to address their entrenched presuppositions.

3. Theories of Refinement. A third class of theories emphasizes continuity between prior conceptions and the target scientific ideas that students are learning. Spelke (1992) argued that normal development involves enrichment of core ideas rather than any fundamental changes in conceptions. For example, ordinary physics knowledge is built on core ideas that are present even in early infancy, such as the idea that objects are solid and do not pass through each other. Spelke did allow, however, that more advanced theories such as quantum mechanics that are learned through formal schooling involve more fundamental changes in conceptions.

Smith, Chi, and Joram (1993/1994) outlined a theory of conceptual change that does not involve fundamental changes in any existing conceptions. Building on diSessa's (1993) theory that naïve scientific knowledge consists of p-prims – fragmentary ideas that are near surface-level generalizations from experiences – J. Smith et al. argued that students' prior conceptions are not incorrect; instead, they are valid conceptions that are misapplied in inappropriate situations. Learning science involves not creating new conceptions from scratch or replacing old conceptions with better ones but rather with learning how fundamentally useful conceptions should be applied to different situations. Conceptual change is predominantly a matter of learning to reconnect conceptions to different situations.

Multiple dimensions of difficulty in conceptual change. Our position combines elements of all three groups of theories in that we think that conceptual change sometimes involves redeploying existing ideas to new situations, often involves learning fundamentally new conceptions, and almost always involves changing beliefs as well as understandings. In contrast to theories of refinement, we think that many important instances of conceptual change do involve fundamental changes in ideas that are not already being deployed in other situations. In contrast to each individual theory of restructuring, we think that there are many kinds of conceptual differences that make conceptual change difficulty across various science topics. Emergent processes (Chi, 2005) are one important kind of conceptual difference that creates serious conceptual difficulties, but so are undifferentiated concepts (Wiser & Amin, 2001) and impoverished causal models of many sorts (Grotzer & Perkins, 2000). We will propose additional specific conceptual difficulties, as well.

In sum, we think that the sources of difficulty in conceptual change are many and multifarious. Our proposal is accordingly termed *multiple-dimensions* theory. Scientific models are extremely diverse in form and content, and any attempt to characterize what makes conceptual change difficult by focusing on one or even several factors is an oversimplification. Learning different kinds of models poses different kinds of difficulty.

Among the theories that we have briefly reviewed, ours is most similar in spirit to that of Grotzer and Perkins (2000). Grotzer and Perkins have attempted to identify a broad range of students' misunderstandings of causal models that can impede conceptual change. Multiple-dimensions theory includes a discussion of a range of difficulties raised by trying to learn causal models, but our conceptualization of difficulties with causal models is somewhat different, and we also identify many other sources of difficulty.

GOALS OF A THEORY OF CONCEPTUAL CHANGE

As educators, our goal in developing a theory of conceptual change is to provide insights into efforts to promote learning in classrooms, particularly learning through model building and other forms of inquiry. To help students achieve conceptual change through inquiry, teachers will benefit from an understanding of the wide range of reasons why conceptual change is difficult. They will be better able to guide discussions and design activities that assist students with the

CONCEPTUAL CHANGE

particular learning difficulties that arise with particular science topics. We will return to this point later in the chapter.

OBSTACLES TO LEARNING MODELS AND THEORIES: MULTIPLE DIMENSIONS OF DIFFICULTY

In this section of the chapter, we will provide an outline of many different dimensions of difficulty in learning new scientific theories. To avoid assuming that students' old or new ideas are theory-like (we think that students' initial ideas are likely to be theory-like in some domains but not in others), we will use the terms *conceptions* and *ideas* to refer to students' old ideas as they attempt to learn scientific models. We do not assume that students always *believe* new conceptions that they construct, although they ordinarily do believe their own *prior conceptions*.

We also make a fundamental distinction between belief and understanding. Students may understand an idea without believing it in any real sense (e.g., students may understand what the teacher says about inertia and force without believing that objects really behave in this way), and they may believe ideas without understanding them (e.g., many supporters of evolutionary theory undoubtedly misunderstand many of its key tenets). Researchers seldom assess both belief in and understanding of an idea, which makes it very difficult to interpret the results of studies designed to assess whether conceptual change has occurred (see Chinn & Samarapungavan, 2001).

The distinction between belief and understanding helps clarify when students' conceptions can be reasonably characterized as *misconceptions*. If students understand a new conception but do not find it to be believable, perhaps because they have everyday experiences that appear to support their own belief, we think it is unreasonable to call the student's own belief a *misconception*. However, if the student is trying to understand a teacher's (or a peer's) explanation, and the student has not understood it as the teacher (or peer) intended, then it is appropriate to say that the student has a *misunderstanding* or *misconception* of the interlocutor's explanation.

In our outline of diverse sources of difficulty in learning scientific theories that begins in the next section, we will consider separately *obstacles to belief* and *obstacles to understanding* because we think that the processes that promote each are at least partly separate; hence the obstacles to belief change and to changes in understanding are overlapping but not identical.

OBSTACLES TO BELIEF CHANGE

Several reasons why students may resist belief change when learning new ideas are presented in Table 1. Some ideas in this table are further elaborated in Chinn and Brewer (1993, 2000). We focus on obstacles to epistemically grounded belief

195

change, that is, belief change that is based on good reasons for belief. Good reasons for belief include fit with empirical evidence and consistency with nonempirical criteria such as parsimony or avoidance of ad-hocness. We think that teachers who are aiming to promote conceptual change in inquiry environments will want to focus on epistemic reasons for belief change. Hence, an analysis of obstacles to epistemically grounded belief change will be particularly useful to them.

Table 1. Sources of Resistance to Epistemically Grounded Belief Change

Source of resistance	Comments
A. Unavailability of an understood alternative theory	When students have available an alternative theory that they understand, belief change is more likely (Posner et al., 1982). Because we hypothesize that understood theories are more likely to be believed, most of the obstacles to understanding presented later in this section will also be obstacles to belief change. Note, however, that this does not require that the students have a *correct understanding* of a new theory. Students may be willing to adopt a new theory that they think they understand but actually misunderstand. The remainder of this table includes additional obstacles to that are especially relevant as obstacles to *belief change*.
B. Characteristics of prior conceptions	
1. Entrenchment of prior conceptions	Some specific ideas in a belief system may be strongly entrenched; instruction that does not specifically address these *entrenched presuppositions* (Vosniadou & Brewer, 1992; Vosniadou et al., 2001) is unlikely to be effective. When nonentrenched beliefs are addressed without addressing entrenched presuppositions, students' beliefs will develop new alternative beliefs that are still be constrained by the entrenched presuppositions (Vosniadou et al., 2001).
a. Entrenchment arising from congruence with known evidence	Students may often believe ideas because they have a strong base of empirical support for them. For example, students have a strong empirical base for believing that things fall downward based on everyday observations, which makes it hard to understand how people could live on the "bottom" of the earth.
b. Entrenchment arising from coherence with other scientific and nonscientific conceptual systems	Preinstructional scientific conceptions may be entrenched because they cohere with other beliefs. For instance, creationist beliefs cohere with other religious beliefs.

c. Entrenchment arising from personal interests	Ideas may be entrenched because abandoning them may be personally threatening. For instance, heavy coffee drinkers show strong motivation to discredit information indicating that heavy caffeine consumption can be harmful to the health (Kunda, 1990).
d. Entrenchment arising from social pressures	Entrenchment may also arise from social pressures. When everyone around a student professes a belief in creationism, it may be hard to change beliefs even if the student finds the evidence for evolution to be strong.
2. Unfalsifiable conceptions	Sometimes conceptions may be structured so that they are virtually unfalsifiable. A student for whom forces behave very differently in different situations and who does not require a parsimonious theory has a theory that is very difficult to dislodge, because there is little evidence that could possibly contradict such a protean theory.
3. Background knowledge related to evaluating evidence.	Background knowledge refers to knowledge systems that are not specifically part of the theory under consideration. Evidence evaluation often relies on background scientific theory. For example, understanding fossil evidence depends on having an understanding of how fossils are created and distributed; the theory of how fossilization occurs is largely independent of the theory being tested.
4. Strategic knowledge of relevant reasoning strategies.	Students must also be able to deploy reasoning strategies that are needed to appreciate relevant evidence. For instance, statistical reasoning strategies are needed to interpret statistical evidence.
C. Epistemological understandings and commitments	Epistemology is complex and multifaceted (Samarapungavan & Chinn, in preparation). Here we mention just a few epistemological commitments that are likely to encourage epistemically grounded belief change.
1. Metaconceptual understanding of theories	Many researchers have argued that science students frequently display an imperfect understanding that theories and evidence are distinct and that different (competing) theories can be advanced to explain the same evidence. Although students often do show an ability to make appropriate theory choices on the basis of evidence (e.g., Samarapungavan, 1992), it appears likely that they do not have a full, explicitly explicable metaawareness of relations between theories and between theories and evidence.
2. Lack of commitment to rigorous coherence	Science students often appear to be satisfied with minimal evidence for a proposition, rather than insisting on rigorous consistency between theories and evidence (Reif & Larkin, 1991). Insisting on rigorous coherence within theories and between theories and evidence should enhance belief change in response to evidence.

	3. Lack of commitment to precision	A feature of scientific research is that scientific theories aim to make increasingly precise predictions; it is likely that students to not appreciate the epistemic value of theories that offer highly precise quantitative predictions and explanations.
	4. Lack of commitment to exposing conceptions to data and to open-mindedness in changing conceptions when warranted	A general open-mindedness about one's own scientific theories will enhance epistemically grounded belief change. Even better is an active disposition to expose one's theories to data.
	5. Lack of commitment to nonempirical criteria for theory choice	Well-grounded theory change can arise from the application of nonempirical criteria such as simplicity, consistency with other theories, and avoidance of ad-hocness.
D. Goals		Educational researchers have documented the general value of a number of goals in school learning, such as mastery or learning goals (as opposed to goals of demonstrating superiority over others) (Pintrich, Marx & Boyle, 1993). However, the interesting question for conceptual change research is which motivational patterns are uniquely beneficial for promoting conceptual change. We think that there is very little research on this issue, but we offer several suggestions.
	1. Lack of goals consistent with the epistemological commitments listed above.	For reasons discussed above, students are unlikely to change beliefs when they do not set goals to (a) evaluate rival theories by comparing each to available data, (b) seek rigorous coherence, (c) choose theories that make accurate and precise predictions, and (d) change beliefs when warranted by the evidence. In general, because instruction often conflates belief with understanding, students may not clearly understand that they must both try to understand new ideas and also (and partly separately) decide whether they believe these ideas.
	2. Goal to come to closure on an issue quickly	Learners may freeze thinking on an issue before they have adequately gathered information (Kruglanski & Webster, 1996; Pintrich et al., 1993). A willingness to hold ideas in abeyance may enable students to examine enough evidence to change their minds.

CONCEPTUAL CHANGE

3. Unwillingness to temporarily suspend belief about theories that are hard to understand.	When new theories are difficult to understand, it may be necessary for students to set aside any attempt to decide whether the new theory is correct or not until they have had a reasonable opportunity to build up an understanding of the theory that is sufficient to permit meaningful theory comparisons. If students attempt to enforce belief too quickly, they may begin to distort the theory so as to make sense of it (Chinn & Samarapungavan, 2001).
4. Goal to achieve personal benefits or preservation of self-image over truth	Sometimes belief change means giving up tangible or psychological benefits, and a willingness to do so is undoubtedly conducive to belief change. In contrast, belief change is probably less likely when learners prefer personal benefits (e.g., for graduate students, believing a theory because it is the basis of the grant that is providing the student's funding) or to preserve self-image (e.g., by believing that their fellow citizens are not responsible for global warming).
E. Characteristics of the data-theory complex	Assuming that students have the strategic and metaconceptual resources to evaluate competing theories with respect to a body of data, epistemically grounded belief change requires that the data actually provide compelling support for the target theory.
1. Characteristics of individual data	One way to examine whether data do in fact support a new theory is to examine individual items of data.
a. Ambiguous	When data are ambiguous with respect to two theories, the data are unlikely to promote belief change. For example, the experiment of dropping a heavy object and a light object to see which falls faster is an ambiguous experiment because it is very difficult to accurately note whether the objects have hit the ground at the same time; as a result, theory change in response to such data is unlikely (Chinn & Malhotra, 2002a).
b. Noncredible	When data are not credible, the data are unlikely to promote belief change. Data are more likely to be credible when (a) the source of information about the data is credible, (b) the studies from which the data are generated employ methods of data collection that are accepted by the learners, (c) there is replication of data, and (d) the data are observed directly (Chinn & Brewer, 1993, 2000).
2. Characteristics of larger data-theory relations	Even when individual data are credible and unambiguous, it may not be the case that the overall body of data provide sufficient support for the new theory.
a. Insufficient data for new theory	One problem is that the data may simply be insufficient in quantity. A single experiment is unlikely to outweigh a lifetime of experiences.

	b. Underdetermination	As we will in a later section, students may have good reason not to change beliefs because the data available to them underdetermine theory choice between their own theory and the theory they are learning.
F.	Misunderstanding of scientific authority	We take believing scientific ideas on the basis of authority to mean trusting the findings and conclusions of other scientists, without undertaking original inquiry on a particular topic oneself. A moment's reflection makes it clear that scientists do and should believe a great deal of what they believe based on authority of this sort (Kitcher, 1993; Norris, 1992). Textbook presentations of central ideas in fields of science are common throughout graduate school training (Giere, 1988), and scientists undoubtedly learn much of what they learn by reading articles in journals, by attending conference presentations, and by being told about colleagues' findings and conclusions in face-to-face conversations.
	1. Overreliance on authority 2. Failure to appreciate epistemic basis for authority	The problem of overreliance on authority arises when the grounds for trusting authority are inadequate. Accepting ideas based on authority should be regarded as nonepistemic only when the grounds for trusting authority are inadequate. Crucially, students should have a deep understanding of how scientific inquiry proceeds and why its results are deemed by scientists to be trustworthy. Once students have this understanding, then they can reasonably judge that findings and conclusions that are generated via scientific inquiry processes can be trusted. However, believing an idea just because a teacher or a textbook says so (when no understanding of the general inquiry processes that lie behind the conclusions asserted by the teacher) is nonepistemic.
G.	Reasoning processes	Assuming that the data warrant changing beliefs, whether students do in fact change beliefs will depend crucially on the reasoning processes that they employ.
	1. Shallow processing	Generally, shallow reasoning processes that involve minimal thinking about the issue are likely to produce either no belief change or shallow belief change (Chinn & Brewer, 1993; Dole & Sinatra, 1998). When shallow processing is present, learners are more likely to use superficial peripheral cues such as the sheer number of arguments given, without reference to the quality of the arguments, or the attractiveness of the source of the new information (Cacioppo, Petty, Kao & Rodriguez, 1986).

2. Lack of needed reasoning strategies	More specifically, students must have available the specific reasoning strategies needed to reach a conclusion. For example, if students fail to understand error of measurement or sampling, then they will interpret a statistically insignificant observed difference as proving that a difference exists. Similarly, learners who believe that variables that are "controlled" by random assignment are not actually controlled because they are not equalized across condition will not be convinced by a study with random assignment (Chinn, 2006).

The taxonomy of obstacles to belief change provided in Table 1 can provide guidance to educators who are developing inquiry curriculum to promote conceptual change. For instance, by analyzing sources of entrenchment, teachers can develop instruction that is more likely to promote belief change. By including readings, activities, and discussions that are directed at productive epistemological commitments and goals, teachers may encourage the development of epistemological commitments and goals that generally support conceptual change. By analyzing the quality of evidence presented in the classroom and assessing honestly whether this evidence is compelling in relation to students' interpretations of their everyday experiences, teachers can evaluate whether classroom inquiry needs to be improved so as to yield an evidentiary base that provides stronger support for targeted scientific models.

OBSTACLES TO CONSTRUCTING UNDERSTANDINGS OF SCIENTIFIC MODELS

Having addressed belief change, we now consider obstacles to developing accurate understandings of difficult new scientific models. Because our analysis is grounded in our understanding of what scientific models are, we begin with a brief overview of our view of scientific models.

Following Giere (1988), we posit that a basic unit of scientific knowledge is a *model*, which is an idealized system, such as the ideal harmonic oscillator. Scientists construct many closely related models that constitute a *family* of related models (e.g., the simple harmonic oscillator is part of a family of models that includes models of pendula, models of harmonic oscillators that include damping through friction, and so on). These families of models are, in turn, clustered into larger *families of families* of models (e.g., Newtonian theory includes the family of harmonic oscillator models, a family of planetary motion models, a family of charged particle interaction models, and so on). Individual models are evaluated by assessing the *extent* to which and the *respects* with which the model is similar to the world.

Building on further work by Machamer, Darden, and Craver (2000) and by Cartwright (1999), we propose that many models include these components: (a)

entities, (b) relevant properties of entities, including capacities that these entities have to act in particular ways, (c) activities that the entities participate in, (d) rules or other constraints governing the activities (sometimes expressed through equations), (e) set-up conditions, including enabling conditions and shields that block off outside influences, (f) intermediate conditions, and (g) final conditions. Importantly, models at different levels of analysis are typically linked. For example, in biological models of disease, many phenomena can be simultaneously modeled at the macro-anatomical level (e.g., asthma results from the constriction of the bronchial airways in response to allergens like pollen), at the cellular level (e.g., asthma is caused by the activation of lymphocytes), and at the molecular level (e.g., the activation of lymphocytes depends on molecular signals from dendritic cells).

Scientific models are representational. They are intended to represent real situations. But real situations are never directly apprehended by people; real situations themselves can only be represented through mental representations. Thus, scientific models are representations that are in turn linked to mental representations of real situations. For example, a scientific model at a microscopic levels of description (e.g., a steel bar represented at the microscopic level as composed of atoms separated by spaces) is mapped onto a mental model of the steel bar as it appears perceptually. Similarly a representation of a moving car as a point mass with properties mass, velocity, acceleration, and a set of force vectors operating on it is linked to the underlying situation of the car moving along a road.

Our account of obstacles to understanding new models refers back to this account of models. Each component of scientific models and each interconnection between models is a potential locus of difficulty as students try to learn the new models, because each component and each interconnection may differ in fundamental ways from students' current understandings.

In general, we propose these general categories of obstacles.

1. *Epistemological understandings* that impede understanding, including views of ideal models that are incompatible with the models to be learned
2. *Goals* for learning that are not well suited for making sense of models that are conceptually different from current ideas
3. *Learning strategies* that are ill suited for conceptual change
4. The actual *conceptual differences between prior conceptions and new conceptions*
5. *Cognitive overload* when running and coordinating models
6. *Difficulties in learning relations among models*

In the following sections, we will briefly discuss each of these sources of difficulty.

Epistemological Understandings

The epistemological ideas that we think promote understanding of new scientific models overlap with the epistemological commitments that promote belief change,

but they are not identical. We conjecture that metaphysical assumptions about the structure of scientific models and the relationship between understanding and beliefs may hinder or facilitate conceptual change. Some examples include:
- *Metaconceptual understanding of theories.* As with belief change, understanding that competing theories can be developed to attempt to explain the same (or overlapping) data is likely to facilitate the ability to successfully understand a new model.
- *Awareness the belief-understanding distinction.* If students explicitly understand that theories cannot be properly evaluated until they are understood, then they are more likely to approach the knowledge construction task in a way that they try not to distort new ideas immediately to fit their beliefs.
- *Understanding the structure of models and theories.* Students who understand the structure of models and theories (as discussed earlier and further elaborated in later sections) will be more likely to construct a successful understanding.
- *Understanding that new ideas may differ in fundamental ways from one's current conceptions.* In a nonscience domain, Schreiner and Chinn (in preparation) found that foreign language students who believe that target language grammar or vocabulary may differ in fundamental ways from their current understandings perform better in their language courses than students who do not have this belief. We think that the same is likely to be true of science learners.
- Understanding the *relations between models and the world.* The learner should understand that in many (but not all) models, most scientists believe that the entities and processes in the model resemble in important respects the entities and processes in the real world. At the same time, the learner understands that models are idealizations and that models that are known to be incorrect may still be useful.

If students lack these metaphysical understandings, or if they have ideas in opposition to these ideas, we think that conceptual change may be impeded.

Goals

Learning goals that focus on rote learning rather than deep understanding are undoubtedly ineffective in conceptual change domains, as they are in most other conceptual domains of learning (Covington, 2000; Pintrich et al., 1993). A theory of conceptual change, however, should focus particularly on learning goals that are uniquely important for learning topics that involve fundamental changes in conception. We think that the goals that learners set when learning conceptual change domains will probably be closely tied to their metaphysical understanding of the structure of scientific models, as outlined above. Effective goals will be closely tied to sound metaphysical understandings. Some candidates for effective goals include:
- Set goals of trying to understand a theory that may be difficult to understand.

- Hold belief in abeyance while trying to understand what proponents of the new theory mean.
- Treat learning a theory as it if were like trying to understand another person's ideas.
- Be alert to differences between current beliefs and the new ideas.
- Look for signs that the new model structures or explanatory structures (including causal structures) may be very different from what one expects; if so, aim to construct a new basic explanatory or model schema.
- Identify one's own beliefs that may be impeding understanding.
- Avoid focusing centrally on equations and procedures (see Hammer, 1994).

Learning Goals

Useful cognitive strategies will be closely tied to goals set for learning. Several examples of learning goals that may be specifically useful in conceptual change topics are:
- Use a selective elaboration strategy that is more cautious than the elaboration that is effective in other domains. That is, be cautious when elaborating and explaining information, because one may fill in gaps with inappropriate knowledge.
- Do not insert mechanisms that are left unspecified by explanations that are read or heard, because scientific mechanisms are often different from intuitive mechanisms.
- Look intently for and focus on words that indicate differences between the target theory and one's own ideas, or between target theory terminology and everyday usages of words.
- Search for cues that indicate that a new high-level model type should be constructed.

Conceptual Differences between Prior Conceptions and New Conceptions

The ultimate source of difficulty in conceptual change domains is that the new ideas are either hard to believe, or hard to understand, or both. We have already discussed reasons why new ideas may be hard to believe. This section examines why they are hard to understand. New scientific models are difficult to understand because the new scientific models are conceptually different from students' existing conceptions in a variety of ways. Here, we outline some of the important kinds of differences that we think exist. These differences are related to the different components of scientific models. Any aspect of a scientific model can be the locus of a fundamental conceptual difference between the learner's ideas and the model to be learned. We will discuss four different categories of differences: (a) differences in general model structure, (b) differences in general causal model structure, (c) differences in specific model components and activities, and (d)

Differences in how models are coordinated with the phenomena that the models represent.

A. Differences in general model structure. One source of difficulty in conceptual change is that the very structure of the model is often unfamiliar to students. Chi's analyses of ontological differences between naïve conceptions and scientific theories (Chi, 1994, 2005; Reiner, Slotta, Chi & Resnick, 2000) overlaps with our analysis, but we think that scientific models take many forms, in addition to the constraint-based interactions (Slotta, Chi & Joram, 1995) and emergent processes (Chi, 2005) that Chi has discussed (see also Collins & Ferguson, 1993).

In addition to *causal models* (discussed in the next section), scientific models may take the form of:
- *Purpose-based models*, including (a) functional models; (b) structure-function models; (c) structure-function models with elaborations to include behaviors, goals, and entity features, which may enable or disable features (Collins & Ferguson, 1993); (d) teleological models, and (e) situation-action models (Collins & Ferguson, 1993).
- *Regularity-based models* that do not make causal claims, including (a) models specifying qualitative regularities, (b) models based on simple quantitative equations, (c) constraint-based systems governed by systems of equations, (d) topographical or spatial models, and (e) conservation schemas.
- *Sampling models* involving random or nonrandom sampling from populations. Sampling models include a different way of looking at entities and events such that causal explanations are eschewed for explanations that appeal only to the results of random sampling. Sampling models can involve sampling of entities or of events.

Many students, especially younger students, may be familiar mainly with functional and perhaps teleological models, in addition to simple causal models. When students learn these other model types, we suspect that they misconstrue them as one of the few types they are familiar with.

B. Differences in general causal model structure. We discuss causal models separately from other models because causal models play such a central role in K-12 education. Although our analysis of causal models has points of similarity to the seminal work of Perkins and Grotzer (2000), there are differences in the specifics of the dimensions we discuss, the range of specific models considered, and characterization of some models. We think that difficulties in learning causal models can arise because students' basic causal conceptions can differ from the causal structure of scientific models along five dimensions discussed below.

Agent-field focus. This refers to whether learners look for causes within individual objects, in the surrounding contextual field, or in a combination of the two. Keller (1983) has identified a bias against more interactionist frameworks as an obstacle in the development of molecular genetics. European-American students' predilection to focus on features of the object (Masuda & Nisbett, 2001) could strongly inhibit understanding of causal models that identify causes in the field or in interactions between the field and foregrounded objects within the field.

Complexity of models. Recently educational scholars have identified many respects in which students' causal thinking tends to be simple rather than complex in the biological (Bell-Basca et al., 2000; Perkins & Grotzer, 2000) and social sciences (DaCosta & Chinn, 2004). Complex causal models have many structural features that students find difficult (Forrester, 1971; Perkins & Grotzer, 2000), including a multiplicity of causes and effects, the presence of mediating steps, the presence of interactions, the presence of stocks and flows, nonlinearity, enabling conditions such as catalysts or activation energies, the presence of shields or barriers to permit desired effects to occur, time delays in effects, amplification or diminution of effects, and a variety of kinds of reciprocal effects.

Level of causal mechanism. The second dimension is the level at which causal mechanisms are identified (cf. Perkins & Grotzer, 2000). We think that students are probably predisposed to look for simple kinds of causes, including simple surface level causes (e.g., a car's large wheels make it go faster), intentional causes (e.g., an animal walks because it wants to go somewhere), and intrinsic causes such as traits or essences (e.g., water flows because it has the nature of flowing). They may also readily posit causes involving straightforward explanations in terms of pieces (e.g., chalk is soft because its molecules are short) or parts (brakes work through the action of constituent parts).

However, students may have much more difficulty with several other kinds of mechanisms, including mechanisms with variables derived from observable variables (e.g., acceleration is a construct derived from speed and time) and constituents (i.e., causal models in which the causal explanation is provided in terms of an underlying mechanism of constituents that may have properties that are very different from the observable entities, such as molecular explanations). Models with relational causes (such as force, which is defined in terms of relations between entities) and models involving higher level causes (such as *culture* treated as a variable in a social science model) may also pose serious difficulties for students.

Types of core causal causes. Mechanisms often "bottom out" in a small subset of fundamental causal processes (Machamer et al., 2000). For instance, a model may bottom out in basic causal processes of pushing and colliding. We think that students' conceptions of possible bottomed-out causes are often inconsistent with the causes that are assumed in actual scientific models. For example, one common core process for children is probably *touch causality*, by which effects are caused by application of force through direct touching (e.g., pushing, pulling, lifting, obstructing). Later, as children gain experience with magnets and electrical charges in school and other settings, their basic stock of core causes may increase to include *electrical* and *magnetic attractions* and *repulsions*. But many scientific models employ core causal processes that are probably unfamiliar to students, such as models with fields.

Difficulties with probabilistic causes and effects. Students may have difficulty understanding probabilistic causal processes (cf. Perkins & Grotzer, 2000). One difficulty is that students may confuse uncertainty with true indeterminism. Students may also have difficulty understanding that a causal process can be operating even if the effects to not occur all the time. Particularly problematic in this regard are probabilistic causes that operate with low probability.

C. *Differences in specific model components and activities.* So far, we have discussed mismatches in very general schemas for models that can obstruct conceptual change. We have argued that when there are mismatches in general model structure, students are likely to fail to construct accurate understandings of the new model.

However, we also think that conceptual change can be difficult *even if the general structure of a model is understood.* For example, even if students have a good general understanding of constraint-based interactions, they may still struggle to understand a particular new model that employs constraint-based interactions. This difficult arises because the particular entities and activities in the model may be difficult for students to understand, even if the basic model structure is understood. We think that those theorists who have focused on general problems of coordinating models with observable phenomena have neglected this source of difficulty.

There are two general sources of difficulty that can arise even when the general model structure is understood. One, discussed in this section, comprises difficulties that arise in understanding the entities and activities of the model considered in isolation, without taking relations of the model to the world into account. The second, discussed in the next section, comprises difficulties that arise in understanding how the entities and activities of the model relate to the entities and activities of the phenomena that the model is meant to explain.

In the account of models we outlined earlier, a model consists of entities (with a variety of properties) and activities that the entities engage in to bring about a variety of effects (Machamer et al., 2000). The activities are governed by a variety of constraints. The entities may play a variety of different roles, such as agents, joint agents, shields, catalysts, enablers, and so on. The entities also can be viewed as having powers or capacities that they carry to other situations, giving them a propensity to behave in certain ways, even if their behavior in a new situation may prove to be different.

We think that as students' efforts to understand entities, their properties, their activities, and the constraints that govern the activities can be undermined by mismatches between beliefs about certain kinds of entities and activities and the specifics of the new model. The difficulties that arise are often ontological difficulties, but they are often ontological difficulties of a sort that is different from what has been discussed by Chi (1992, 2005). Chi has focused on problems such as misunderstanding processes as substances or misunderstanding indirect processes as direct processes. But we think there are other kinds of ontological difficulties that can arise. Consider entities that are substances but are very new and unusual kinds of substances, which lack some ordinary properties of substances (e.g., color is not a property that can be applied to the new entity) and which interact with other entities quite differently from ordinary objects in the real world). It seems to us that these differences do not cross the substance/process boundary that Chi has discussed, but they nonetheless pose ontological difficulties for students trying to understand how something can be a material entity and yet not have the ordinary properties of materials. Another example is an entity such as a photon, which is massless and hence not in that sense a substance and yet has some properties ordinarily attributed to substances. These differences are differences in how entities are conceived that seem to us to be unrelated to the substance/process distinction.

Constraints that operate in models may also be counter to current conceptions of possible constraints. One example is the process of electron tunneling, in which electrons can spontaneously appear on the other side of a barrier while having zero probability of ever appearing at points within the barrier. Another example comes from the shell-filling model of chemical reactions. Engendering stability by filling electron shells is likely to seem a very new kind of core process for students. Models often incorporate entities and activities that are difficult to understand, because they lack the ordinary kinds of properties of familiar entities and activities.

D. Difficulties coordinating models with the phenomena that the models represent.
As we have noted, scientific models are linked to mental models of situations (Giere, 1988). The scientific model provides a theoretical redescription or explanation of one's more prosaic mental model of the situation.

In Table 4, we have outlined some difficulties that can arise as students try to understand the relationship between the scientific model and the more prosaic model of the situation. To simplify the terminology, we refer to scientific models simply as *models* and the mental models of situations as *phenomena,* keeping in mind that phenomena are themselves a mental construction. We refer to connections across models and phenomena as *mappings*, and we will also use the verb *coordinate* to refer to efforts to map from models to phenomena. Although models can be connected to multiple other representations (visual observations, graphs, models in different formats), we will focus here on the relationship between the model and the real-world phenomenon that the model is intended to represent.

Chi (2005) has argued that conceptual change is difficult principally in cases of emergent processes, in which spontaneous, random, nonpurposeful collisions among molecules can create emergent patterns at the macroscopic level. For instance, random collisions among molecules are sufficient to explain the process of diffusion. We certainly agree that it is extremely difficult for students to understand how such indirect molecular processes can yield emergent processes at a higher, macroscopic level of phenomena (see also Penner, 2000; Resnick, 1996).

But we think that there are other kinds of mappings between models and phenomena that are difficult. Table 2 lists several of these mapping difficulties. All of these potential differences between models and phenomena can be sources of difficulty as students attempt to understand new models. We think that many of the differences in Table 2 apply to

Table 2. Difficulties in Mapping Models to Phenomena

Mismatch type	Comments
Differences in entities across models and phenomena	
Number of entities	It may be easier to make one-to-one mappings (e.g., from an idealized object in force diagram to the object it represents) than to make one-to-many mappings (e.g., from a liquid in a cup to a molecule representation of the liquid and cup). Different kinds of one-to-many or many-to-one mappings may also differ in difficulty.
Mismatches in properties of entities	It may be difficult for learners to attend to properties that are emphasized in the model but that are absent or not salient the student's understanding of the phenomenon. An example is understanding variation as a feature of entities in a population in evolutionary models. Even if students encode variation as a relevant property of populations, it is unlikely to be salient to them.
Property salience	It may be difficult for learners to attend to properties that are emphasized in the model and are present but *nonsalient* in the student's understanding of the phenomenon. An example would be understanding variation as a feature of entities in a population.
Unmapped components	Models often abstract away from phenomena, leaving out entities, properties, or activities that are important in the everyday model of the phenomena. For instance, most aspects of human motivation and cognition are omitted in some economic models of the behavior of economic actors; the omission of such components may be difficult for many learners.

Property-entity mismatches	Sometimes *properties* of entities in scientific models correspond to *entities* in students' naïve models of phenomena. For instance, heat is an extensive property or characteristic of substances in the scientific model but is often treated as an entity (a substance) by students (Wiser & Amin, 2001).
Entity-activity mismatches	This category concerns *activities that appear to be entities*. An example is *sound*, which appears to be an entity at first glance but actually describes the outcome of a complex activity. This problem is similar to what was described in Chi's earlier ontological work.
Differences in constraints across models and mapped phenomena	
Constraint mismatches	Some constraints in models may be strongly inconsistent with constraints believed to operate in the phenomenon. An example is the existence of reaction forces that are equal to forces applied to an object. Students' models of phenomena do not include any such constraints on interactions.
Constraints with no parallel in the everyday world of phenomena	An example is the Pauli Principle that constrains many causal processes at the atomic level. The constraint that no two electrons in an atom can have identical quantum numbers has no parallel in the everyday world and may even conflict with everyday world constraints; for instance, there is no physical constraint that prohibits two soccer balls on a field from having the same spin.
Other coordination difficulties	
Causes crossing levels	One possible error is to causally connect a cause at one level. For instance, saying that a high temperature *causes* molecules to move faster confuses causes at different levels.
Random-nonrandom mismatches	This category is meant to capture part of the particularly important situation that has been discussed by Chi (2005) as the emergent processes schema. When seemingly nonrandom processes at the level phenomena (diffusion) are modeled by processes involving random, uniformly acting entities (randomly moving molecules), students are likely to have difficulties.

Expected size of effects	Learners may have difficulty comprehending how small entities can have large effects. When trying to introduce germ theory in Peru (as a part of an educational effort to encourage people to boil water), developers met with difficulty because villagers could not understand how anything as very tiny as viruses and bacteria could cause any harm to people, who are so much larger (Rogers, 1995).
Arbitrary mappings	Prior to more advanced instruction in chemistry, models of chemistry do not account for why it makes sense to think that different chemical substances are defined by different combinations of atoms in molecules. Seemingly arbitrary mappings may be problematic for students (and may contribute to lack of belief as well as to misunderstandings).
Quantitative relations	The process of mapping detailed quantitative relations from models to phenomena can be very challenging. Learning Mendelian genetics, for example, requires not only understanding the range of offspring resulting from a particular cross but working out the specific quantitative pattern. We think that this is a very important source of difficulty that has been little studied to this point.

Summary. Throughout this section, we have provided an outline of differences between prior conceptions and the various components of scientific models to be learned. In accordance with the central claim of multiple-dimensions theory, we think there are many kinds of mismatches between models and phenomena that can pose serious difficulties for learners. Different topics will pose different difficulties because they involve different specific mismatches. We think that instruction is likely to be most effective when instructional designers and teachers are aware of the particular mismatches that pose difficulties for a particular topic. Hence, the outline of potential mismatches in this chapter can provide a framework for identifying crucial differences between old and new conceptions that should be addressed in some way during instruction.

Cognitive Load Issues in Running and Coordinating Models

Even when such mismatches do not exist or are overcome, conceptual change may fall short because of cognitive load issues (Sweller, 1994). Scientific models are often very complex. Such complexity may make it very difficult for learners to overcome working memory limitations so as to successfully learn the model. This complexity expands when learners try to coordinate the model with phenomena or with other models. Any model feature that increases the number of elements of information that must be held in working memory at one time will inhibit learning. For instance, models whose actions can be understood only by mentally running multiple entities to see how they interact will be harder to understand than models whose actions can be understood by running one or two entities.

Learning Relations among Models

As we have noted, theories comprise families of families of models (Giere, 1988). Even when individual models are well understood, conceptual change may be limited by failure to make needed interconnections among related models. Several particularly important types of relations that should be learned are:

- *Contrastive models.* We think that contrastive models are likely to be particularly potent in promoting understanding of processes that might not otherwise be salient to learners. For instance, to understand the effects of intermolecular bond strength on the rate of evaporation, the learner can contrast this model of evaporation with a model of a substance with a stronger or weaker bond strength, such as isopropyl (rubbing alcohol), which allows the molecules escape to the air occurs much more quickly. By contrasting the explanatory model of isopropyl with the explanatory model of water, the learner apprehends the causal role of bond strength. We think that students often fail to construct and meticulously compare contrasts.
- *Range of instances or phenomena to which the model applies.* Another problem in conceptual change is to learn the appropriate range of instances or phenomena to which models and families of models apply. This is a central issue raised by theories of refinement.
- *Model-data relations.* The relationship between models and data that provide evidentiary support for the data are often more complicated than the relationship between models and the prototypical situations that they provide models of (Chinn & Brewer, 2001). It can be easier to understand model-instance relations than model-data relations. An example comes from evolutionary theory. Experimental manipulations have been carried out that can stimulate the growth of teeth in hens. Evolutionary theory can readily explain this phenomenon, but the experiment does not represent an instance of evolutionary change.
- *Learning relations among models within and between families.* Models within a family deal with a range of modifications of a core model. If a model of a simple oscillator composed of a weight attached to spring is taken as the core model, models within the family include spring models that incorporate damping effects, closely related models of pendula, and so on (Giere, 1988). Learning the interconnections among families of models requires recognizing common entities and capacities across these situations, abstracting common processes, and understanding the conditions under which different models are and are not applicable. As students connect different models, students must

construct abstract variables, which we think is difficult for many. We think that many students persist in thinking of variables as specific causes or a bipolar quantity rather than as a general quantity that can take a continuous range of values.

Even more difficult is the interconnection of models within a theory that belong to different families. The connections between these models may be at a very abstract level. For example, what is shared by the concept of force in gravitational models and the concept of force in spring models is something that is very abstract, with few if any superficial similarities. The challenge for students is construct abstract representations that can unify entities and activities that are shared across these diverse models.

Summary

Throughout this section, we have argued that there are not just one or several but many varied sources of difficulty when learning new scientific models. We supported this claim by cataloguing numerous, diverse sources of difficulty. There are researchers who have identified some of these sources of difficulty as particularly critical, and while we agree that there are some that reliably pose significant difficulties, we think that there is no one particular cluster of difficulties that can claim to account for why conceptual change is difficult.

One goal of our work in cataloging a broad range of sources of difficulty has been to bring together many important sources of difficulty that are currently scattered across the various literatures that deal with conceptual change. A second goal is to provide a framework of issues that need to be considered when designing instruction for conceptual change. We think effective instruction requires a nuanced, detailed understanding of the wide range of difficulties that can arise in conceptual change domains. Instruction that enables students to learn to use scientific models effectively will likely require teachers to have a deep understanding of the sources of difficulty that are salient for each scientific topic. The taxonomy presented in this chapter can provide a guide for the range of issues to consider when developing instruction.

INSTRUCTION TO PROMOTE BELIEF IN MODELS, UNDERSTANDING OF MODELS, AND PROFICIENT USE OF MODELS

Our analysis of conceptual change suggests many complexities and difficulties that need to be surmounted to devise effective instruction to promote conceptual change. In this final section, we will discuss several instructional issues that we think are important for the design of effective conceptual change instruction.

Three basic instructional activities to promote conceptual change are: (a) instructors explain models to students, (b) instructors present and explain evidence that supports models, (c) students engage in a variety of model use activities (prediction, modification, simple design, explanation, revision, testing, and complex design). In this discussion, we will discuss some instructional issues raised by these basic methods.

Explanations Provided by Instructors

Explanations of scientific models can be given in lectures, in videos or on television, in textbooks, in trade books, on websites, or in explanations provided in computer inquiry systems. Instructors may provide lengthy explanations, or they may give bits of explanations when appropriate. For instance, a class may collectively generate most a scientific model in a class discussion, but the teacher may suggest a crucial model component at a key juncture. Explanations can also be provided by students who have learned these explanations from trade books, television, or other classes.

Although some researchers have discounted explanations as a viable form of promoting conceptual change, there seems little doubt that practicing scientists have learned much of what they know from textbooks and lectures (Brewer & Mishra, 1998). In addition, a large body of research has shown that appropriately written explanatory texts are effective at promoting conceptual change (e.g., Guzzetti, Snyder, Glass & Gamas, 1993). Other researchers have suggested that teacher explanations provided during discussions contribute to conceptual change (e.g., Vosniadou et al., 2001; Wiser & Amin, 2001).

We think that explanations that contribute to successful understanding have one crucial characteristic: They block importations from prior knowledge. Importations occur when students encounter explanations that are sketchy, so that gaps have to be filled to make sense of them. Students will then import prior conceptions to fill these gaps. For instance, if a middle-school student believes that water is composed of tiny drops of water, when that student reads that "molecules are small particles that make up water," the student will likely import prior ideas about water being drops of water because the explanation has left a gap that can readily be filled in by prior knowledge.

Blocking importations can be accomplished in four main ways. First, texts that incorporate *refutations* have been found to encourage conceptual change (Guzzetti et al., 1993). Refutations are explicit statements in the text that alert readers to points at which their prior conceptions may be in conflict with the theory being explained. Refutations have clear potential to block importations from prior conceptions because learners have been explicitly warned that such importations will lead to misunderstandings of the scientific theory.

Second, when explanations are *highly explicit*, explaining clearly all relevant model entities and activities, students are less likely to import prior knowledge to fill gaps in explanations. Chinn (2006) found that explicit texts were as effective or even more effective than refutation texts at promoting learning of conceptual change topics. Explicit texts, like refutation texts, cued readers to points at which they should avoid importing prior knowledge.

Third, when gaps exist in the scientific models because scientists do not yet know how to fill the gaps, explanations should be very explicit about these gaps. When causal mechanisms are not understood by scientists (or when causal mechanisms do not exist because the model is noncausal), texts should explicitly say so, so as to encourage students not to import naive causal mechanisms from prior knowledge.

Fourth, explicitly explaining higher-level abstract schemas (such as more advanced causal schemas or the emergent) appears to promote conceptual change. Examples include explicitly teaching students more sophisticated causal schemas that they can apply to the various scientific models that they are learning (e.g., Bell-Basca et al., 2000) and teaching students the emergent process schema (Slotta & Chi, 1995).

Despite strong evidence for the effectiveness of instructor-provided explanations in promoting conceptual change, our analysis of the obstacles to conceptual change strongly suggests that providing explanations will not suffice to produce robust conceptual change in many students. Many different model components may pose difficulty to students. Thus, even a student who actively attends to and reflects upon provided explanations will very likely end up with a misunderstanding of the model; there are too many different directions in which understanding can go astray.

We think that students' misunderstandings are most likely to be uncovered only when students attempt to use the model, and they find (or the instructor finds) that they do not understand it well or accurately. With many potential sources of misunderstanding of any model, extended opportunities for model use and model repair are probably essential to develop an accurate understanding of the scientific model. Thus, although we think that well-crafted explanations can play a key role in promoting conceptual change, they have a limited role. Extensive opportunities to use and/or test models are essential.

Providing Evidence for Theories: Student Inquiry and Instructor-Provided Evidence

Students may also encounter evidence for scientific models through inquiry or through evidence provided by teachers, websites, videos, or other sources. Although we fully concur with the science education community that student should engage in inquiry and evidence evaluation, our reasons may be somewhat different from many researchers'. We believe that classroom investigations can provide a sound evidentiary basis for believing only some (not all) components of their models. That is, we believe that classroom inquiry cannot provide strong evidence for many important aspects of the theories students learn in school. In particular, we think that classroom inquiry does not provide decisive evidence against many of the alternative frameworks held by students prior to instruction.

Underdetermination of theories in classrooms. In contrast to some philosophers (notably Kuhn, 1962), we believe that choices between competing theories in the history of science are typically not underdetermined by data (Samarapungavan and Chinn, in preparation). But we think that choices between broad alternative frameworks in science classrooms are typically underdetermined by evidence, and that beliefs about many of the features of models learned in school are also underdetermined by evidence.

In our view, compelling classroom evidence can be provided for many scientific *phenomena* that are explained by models. For instance, many students may become convinced of the law of the conservation of mass on the basis of classroom inquiry showing that mass does not change through a variety of chemical and nonchemical changes. Similarly, classroom experiments can establish that water "evaporates" (whatever that means) when left in an open (but not a closed) container. The belief that sweaters make things hotter on the inside, regardless of whether there is a heat source inside the sweater, can eventually be dislodged through an extended series of experiments (Watson & Konicek, 1990). Thus, although it may not be easy (Chinn & Brewer, 1993), classroom experiments may lead students to construct more accurate knowledge about some of the phenomena to be explained by scientific theories.

But many components of the models that explain these phenomena are *not* uniquely supported by data that can be gathered in the classroom. For example, classroom inquiry is unlikely to provide compelling evidence for many aspects of molecular models. What experiments done in class demonstrate that individual molecules should not be regarded as having have temperature or color? A student could reasonably adopt a theory that was molecular theory in all respects except that individual molecules do have temperature and color and still explain all data that can be gathered in the classroom. Similarly, no classroom experiment rules out the existence of ether.

Can evidence described by instructors (or by videos or texts) fill the evidence gaps left by hands-on inquiry in the classroom? An example of evidence that is often presented in support of evolutionary theory is the famous study of the dark and light moths in England (Allchin, 2003), in which industrial pollution darkened the bark of trees on which the moths lived, and in consequence the distribution of light and dark moths in the overall population shifted in favor of the dark moths, which were better camouflaged by the darker bark. But this evidence and most other evidence for evolution found in textbooks (including the distribution of finches on the Galapagos Islands) does *not* establish that radically new species can develop through evolutionary processes. Many creationists accept that minor species changes can occur via natural selection but deny that large shifts can occur, as from reptiles to birds, from songbirds to hawks, or even from robins to canaries. Even if a ten or twenty items of evidence were made available to students, we think it would still not properly demonstrate that evolutionary theory is superior to its rivals. The evidence for speciation through evolutionary processes is a grand tapestry, any several items of which may not be sufficient to propel belief change in a student who is a skeptic. It is the coherence of the overall fabric that is decisive.

In many cases of textbook science, the evidence that decisively favors one theory over another is highly technical and may involve advanced mathematics that is not accessible to students. The evidence that is conceptually accessible to students is often the very same evidence that many scientists did not find convincing. For example, evidence for continental drift in the first half of the twentieth century included the striking similarities between geologic formations in South America and Africa and similarities in fossil flora and fauna in these two locations. Such data can be readily understood by middle school students, but evidence of this sort (and much more) did not before 1958 persuade a clear majority of scientists that continental drift (Laudan, 1996; Solomon, 2001). Scientists did not achieve consensus on plate tectonics theory until they had generated more than a decade of research on paleomagnetism, seismic activity, linear magnetic anomalies on ocean floors, and transform faults (Solomon, 2001). The evidence that was persuasive to scientists is highly technical and is much less readily understood by science students. The evidence accessible to students is often insufficient to decisively favor one theory over alternate theories that the student may advance.

We do think that some aspects of model *revision* can be reasonably based on evidence, and choices between some alternate versions of models may not be underdetermined by evidence. For instance, Etkina, Matilsky, and Lawrence (2003) have impressively demonstrated that groups of high school students can develop successful models of astrophysics data. Similarly, some students who work with data on population changes in bird species are able to develop appropriate natural-selection-based models of the data (Reiser, Tabak, Sandoval, Smith, Steinmuller & Leone, 2001). However, we think that in these cases, students have typically accepted without proof the broad outlines of the theories that they are working with, and they are adjusting assumptions of the models within the broad framework. The data that they are working with do not establish the validity of the overall framework, but only that *within this framework,* one particular model is more consistent with the data than another.

Our conclusion is that classroom inquiry can support the construction of more accurate beliefs about phenomena, but the choices of broader models to explain these phenomena will frequently be underdetermined by data. In addition, classroom inquiry can reasonably support choices between some competing versions of scientific models, but basic choices between fundamentally different models will often be underdetermined by evidence available in classrooms.

Thus, although classroom inquiry can help support rational belief, it has clear limitations. But we think that classroom inquiry has other benefits apart from providing rational grounding for belief, as we discuss below.

Promoting understanding of theories. Connecting models to data is a powerful way to learn about the theories being tested. A student who is modifying models of gene recombination based on computer simulated data has the potential to learn a great deal about how genetics models is linked to patterns of offspring. This is likely to enhance students' understanding of the workings of the model, as well as to help them see how to map the model to real-world phenomena (Stewart, Hafner, Johnson & Finkel, 1992).

Establishing the plausibility of theories. Even if the data available to students does not compel rational belief, extensive work with theories − including applying theories to a variety of situations as well as more directly testing and revising theories − can show to students that the new theory can account for an impressive array of phenomena in the world. This can make the new theories highly plausible. Many students may be willing to believe theories if they are sufficiently plausible, even in the absence of decisive evidence.

Engendering an appreciation of the epistemology and methods of science. Even if classroom inquiry cannot provide complete data in support of scientific theories, it can help students learn about processes of science, and it can engender an appreciation of the methods of science and the overall epistemic commitments that motivate scientists. When students revise their models on the basis of inquiry, and when they develop a sounder understanding of phenomena in the world through experiments in the classroom, they are learning a great deal about scientific inquiry. This is a valuable outcome in its own right, and it can also provide students with a basis for sound trust in the authority of scientists. If students appreciate that the inquiry that they have carried out in the classroom has led them to form more accurate beliefs about phenomena in the world, they may legitimately appreciate that scientists who use the same or similar methods, and who are animated by the same epistemic values that they have put into practice in the classroom, are likely to generate results that can reasonably be trusted.

Students' Active Engagement with Model Use

We have argued that active student work with model revision and testing can provide some grounds (though not always conclusive) grounds for believing the scientific models. We also think that active student work with the many models that make up a theory is also crucial to developing a deep understanding of a theory. Active use of models will give students many opportunities to work out interrelations among entities, their properties, activities, and the constraints governing activities. In the final section of the paper, we address learning models by using them.

CONCEPTUAL CHANGE

LEARNING MODELS BY USING MODELS

A central way to promote belief in and understanding of scientific models is to engage students in using models so that they can see how they work, repair their misunderstandings as they go along, and appreciate some of the reasons why scientists find them useful. In this section, we first seven basic ways in which models can be used, and then we discuss several instructional issues.

Types of Model Use

We propose that there are seven basic ways in which models can be used: prediction, modification, simple design, explanation, revision, testing, and complex design. Our goal in discussing these seven basic forms of model use is to provide a taxonomy of templates for tasks that can be used when teachers or instructional designers develop inquiry instruction.

Typically, using models requires coordination of a model and the phenomena that the model represents. To illustrate, we will discuss the example of the different ways in which students can use the molecular model of evaporation. The microscopic model and the macroscopic that is represented by the microscopic model can be diagrammed as shown below:

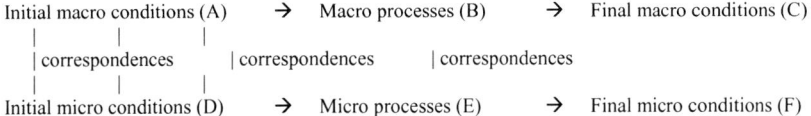

For example, in a model of evaporation from a bowl of water, the initial macroscopic conditions are the bowl filled with water (A). The water steadily but very slowly decreases in amount (B) until it is gone (C). The microscopic explanation accounts for behaviors and movements of the molecules of the system throughout this process.

Prediction. With prediction, the learner is asked to tell what would happen given initial phenomena conditions. For instance, the learner is asked to say what if anything would change if a bowl of water were set in a hot dry room vs. a cool dry room. This task requires the learner to retrieve the relevant model, map from A to D, run the model through E and F, and then to map back onto C.

Modification. The learner is shown a model and asked to modify it so as to produce a different outcome. This model modification is to be done within the rules for the model by revising model parameters; this is not a theory revision task. For instance, the learner is asked to say how the evaporation rate for a given bowl

219

of water could be increased by 50%. This requires students to evaluate the current model D-E-F and modify D or E so as to produce a new F.

Simple design. Design tasks involve asking learners to design a situation (from scratch, or largely from scratch) that will produce a desired outcome. The simple design task involves setting up a situation that will produce a desired outcome, but there is no requirement to depart much from known models. An example would be to set up a room that would maximize the evaporation rate of water.

Explanation. With the task of explanation, the learner can be asked to explain (a) a particular outcome or process, such as why a particular liquid on the table evaporated very quickly; or (b) a general phenomenon, such as why isopropyl in general evaporates faster than water or why a particular graphed relationship is obtained between ambient temperature and evaporation rate. As above, the task is to use the known model rules and parameters to come up with an explanation. The task of explanation is of course much harder when learners are not provided with cues that they should use a particular model. In such cases, the learners must both decide which models to use and then work out how the model parameters can be filled in so as to provide the needed explanation.

The explanation task includes the following processes: (a) The learner is provided with outcome C, or perhaps C with A, or C with A and B. (b) The learner selects an appropriate model D-E-F to use to create the explanation. (c) D-E-F must be constructed in a way so as to correspond to A-B-C at each step of the process.

Revision. Unlike the four previous tasks, the revision task requires the learner to substantively alter the model, not just to modify values of parameters. In the revision task, the student sees an outcome that is not predicted by the model and revises the model itself (the entities, invariant properties of entities, invariant constraints, and so on) so that the unexpected outcome can be explained.

Testing. Experimental testing of a model often involves instantiating models in experimental conditions to investigate what happens. However, testing typically requires elaborating the model in order to yield measures. Some measures may interfere with the operation of the model, which further complicates efforts to test the model (Chinn & Malhotra, 2002b).

Nonexperimental testing involves looking for phenomena in the world that can be regarded as instantiations of the model (often comparing these with phenomena that vary the model in some way or are noninstances of the phenomena). As with experimental testing, this form of testing often requires elaborations to link the model with observations. For instance, understanding that a particular fossil find supports the idea that dinosaurs were warm-blooded may require an understanding of the fossilization process to judge the plausibility of the connection between the evidence and the theory (Chinn & Brewer, 2001).

Complex design. We distinguish between complex design and simple design on the basis of how "distant" the design situation is from the exemplary models. As Cartwright (1999) has argued, the further scientists depart from studied situations in which reliable results can be produced, the more likely it is that the model will not work properly (see also Guala, 2003). In designing new systems, scientists must take their knowledge of entities and their capacities, together with known interactions in other situations, and make their best guesses about how to create a dependable system that will reliably yield the desired range of outcomes. Complex design requires designers to bring in information from many different models and then to try to create a system of entities and activities that will yield the desired outcomes. But because new situations may be fundamentally unpredictable and certainly cannot be routinely extrapolated from past situations that are fairly different, the process is error-prone.

An example of a complex design task involving evaporation would be the assignment to design a system to evaporate seawater commercially. This would require integration of other models (e.g., condensation models, economic models) with evaporation models. Attempting to solve this problem will quickly yield many unanswered questions. For examples, students working on a particular design might find that they do not know how rapidly water will condense as it passes through pipes to collection bins. There will likely be many such questions that go beyond models taught in K-12 education, and even beyond models known by scientists.

Implicit in our discussion throughout this section is that model use involves mapping back and forth from models to the phenomena that they represent. It is possible to map models onto other models as well as onto phenomena (e.g., mapping chemical reaction equation representations to molecular representations). We believe that tasks that involve such mapping will be the most productive for learning (see Kozma, 2003), because mapping is a very difficult process that is unlikely to be mastered without extensive occasions to engage in mapping.

It is also possible, however, to provide useful tasks that do not involve mapping. An example of such a task would be a prediction task involving the microscopic model (only) of evaporation: "Suppose that there is no variation in speed of the molecules; what will happen to the molecules in the model?" This question can be answered using the scientific model without making any connections to real-world phenomena of evaporation at all, and it seems useful for students to have opportunities to think with models in this way. But without extensive opportunities to engage in mapping, as well, we think students are unlikely to master the use of models.

Instructional designers and researchers can use these task formats as a guide for creating tasks to promote inquiry. We think that it would be valuable for researchers to investigate the relative effects of these various tasks. Researchers and educators can also use this taxonomy of tasks as a guide for developing different types of measures to assess understanding of models.

Instructional Issues in Learning through Model Use

To be maximally effective, active engagement with models should aim to accomplish these key goals, which correspond to the dimensions along which we have argued conceptual change is difficult:
- promote productive epistemological commitments, learning goals, learning strategies, and reasoning strategies;
- block inappropriate importations from prior knowledge or correct them when they occur;
- fill in the details of all the entities and activities that occur within models;
- work out the mappings between models and phenomena;
- establish interconnections within and between models; and
- link models with evidence that is as convincing as possible.

Active engagement alone does not ensure that these goals can be accomplished. Indeed, because importation is rampant in conceptual change domains, it is essential to structure instruction so that students' misunderstandings are detected and addressed and so that students elaborate models sufficiently. We will consider three instructional approaches that can meet these goals: constructive argumentation, focusing of attention, and reducing cognitive load.

Constructive argumentation. Peer interaction during group work is one common forum in which students can articulate ideas about models and how models are connected to phenomena. What kind of peer interaction will promote belief and understanding of scientific models and theories? An exemplar of an ideal kind of interaction comes from a description by Wilensky and Resnick (1999) of two computer scientists discussing a traffic jam simulation that showed the traffic jam moving backward as the cars themselves moved forward:

> They were not at all surprised that the traffic jams were moving backwards. They were well aware of the phenomenon. But then one of the researchers said: "You know, I've heard that's why there are so many accidents on the freeways in Los Angeles. The traffic jams are moving backwards and the cars are rushing forward, so there are lots of accidents." The other researcher thought for a moment, then replied: "Wait a minute. Cars crash into other cars, not into traffic jams." In short, he believed that the first researcher had confused levels, mixing cars and jams inappropriately. The researchers then spent half an hour trying to sort out the problem. (Wilensky & Resnick, 1999, p. 4)

The first scientist made, in our taxonomy, a basic error in coordinating models with phenomena, which we have called a causes-crossing-levels error, in having backward-moving traffic jams, an entity at the phenomenon level, causing accidents, an event at the level of the individual cars. We view this interaction as exemplary for three reasons:

1. One participant articulated a mistaken view that involved a fundamental but common error in understanding models (this error could be viewed both as a mistaken belief and as a misunderstanding of the expert scientific model). In conceptual change domains, it is almost assured that mistaken ideas and beliefs that do not fit available evidence will be rampant. Clearly articulating ideas is essential for eliciting possible challenges.
2. A second participant challenged the view, implying a contrary position ("wait a minute" implies disagreement and giving a reason ("cars crash into other cars, not into traffic jams"). Without challenges of this sort, there is no reason for either participant to build a better understanding. Unless students challenge each others articulated ideas, misunderstandings will persist indefinitely.
3. The participants then worked together for an extended time to resolve the disagreement. Although the details are not presented, we would suppose that the computer scientists engaged in extensive articulation of ideas about the models at different levels and how they should be coordinated, together with challenges to ideas (one's own and one's partner) where appropriate. The fact that the issue was not easily resolved suggests that the second participant had a crucial insight but may not have worked out all the details of model coordination, making extended joint dialogue necessary. We would presume that the dialog would include co-constructions of knowledge (e.g., one participant elaborating an idea expressed by the other or giving reasons in support of something said by the other) as well as challenges.

Effective peer interaction within groups will require interchanges analogous to this one (see also Duschl & Osborne, 2002). We call discourse such as this *constructive argumentation*. It is *argumentation* because the participants make assertions and give reasons and evidence for their assertions, and because the participants are willing to challenge each other when necessary. It is *constructive* because the goal is to construct valid knowledge, not to win arguments, and the participants have a mutual goal to achieve understanding.

The goal of classroom argumentation should be to encourage articulation of all the ideas that can lead to difficulty when learning science and to encourage students to challenge each other's beliefs and understandings constructively. Specifically, argumentation should be structured so that students extensively *elaborate* model components (entities, activities, etc.), detect each others' misunderstandings, connect models to evidence and models to other models, and notice gaps in understanding.

Teacher feedback during constructive argumentation can help students detect and correct misunderstandings. However, excessive use of teacher-provided feedback has several potential disadvantages. It can discourage students from using more active self-evaluation strategies that can help them spot faulty understandings on their own. Moreover, teacher-provided feedback may promote only partial understanding at the expense of belief. For example, if a teacher provides feedback to a group during student inquiry, the students in that group may stop conducting their investigation because they now know the "answer." If they had completed the investigation, their continued active engagement might have led to more elaborated understanding of the model, and they might have generated data that would provide epistemic reasons to believe a particular version of a model. Thus, it is highly desirable to structure tasks in ways that enable students within the group will detect and correct their own and each other's misunderstandings.

An important feature of productive argumentation is that the collective argumentation is deep, in that students give not only give reasons for their ideas but also elaborate on these reasons in greater depth. Chinn, O'Donnell, and Jinks (2000) found that students who participated in discussions in which reasons were elaborated and/or debated in depth learned more than students who gave reasons without such elaboration.

Focusing students' attention on key aspects of models. Many efforts to support students' understanding of scientific models provide scaffolding that focuses learners' attention on key aspects of the model. Common forms of such scaffolding are:
– *Careful selection of tasks.* For example, if the goal is to help students notice the role of intermolecular bonding in evaporation, one form of scaffolding is to have students focus on a task that highlights this distinction, as by posing the task of explaining the differing evaporation rates of two different liquids. In conceptual change domains, it is generally important to be sure that tasks focus students' attention on coordinating models with phenomena and with other representations (Kozma, 2003).
– Providing *representations that highlight key points.* An animation could focus students' attention on the role of intermolecular bonds by varying the intensity of the color of the molecules depending on how strong the bonds were (cf. Kozma, Chin, Russell & Marx, 2000).
– *Asking questions or prompting* students. Students could be prompted to consider whether they had considered bond strength in their explanations.
– Providing *schemas to fill in.* Students could be provided with explanatory schemas that provide outlines of features to include in the explanations, such as specifying what kind of molecules are involved, what the range of velocities is, what the intermolecular bond strength is, and so on (cf. Reiser et al., 2001).
– Having students *self-evaluate* their collaboration. Students could be asked to self-evaluate their group work based on whether their explanations had explicitly included these same features (cf. White & Frederiksen, 1998).
Self-evaluation may be particularly powerful in helping students learn the general metaconceptual criteria that apply to modeling tasks (White & Frederiksen, 1998).

Reducing cognitive load. A crucial feature of many simulations and animations is that they have the potential to reduce cognitive load. Students can alter parameters of a model and observe what happens when they run the model, precluding the need to run the model in their own working memories, which is much more prone to error due to overload of working memory. The scaffolding techniques discussed above also reduce cognitive load by precluding the need for students to hold all the needed information in memory all at once until relevant schemas are built up. When cognitive load is reduced, students should be more likely to detect and correct misunderstandings, to elaborate key components of models, to connect models with phenomena, and to relate models to each other.

CONCLUSION

Scientific models often differ in fundamental ways from students' prior conceptions. Our core position in this paper is that there are many ways in which scientific models can differ from students' prior conceptions, and, as a result, conceptual change is difficult for manifold reasons. To support our position, we have catalogued many of the dimensions along which conceptual change is difficult. We think that teachers and instructional designers should take account of all these dimensions when developing instruction. Instruction should engage students in thinking about their epistemological commitments, goals, and cognitive strategies; their general understanding of causal and noncausal models; the specific entities and activities of the models they are learning, together with how these models link to phenomena; and a variety of relations with other models. Instruction should also consider processes of belief change as well as processes in constructing new understandings. The framework provided in this paper can guide instructional designers as they identify the sources of difficulty that are likely to be prominent with any given scientific topic. The framework also points to general epistemological commitments, goals, and learning and reasoning strategies that are of general benefit in promoting conceptual change.

Although instructor-provided explanations have a clear role to play in helping students learn to use models, we think that extensive model use by students is probably the most effective means of promoting belief in and understanding of models. Students will learn most if the model use tasks are designed to facilitate extensive elaboration of model components, detection and correction of model misunderstandings, and rich interconnections among related models. Instructional formats such as constructive argumentation need to be set up to support these learning processes.

Clark A. Chinn
Department of Educational Psychology
Rutgers University

Ala Samarapungavan
Department of Educational Studies
Purdue University

JOSEPH KRAJCIK

COMMENTARY ON CHINN'S AND SAMARAPUNGAVAN'S PAPER

Chinn's and Samarapungavan's paper explores an important topic in science education: how to bring about conceptual change in learners. Their premise is that learning to use scientific knowledge is difficult because it involves conceptual change. One interesting point that they make is that conceptual change involves both changing beliefs and changing ideas and understanding. For instance, for various reasons a student could believe in a scientific idea (evolution) without really understanding it. Or, a student could understand an idea (the smallest building blocks of matter are atoms) without believing in the idea.

I concur with Chinn and Samarapungavan – learning science is challenging. Students do have challenges learning the various major ideas in the disciple like energy, evolution, and function/structure aspects of biology as well as various scientific practices such as scientific modeling and argumentation. I like to refer to these as the "big ideas" of science – those ideas that give explanatory power within and across disciplines that allows an individual to understand a variety of different ideas within and/or between science disciplines that explain a variety of phenomena (see Smith, et al., 2005, for further commentary on "big ideas"). When a learner has understanding of big ideas, it gives the individual powerful ways of thinking about the world. In their paper, Chinn and Samarapungavan discuss the particulate nature of matter. Certainly, this would qualify as a "big idea." Unfortunately, science instruction and most curriculum materials in schools deal with big ideas in a cursory manner. Most science classrooms present many topics at a superficial level, focus on technical vocabulary, fail to consider students' prior knowledge, lack coherent explanations of real-world phenomena, and provide students with few opportunities to develop explanations of phenomena (Kesidou & Roseman, 2002). Big ideas take time to develop and mature. Students need to work with these ideas in rich and varied contexts if learning that is to last will occur. Developing expertise of challenging content like the particulate nature of matter or of challenging scientific practices like modeling and argumentation takes a long time. Science instruction is not structured in a manner to promote such understanding.

Now that we have established that learning (conceptual change) is difficult, I interpret Chinn and Samarapungavan as making three bold claims: First, they claim it is okay to learn from reading or from hearing someone else tell you about various models used to explain phenomena. They refer to this as trusting authority of established scientists. Their evidence for this claim is that many scientists, graduate students, etc. have learned through reading textbooks. Certainly, all of us attending this conference would fit into this category, but an important question we need to consider is if this is true of all learners (or even most, particularly children or other novices). We need to consider that the development of understanding is a continuous process that requires students to construct and reconstruct what they know from new experiences and ideas, and prior knowledge and experiences. Learners actively build knowledge as they explore the surrounding world, observe and interact with phenomena, take in new ideas, make connections between new and old ideas, and discuss and interact with others. Certainly, reading and listening to others is one way to acquire new ideas; however, for many learners it is an impoverished mechanism.

The second bold statement that they make is that teachers and textbooks simply cannot provide the evidence to build a case for most scientific models we present in school to explain phenomena. As they point out, we can provide evidence for macroscopic models, but these, in some respect, do not get at the underlying mechanisms. They argue that the evidence for most models is just too complex for learners because learners do not have the tools to understand all the underlying supporting information and evidence. Is the solution not to present such models to students or to realize we can never present a complete evidence-based justification for the use of the model?

If we reflect on this idea, I think we will realize that Chinn and Samarapungavan make an extremely strong and important point. For instance, to understand quantum mechanics takes sophisticated mathematics that is beyond the reach of even the most capable learners. Does this mean we do not present the current atomic model to learners or do we present the model so students can work with the ideas and see the importance of the idea in explaining phenomena? My strongest disciplinary knowledge is chemistry. Now, do I really understand all the evidence for believing in the particulate nature of matter? I can look at a scanning tunneling micrograph and see what I would call "atoms", but if I did not have this model of matter made up of atoms in my mind, would I really see particles? I did study quantum mechanics, but do I really understand the evidence or do I believe in the model because it allows me to explain a large array of phenomenon that I encounter in my daily life? To be honest, my understanding of quantum mechanics and how it provides a justification for the particulate nature of matter is fuzzy at best. But, I do understand that the particulate nature of matter is a good model for explaining a wide range of phenomena that occurs. Chinn and Samarapungavan argue that we need to show that the model/theory can explain a range of phenomena and that learners need to apply this model for them to believe and understand the model.

Given the emphasis about scientific inquiry stressed by *National Science Education Standards*, these are indeed two bold statements. The *Science Education Standards* (*NRC*, 1996) proclaim that inquiry should be the preferred way of teaching science in school. Now as I will discuss below, I do not really believe that statements made by Chinn and Samarapungavan and the ideas put forth in the *National Standards* are contradictory to each other. I also think that these two statements are consistent with the new science education learning environments that Duschl and Grandy (this volume) refer to as new inquiry curriculum or as immersion units.

Chinn and Samarapungavan make their third bold statement that proposes a new way to bring about conceptual change (or the development of understanding): 1) teachers explain models to students, 2) teachers present and explain evidence that support models, 3) students engage in model use activities, 4) students engage in scientific inquiry as in carrying-out model testing activities.

Overall, what this last statement says is that a central way to promote belief in and understanding of scientific models is to engage students in using models so that they can see how models work, build their understanding (make connection) as their ideas mature and appreciate some of the reasons scientists find models useful. Although this is a strong claim and we do not have empirical evidence to back it, it has merit when we think of students doing science in school.

In describing their instructional sequence (I refer to this as an instructional sequence rather than model to avoid confusion between the use of the word model) to bring about conceptual change, they discuss several important strategies. First, they discuss the importance of showing and allowing learners to map back and forth from models to the phenomena they represent. This is an important point and seldom stressed in schools or science learning materials. Learners have a difficult time separating out what the model is and what are the phenomena that the model explains – often because in school we do not help learners understand the difference between the two. Second, they stress that in explaining models, teachers need to 1) make explicit statements regarding how students current ideas might be in conflict with the ideas of the model, 2) provide rich information necessary to understanding the model, 3) explain when gaps in the model are present, 4) provide higher-level schemes such as sophisticated causal schemas that students can apply to the various scientific models. Third, students need to engage in building and testing models for conceptual change to occur. Building and testing models allow students to engage in argument and debate. External models allow students' understanding to be visible. As such, models allow learners to engage in productive –back and forth dialog. Chinn and Samarapungavan refer to this process as constructive argumentation. Finally, they discuss how learners need to use models to predict phenomena, modify and create simple designs to test features of the model, develop explanations, and revise and test models.

So where do I stand on these idea? For the most part, I agree with a few caveats. I agree that it is too difficult to present all of the evidence behind models and theories but not to present rich explanatory models to learners will leave learners blind. What we need to do in school is show the applicability of these models to explain phenomena. I support students building, testing and revising models as a way to make connections between ideas. Moreover, students engaged in building, testing and revising models is an important part of inquiry.

However, what does concern me is their first step in the process – teachers presenting models to learners. I do not think teachers can "just present models" and expect students to understand, no matter how explicit the description. For models to be of value, learners need to realize the value of the model. They need to see how the model has value and meaning in understanding situations that are important to them and beyond the realm of the classroom. For learners to develop understandings that they can use in new situations, they need to apply their emerging understandings in situations that are meaningful and important to them. Students need to acquire information (i.e. the new model) in a meaningful context (Blumenfield, Soloway, Marx, Krajcik, Guzdial & Palincsar, 1992;) and relate it to their prior knowledge and experiences. I like to refer to this as contextualization (Krajcik, Blumenfeld, Marx & Soloway, 2000). In our work, we contextualize by have students engaged in finding solutions to driving questions that are real world, non-trivial and important.

Reiser (2004) takes a slightly different cut on contextualization and refers to it as problematizing the context. He discusses problematizing in relation to new learning technologies, but tasks that don't involve the use of learning technologies can problematize as well. A task that is problematized leads students to encounter and grapple with important ideas or processes. Reiser discusses three conditions necessary for problematizing: Problematizing involves focusing students' on a situation that needs resolution. Second, problematizing involves engaging students in reasoning about an aspect of a problem that may involve creating a sense of dissonance or curiosity. Finally, it may involve creating interest in some aspect of a problem or getting students to care about understanding or resolving the problem. (I think it is important to consider how different problematizing is from the dissatisfaction phase of Posner's, Strike's, Hewson's and Gertzog's (1992) seminal work.)

So, while I agree that students might not 'discover' the principles of kinetic molecular theory or evolution, we need to put them in meaningful situations in which they are wrestling to make sense of data (phenomena) so that they can see the applicability of the model. In such situations, learners can form connections between the new information and their prior knowledge to develop better, larger, and more linked conceptual understanding. In addition, learners need to apply models to different situations for their new understandings to be useful in a variety of situations (Bransford, Brown & Coxing, 1999).

Now, perhaps Chinn and Samarapungavan have this as an implicit component of their instructional sequence, but I do not get any sense of it. We need to introduce models in meaningful, problematized contexts. Chinn and Samarapungavan do present one potentially meaningful context or problematized scenario – the evaporation of seawater. As fresh water becomes more limited, particularly in certain parts of the world like Southern California or the Middle East, this situation can serve as a rich context in which students can apply a variety of models and theories. However, they raise this as an example in their third component: students engage in their model-use activities, rather than a way to initially contextualize students in the understanding the model.

The second point I would like raise is that a learner's understanding of a model and of modeling will grow over time. We cannot present the complete evidence-based model to learners, but rather need to introduce parts and revisit the model as students explore new phenomena. One of the major concerns I have is that in science education, we do not progressively allow students to develop their understandings of the 'big ideas' of science. We do need, however, to make certain that the models we present are developmentally appropriate for the students. For instance, we can introduce some basic components of kinetic molecular theory to children in early middle school to explain the compression and expansion of gases. However, I would suggest that it be introduced as only one alternative. Other alternatives also need to be considered and children need to discuss the pluses and minuses of the various models. However, the kinetic molecular model can also grow to explain phenomena like evaporation, melting and heating. Learners need to understand the advantages of the kinetic molecular theory in explaining and predicting these phenomena. As learners grow and develop, the model can also grow and develop, showing how the model helps to explain new and more complex phenomena, like how old substances can form new substances. Developing expertise of challenging content like the particulate nature of matter or of challenging scientific practices like modeling and argumentation takes a long time. Now, the evidence for students believing in the model and understanding the model is the applicability of the model to explain an array of phenomena they encounter. In some respect, this is like the fruitfulness component of the Posner et al. model of conceptual change (1982). The point here is the model must be developed over time and must explain new and more complex phenomena. As students develop understanding, more and more evidence needs to be collected regarding the model and students need to continually use the model to develop rich understanding of the model and modeling. Of course, one area we need to work on is creating science learning environments that sequence students understanding of big ideas over time in a developmentally appropriate manner.

As a parent of three children who all played sports in elementary and high school I would like to present an analogy of how children learn to play sports. Because two of my children as adults and my spouse continue to play soccer, let me use soccer as the example. I don't know how many of you have ever watched children in first or second grade play a soccer game, but it only has a semi-resemblance to what my daughter's college games look like. Rather than the children playing positions, they all run in a "herd", kicking up dust. They all want to run to the ball and do their part without understanding a series of progressive actions between several players can work in concert to achieve a particular goal (no pun intended!). Some of the children kick the ball once and don't realize the importance of following the ball. One or two children might be just standing in a random area, not aware of where the ball is or what they should be doing. There is a child in the goalie box, but often this child is picking grass and throwing it in the air or doing cartwheels. One or two of the children run a little faster than the other children and often break out of the herd and score to the delight of their parents.

Any of us who have watched young children attempt to play "soccer" have these images in our minds. Yet, these children are playing or "doing" soccer. As parents, grandparents and friends we cheer and support these children in their efforts. As the children mature to middle school, the game resembles more of what we know as soccer. Most of the children now know to keep their positions, understand defense and offense, and which side is their goal. Some of the children have developed excellent passing skills and realize that by kicking the ball to another player it is more efficient and effective than dribbling the ball through players. Yet, we still might see a child picking up and tossing grass. By eighth grade, more children understand the importance of passing and although the confusing rule of "off sides" is in full use, most of the children as well as the parents on the sideline only have a mild grasp of what this means. These children too are involved in playing or "doing" soccer and as parents we applaud and enjoy the game. By high school varsity games, the players' passes are crisp and sharp. Games can be won in the last second of a game if a break away occurs. The players (and the parents!) have developed an understanding of the finer aspects of the game such as off sides, penalty kicks, yellow and red cards and the importance of keeping the ball out of the goalie box. Games are exciting and intense. These children too are playing and "doing" soccer. As parents we now understand the unlikelihood of our children becoming professional soccer players, but we encourage and reward their playing by going to their games because we understand the importance and benefits of life long habits of exercise. As these children matured their understanding and ability to play soccer grew. Soccer is a complex game and we don't expect children to learn how to play it overnight. It takes time and developmental progression to become better at soccer. As parents, we seem to understand this. Similar to children doing soccer in second grade, children can also do science in second grade. It might only have a vague resemblance to what scientists do, but still the children are doing science. If we can continue this developmental progression, by the time they get to high school, they can be engaged in some pretty exciting and challenging science! Now, I know this analogy is not 100% perfect; not all kids play soccer but we want all children to be scientifically literate; however, we also realize that all analogies breakdown at some level.

This vision of science education and inquiry that I have tired to outline above is consistent with what occurs in many of the new immersion units that Duschl and Grandy (this volume) refer to in their manuscript. Many of these units introduce complex ideas (the models/theories) in an engaging context and then let learners apply and use these ideas to explain challenging phenomena. The goal is not to discover the scientific ideas. For instance, in the Biology Guided Inquiry Learning Environment (BGuILE) (Reiser, Tabak, Sandoval, Smith, Steinmuller & Leone, 2001) students explore similar problems that biologist explore – why some birds survived and others did not on the Galapagos Islands, and apply the ideas of natural selection to understand the problems by designing investigations and constructing explanations. Some new inversion units also involve new computer-based modeling programs. Students can use modeling tools like Netlogo (Wilensky, 1999) or Model-It (Jackson, Krajcik & Soloway, 2000) that allow them to build, test, revise models and various visualization tools that allow them to explore complex data patterns (Edelson, Gordin & Pea, 1999).

Inquiry then is not to discover the "big ideas" of science. It took great minds to do this building using the work of other great minds; we cannot expect students to do this in school. Rather, one way to envision inquiry is students using and building models to explain an array of increasingly complex phenomena that they really care about and find meaningful. Learners need to see the applicability of the model to solve various problems, if they will both understand and believe it. We must also remember that it will take time for students' models to develop, but we need to give them important developmental experiences, if we hope they develop rich understanding and the ability to "do" science. This approach will foster students to develop fundamental understandings of the "big ideas" so that they have more powerful ways to think about their world.

Joseph Krajcik
University of Michigan
Weston Visiting Professor of Science Education
Weizmann Institute of Science

STEPHEN P. NORRIS AND LINDA M. PHILLIPS

READING AS INQUIRY

Scientists read and write text a great deal of the time. They ponder phrasing that will capture what they mean, and, in so doing often construct what they mean while they write. They search for expressions that will carry the level of exactness they intend. They choose words carefully to distinguish between degrees of certainty they wish to express. They select genres to describe what they did to collect their data and to provide justifications for their methods. They puzzle over the meanings of what other scientists have written, question their own and other scientists' interpretations of text, and they sometimes challenge and at other times endorse what is written. They make choices about what to read, and, once a choice is made, make further choices about how closely and critically to read, and about what to seek in their selections. These activities of constructing, interpreting, selecting, and critiquing texts are, we contend, as much a part of what scientists do as are collecting, interpreting, and challenging data. These activities with text are, we shall show, as much a part of scientific inquiry as are observation, measurement, and calculation.

This paper will concentrate on reading, although much that we have to say applies equally to writing. We shall show how reading scientific text is an inquiry activity and how reading as inquiry could become a part of school science instruction. In section one, we shall outline some of what is known about the nature of scientific text and about the inquiry that scientists engage when they read and write it. By way of contrast, the second section will show how the language of science teaching, the language of school science, differs dramatically from the language of science. This section will illustrate some of the interpretation pitfalls that students who have excelled in school science encounter when they are asked to read text that resembles the language of science. Section three will offer an explanatory account of students' performance in reading scientific text postulating that the problem lies fundamentally in the theory of reading that they hold (what we shall call "a simple view of reading") and that they have been taught in school. Finally, in section four we shall offer an alternative to the simple view of reading that fits well with the inquiry goals of science education, namely, a theory of reading as inquiry.

As we discuss more fully in the fourth section, we include under the concept of text not only printed words, but also graphs, charts, data tables, diagrams, figures, mathematical equations – whatever appears on the printed page or other medium and is meant to be read. Recently, the word "inscription" has been used instead of or in addition to "text". Some uses of the term "inscription" contrast with our use of "text", in that they include everything but words (except when these appear as labels in diagrams, charts, graphs, figures, etc.) (e.g., Bowen & Roth, 2002). Other uses of "inscription" seem to coincide with our use of "text", but then take text (on their sense) to be one form of inscription (e.g., Forman & Ansell, 2002). There is no standard usage, which can lead to confusion. Therefore, we take this time to make clear our usage in this paper.

I. THE LANGUAGE OF SCIENCE

Suppe (1998) examined more than 1000 data-based papers in science. He concluded that the papers perform the following speech acts:
- present the data arising from an observation or series of observations;
- make a case for the relevance of the observations in addressing some scientific problem;
- provide detail on the methods used to collect the data and to analyze it;
- provide and justify an interpretation of the data; and
- identify, acknowledge, and possibly impeach specific doubts that can underpin alternative interpretations of the data.

These speech acts create the argumentative structure of the paper. Although it is not particularly relevant to our task here, Suppe concluded that the argumentative structure of such papers is not captured by any of the major philosophical theories of scientific reasoning – not by hypothetico-deductivism, not by Popper's falsificationism, not by Bayesian induction, and not by inference to the best explanation. Rather, the central structure of such papers is arguing for a favoured interpretation of the results and against the alternatives. As an example, Suppe chose W. Jason Morgan's (1968) famous paper credited with laying the foundation for current Plate Tectonics. On Suppe's account, the paper is paradigmatic of how scientists argue their claims by interpreting data and by impeaching alternative interpretations.

We do not have to accept any of Suppe's more controversial and contested claims about the failures of major philosophical theories of scientific reasoning in order to derive lessons about the language of science. The first lesson is that scientists engage a variety of speech acts when they write and read scientific text, including: presenting data, making relevance cases, describing methods, justifying interpretations, and impeaching interpretations. The second lesson is that these speech acts are marshalled into an overall argumentative structure that defines the nature of the text.

We conducted our own examination of data-based research reports by some colleagues in physics. One paper reported a study of hysteresis, a phenomenon whereby a system undergoes a change of state in response to changes in its environment that is not reversible via the same route when the environmental changes are run backwards. For example, glass thermometers, once heated and having undergone expansion, do not return to their original size when they are cooled to the original temperature by reversing the path to expansion. If we magnetise a piece of iron by placing it in the magnetic field of an electrical coil, decreasing the intensity of the magnetic field produced by the coil does not lead to the piece of iron losing its magnetism by simply reversing the route of its magnetization. The states of some physical systems are thus dependent not only upon their current environments but upon their histories. So it is, apparently, with the adsorption and desorption of helium in silica aerogels, a type of porous media studied in condensed matter physics (Beamish & Herman, 2003).

Beamish's and Herman's paper of just two pages exhibits much of the argumentative structure identified by Suppe. What interests us in addition, however, is the range of speech acts present in the paper. Beamish and Herman:

1. **motivated** their study – "Silica aerogels ... provide a unique opportunity to study the effects of disorder on phase transitions" (p. 340).
2. **reported** relevant past results – "Also, recent experiments ... showed features characteristic of capillary condensation rather than true two-phase coexistence ..." (p. 340).
3. **reported** limitations of past research – "... although long thermal time constants made it difficult to determine the equilibrium behavior" (p. 340).
4. **described** what was done – "... a thin (0.6 mm) disc was cut ... Copper electrodes (9 mm diameter) were evaporated ... This was sealed into a copper cell . . . Temperatures were measured and controlled ... A room temperature gas handling system and flow controller allowed us to admit or remove helium at controlled rates. . ." (p. 340).
5. **argued** for the suitability of techniques – "This effect [thermal lags when gas was admitted or removed] appears as a rate dependent hysteresis, so it was essential to check directly for equilibrium" (p. 341).
6. **explained** observations – This behavior [The equilibrium isotherm ... had a small but reproducible hysteresis loop and did not exhibit the sharp vertical step characteristic of two phase coexistence] is characteristic of surface tension driven capillary condensation in a porous medium with a narrow range of pore sizes" (p. 341).
7. **conjectured** what might be happening – "At low temperatures we observed hysteresis between filling and emptying, as expected for capillary condensation; this hysteresis disappeared above 5.155 K. The vanishing of the hysteresis may correspond to the critical temperature of the confined fluid" (p. 341).
8. **challenged** interpretations – "However, this identification [with the critical temperature] is not clear, since all our isotherms have finite slopes [while theory would suggest an infinite slope at this point]" (p. 341).

The other paper we examined studied the transition from crystalline to amorphous solids in certain nanocrystals under the influence of radiation (Meldrum, Boatner & Ewing, 2002). In comparison to crystalline solids, amorphous solids (a common example is glass) do not have their molecules arranged in an orderly geometric pattern. They have no definite melting or freezing points, usually have high viscosities (glass runs very slowly), and fracture along irregular patterns rather than regular cleavage lines. This paper is about twice as long as the previous one and thus has the space to exhibit more extended structures. As with the previous paper, the overall pattern is the argumentative one identified by Suppe. The paper contained many of the speech acts described for the Beamish and Herman paper, and we shall not include further examples of these. However, this paper offered many more and much more extended arguments, so we shall exemplify and discuss these.

1. As part of the motivation of the study, whose aim was to provide evidence for the amorphization of nanocrystalline zirconia by ion irradiation, the authors provided a full-paragraph case for the proposition that zirconia is one of the most radiation resistant ceramics known. The point of the argument was to signal that the results of this study were not to be expected, that previous conclusions will have to be modified, and that therefore this study was significant.
2. In addition to describing what was done to collect data, two justifications were provided for the methods chosen, which were deemed particularly appropriate in that they avoided certain pitfalls that would have led to misleading results.
3. Imbedded in the description of the results, a case was made that the cystals of zirconia that reached the amorphous state were sufficiently small to allow the formation of tetragonal zirconia. However, in the case under study, there was incomplete information for calculating the critical size for such formation. By comparing two related cases where the relevant calculations could be made, the authors argued that the size of their crystals was smaller than either of these two cases and that therefore they should be adequately small to support the observations made.
4. As was the case in the work by Morgan analyzed by Suppe, these authors spent considerable effort arguing against alternative interpretations of the observations they made. They carefully spelled out how various steps of their method made other possible mechanisms implausible.

We have argued previously that scientific language is textured and structured in ways that are significant for science education (Norris & Phillips, 1994b). Scientific language is *textured* in that not all of its statements have the same reported or implied truth status. We see this texture in the papers examined. There are statements offered as truths, for example, descriptions of what has been found in the past, of what was done in the current study, and of what was observed. There are statements offered as probable or as having uncertain truth status, for example, conjectured explanations. There are statements offered as false, for example, explanations that the authors aim to impeach. Scientific language is *structured* in that not all of its statements have the same epistemic status and role. Scientific language is structured largely using a metalanguage including such terms as 'cause', 'effect', 'observe', 'hypothesis', 'data', 'results', 'explanation', 'prediction', but also using argumentation both in the form of justifications for actions taken and of reasons for conclusions drawn. We saw this metalanguage and these argumentative forms in the papers examined. Each of the metalanguage terms and each of the argumentative forms imply a relationship. Taken together, these relationships hold the pieces together.

Myers, who has done an enormous amount of research into the nature of scientific text, makes a very interesting point that relates to our ideas of texture and structure. Nonspecialists faced with scientific texts, he says, often interpret their difficulty in understanding as one of not knowing the vocabulary, which, it is true, can be daunting. However, argues Myers, the problem will not go away by using a dictionary, even a very good one, because much of the difficulty interpreting scientific text lies in grasping the connections of one statement to another (Myers, 1991). Although our approach to understanding scientific texts differs from Myers, we agree with his view that "scientific texts must be understood as part of the larger processes of argument and alliances within scientific communities" (1991, p. 3). This conclusion seems in accord with Bazerman's (1985) report on how physicists read physics. He reports their being impressed with scientists who admit experimental or methodological difficulties, who present the methods clearly, and who take pains to point out artifacts and difficulties in the results that might otherwise be overlooked. The question of scientific vocabulary and its role in impeding the understanding of scientific text will be raised again in section three, where we associate the concern with teaching vocabulary with a simple and particularly narrow view of reading.

In summary, this section has shown that the language of science at the macrolevel is argumentative, in the sense that it is structured so as to support conclusions on the basis of reasons. At the microlevel, scientific language is usefully characterized as performing a variety of speech acts, which, when taken in combination, create the argumentative structure seen at the larger scale. Scientific language is understood at the macrolevel by one who grasps the connections that are made and implied among its microlevel parts.

II. THE LANGUAGE OF SCHOOL SCIENCE

The language of science teaching is manifested in various forms, including the language of science textbooks and the language of instruction in science classrooms. The language of science teaching is also indirectly revealed in the language abilities of students who have excelled in school science.

Studies of science textbooks have come to several conclusions. One conclusion by Myers (1997), drawn on the basis of his analysis of a major university-level textbook in molecular genetics, sets the tone for many of the conclusions reached by others. In speaking of what the pictures in the textbook do *not* do, he says: "They never provide proof; the **show** that is so common here always means **illustrate**, not **demonstrate**" (p. 98). This conclusion is telling, because it points to the general divergence in the reasoning found in textbooks from the reasoning of science. Whereas science is fundamentally concerned with using language to demonstrate conclusions, in the sense of providing evidence and argument for them, science textbooks are concerned with illustrating, in the sense of showing, summarizing, and defining.

In a study comparing the language of scientific journal articles to the language of science textbooks, Myers (1992) reached similar conclusions. For example, in contrast to journal articles, in which statements frequently are qualified with hedges, textbooks tend to present statements as accredited facts that require no hedging. In journal articles, photographs, diagrams, graphs, and other illustrations serve the function of proof by providing experimental results. In contrast, illustrations in textbooks generally are used to picture and have no argumentative function. Myers concludes that "the textbook generally makes it easier for the student reader than the article, but in doing so it makes is harder for those students who later encounter (or write) research articles" (p. 12). He contrasts the two genres by imagining the activities of a reader in each case (p. 13):

The reader of the article:	*The reader of the textbook:*
– sorts out the new knowledge from the old,	– arranges facts in order,
– attributes credit to researchers,	– separates facts from researchers,
– assesses the certainty of statements,	– takes most knowledge as accepted,
– infers cohesive links using knowledge,	– infers knowledge using cohesive links,
– traces the relations to other texts,	– uses the array of sections in the book,
– evaluates the illustrations.	– uses illustrations to clarify the text.

The contrast seems to capture the frame of mind, in the case of the article, of a reader dealing with an argument, and, in the case of the textbook, of a reader dealing with an exposition.

Our own research on the anatomy of school science textbooks (e.g., Penney, Norris, Phillips & Clark, 2003) supports the pattern found by others. We compared textbooks to media reports of science, rather than to scientific research articles, because it is such reports that serve as an important source of science for lifelong learning, and because school science teaching takes as one of its goals preparation for lifelong scientific learning. We found that between 90% and 99% of the statements in the textbooks sampled presented science as truths. By comparison, between 57% and 92% of the statements in the media reports sampled presented science as truths, with on average 24% of statements presented as uncertain of truth status. The media reports were therefore considerably more hedged than the textbooks and more closely resembled science journal articles in this respect. The type of text most frequently found was expository in both the textbooks (75% to 100%, averaging 92%) and media reports (57% to 100%, averaging 83%). What was not expository in the textbooks was narrative, marking a difference from the media reports where the non-expository text was argumentative. Examining the role in scientific reasoning, between 51% and 77% of the statements in textbooks provided facts or conclusions. Given that only 2% of statements provided reasons, most of the facts and conclusions were presented without providing either insight into their origin or into how they were related to other statements. Striking differences were noted between these findings and the media reports. Media reports devoted 13% of their space to what prompted the research to be done and another 13% to how the research was done, both proportions being four to six times those found in textbooks. The media reports also devoted between 2% to 43% (13% on average) of their space to providing reasons, over six times the proportion devoted to reasons in the textbooks. Again, as in findings by others, textbooks expose and do not argue. Given that "Science textbooks are the ultimate source of science knowledge in many science classrooms . . . to the extent that, in many ways, they become the embodiment of science for students" (Penney, Norris, Phillips & Clark, 2003, p. 418), it would not be surprising that the form of discourse found in textbooks becomes a model that students try to emulate. We thus find an explanation for why students tend not to look for reasons for views and to accept rather than to think critically.

Not surprisingly, perhaps, the language of science teaching resembles the language of science textbooks. Martin (1993) has argued that literacy in science has to be considered from the point of view both of the field (the substantive content under focus) and the genre (the patterns of text that package the field). We have made a related distinction between the derived sense of scientific literacy (having knowledge of the substantive content) and the fundamental sense of scientific literacy (being able to read and write when the content is science) (Norris & Phillips, 2003). Science curricula and, consequently, science teachers, tend to focus on the field, on the derived sense of scientific literacy. The main aim of reading and writing in this context is the "student's understanding of relatively isolated technical terms. When students are asked to write at length . . . for the most part they end up copying . . . passages of text from research materials they have found (Wignell, 1987)" (Martin, 1993, p. 201). This conjecture coincides with the empirical findings of Ebbers and Rowell (2002) and Rowell and Ebbers (2004). Newton and Newton and their colleagues found a similar phenomenon among 76 science tradebooks for children: "All the books provided a significant body of facts and descriptions . . . Many books, however, did not go beyond that and did not supply much in the way of reasons for why things happen as they do" (2002, p. 236). Their study of discourse in primary school science classrooms reflects the same thing: "teachers discourse was often largely confined to developing vocabulary and descriptive understandings of phenomena and situations ... [with] little evidence of an oral press for causal understanding ... with its persistent emphasis on reasoning, argument and explanation (Newton & Newton, 2000, p. 607). Generally, the instruction that students receive when learning to read does not address the genres found in scientific writing and thus students do not acquire scientific literacy in its fundamental sense.

Hence, although students are expected to be able to read scientific and other non-fiction texts by the time they leave elementary school, they usually have great difficulty doing so (Peacock & Weedon, 2002). In our research, we have been identifying and quantifying the nature of some of these shortcomings in students' reading ability. In one study (Norris & Phillips, 1994b), we asked top-performing high school science students to interpret various aspects of the pragmatic meaning of media reports of science: the expressed degree of certainty with which statements were reported; the scientific status of statements (e.g., whether the statements were causal generalizations, observations, or descriptions of method); and the role of statements in the scientific reasoning (e.g., whether the statements were justifications for procedures, evidence for conclusions, or conclusions). There were several salient results. First, students demonstrated a certainty bias that skewed their interpretations of the expressed degree of certainty of statements towards being more certain than their authors had written them. Second, students were less able to interpret the role of statements in the scientific reasoning of the reports than they were able to interpret the scientific status of statements taken one at a time. The difference was quite large, with fewer than one-half able to interpret the role of statements and about 90% able to identify the nature of statements that could be assessed independently of others. The difference seemed to be due to a weaker ability to interpret statements whose role could only be inferred by recognizing the connections implied to other statements. When literal interpretation alone provided significant cues, such as frequently is the case with observation statements ("We observed that ..."; We saw that ..."; "We noticed that ...") and reports of method ("We measured the ..."; We attached the probe to ..."), their performance was significantly improved.

The ability to interpret degrees of expressed certainty and the ability to interpret the relationships among statements are central to understanding science. As we have seen, scientific writings express a range of epistemic attitudes towards statements and they are fundamentally argumentative prose. If a reader does not grasp these meanings, then the reader fails to grasp the science. For example, reexamining the Beamish and Herman article, a reader who interpreted them as saying that the vanishing of the hysteresis at 5.115K corresponds to the critical temperature of the fluid would be ascribing more certainty to this statement than the authors intended. In this case, the science is a conjecture and Beamish and Herman were careful to qualify it as such. To interpret them as offering something more certain than a conjecture is to miss an important aspect of the science. Beamish and Herman even offer reasons why their conjecture might be false. They are involved in a dialectic with themselves and, in effect, are inviting their readers to take part. This form of reasoning and its associated epistemic stance is also part of the science that a misinterpretation might miss or even distort.

We conducted a similar study with undergraduate university students (Norris, Phillips & Korpan, 2003) who on average had taken eight more single terms of science (11.8 v. 3.8) than the high school students in the study just discussed. As such, these university students had an education in science that was far in excess of the average nonscientist. We found, however, that these students performed almost identically to the high school students. They showed the same certainty bias, the same strength in identifying observation and method statements, and the same weakness interpreting the role of statements in the reports' reasoning. Their increased science education did not help them on these tasks.

We asked these university students two additional questions that helped us better understand their performance. First, for each question that required them to make an interpretation, we asked them to identify where in the reports they found the information they needed to answer it. Whereas overall they answered only about one-half of the interpretive questions correctly, they correctly identified the place in the report with the needed information about three-quarters of the time. We also asked the students how difficult they found the reports to read. At most, only 5 percent judged that any report was very difficult to read, more than one-half found the reports easy or very easy, and more than 90 percent found the reading difficulty to be at least about right. That is, their self-assessments of the reading difficulty of the reports underestimated dramatically the difficulty they actually experienced reading them. We believe that these results are key to understanding the difficulty that students experience reading scientific text. We will take up this issue in section three.

In the final study we shall discuss (Phillips & Norris, 1999), high school students were again asked to read media reports of science. This time, however, they were asked questions that probed the relationship between the content of the reports and their beliefs. Before reading each report, students were asked to respond to a question that gauged their attitude toward the topic of the report. For instance, before a report on the relationship between weather and sickness, they were asked whether they believed weather can make one sick and why they believed what they did. Before a report on the possible discovery of a new animal species, they were asked whether they believed new animal species were still being discovered and why they believed what they did. After reading each report, they were asked whether they now were more, less, or equally certain of their previous view and what had made up their minds.

The results are interesting. For every report, more than two-thirds of students adopted what we called a text-based position in supporting their degrees of certainty. They either deferred absolutely to what the reports said, simply paraphrased parts of the reports to support their positions, or agreed with the reports on the grounds that the text and their own beliefs coincided. A significant minority, less than 20 percent for each report, adopted a background-belief-based position by imposing interpretations on the reports that accorded with their pre-existing beliefs. A smaller minority, around 10 percent, adopted critical positions with respect to the reports by giving reasons either for or against the views taken in the reports. Overall, the data tell us that students' positions do not follow from a critical assessment of reasons. Contrasting their responses to three reports demonstrates this point. The response patterns to the New Animal Species, Breakfast, and Cows' Milk and Diabetes reports were almost identical. Almost all the students responded "yes" to the initial questions: whether new animal species were still being found, whether breakfast is important to health, and whether mothers should breast feed their babies. After reading the reports, fewer than 10 percent of students were less certain of their position, about one-third were equally certain, and about 60 percent were more certain. As it stood, only the Breakfast report supported an affirmative answer to the question asked. The other two reports supported neither an affirmative nor a negative response. Thus, for the Breakfast report it was reasonable to be equally certain after reading it, or even to be more certain, that breakfast is important to health. However, the reasonableness of the students' responses to this report is called into question by their responses to the New Animal Species and Cows' Milk and Diabetes reports where it was not reasonable to be more certain on the basis of reading the report, but 60 percent of the students nevertheless were. It thus seems that there was very little critical evaluation underlying students' responses, and that their degrees of certainty were influenced by factors that frequently were unreasonable.

In sum, we are confident in concluding that the language of school science – be it revealed in textbooks, instructional language, or the language abilities of students who have excelled in school science – stands in marked contrast to the language of science itself. The former is characterized by attention to word meanings, facts, illustrations, and belief formation. The latter, although attentive to all of the elements of the former, is more concerned with relationships that create holistic meanings, logical argument and probabilistic demonstrations, and the avoidance of credulity. In school science, belief comes easily to those prepared to defer. In science, belief is hard won, and is never immune to challenge. The two enterprises are indeed very different, perhaps antagonistically so. Much of the antagonism, we believe, is revealed in the view of reading that underlies school science.

III. A SIMPLE VIEW OF READING

Science teachers, and content area teachers generally, believe that it falls to other teachers to teach reading and writing. "Traditionally science teachers have had little concern for text . . . [R]eading is not seen as an important part of science education" (Wellington & Osborne, 2001, pp. 41-42). Indeed, criticism of undue attention paid to science text and science reading has "focused the science education community's effort on eliminating text from science instruction", because of a "perception that science reading [is] a passive, text-driven, meaning-taking process" (Yore, Craig & Maguire, 1998, p. 28). Reading can seem a simple process and text sometimes can seem transparent. Indeed, for accomplished readers, reading frequently appears automatic and little more than recognizing the words and locating information in the text. This simple view of reading penetrates education very deeply. Spiro and Myers (1984) described the view well:

> According to this view, readers go through a series of processing stages that progress from smaller units of analysis in text to larger ones. Roughly, features of letters are detected, letters are recognized, strings of letters are identified as words, concatenated words are analyzed to determine sentence meaning, and, finally, sets of sentences are considered together to produce the meaning of a connected discourse. (p. 478)

Haas and Flower (1988) reported that many students are good readers according to the simple view of reading: they know the words, read flexibly, identify and locate information, and recall content. Yet, these same students paraphrase when asked to analyze, summarize when asked to criticize, and retell when asked to interpret. What is missing from the simple view of reading that leads to such shallow responses?

In children's minds, the simple view engenders a belief that "reading is being able to say the words correctly, a passage of unrelated words [seems] just as readable as an intact passage" (Baker & Brown, 1984, p. 359). There is strong reason to believe that even today teachers are unwittingly fostering this simple view of reading, despite over five decades of research showing that skilled word recognition is not reading. "Although . . . skilled decoding is necessary for skilled comprehension . . . decoding is not sufficient . . . Despite the plethora of research establishing the efficacy of comprehension strategies instruction, very little comprehension strategies instruction occurs in elementary schools" (Collins Block & Pressley, 2002, pp. 384-385). Indeed, as few as seven years ago, Pressley and Wharton-McDonald found it necessary to challenge "the myth that children will be able to comprehend a text simply because they can decode words in it" (1997, p. 448).

We find a reliance on this simple, word-recognition-and-information-location, view of reading in many science education papers. For example, Miller (1998) wondered whether "scientific literacy might be defined as the ability to read and write about science and technology" (pp. 203-204). He proposed a concept of civic scientific literacy with two dimensions: a vocabulary dimension, referring to "a vocabulary of basic scientific constructs sufficient to read competing views in a newspaper or magazine" (1998, p. 205); and a process of inquiry dimension, referring to "understanding and competence to comprehend and follow arguments about science and technology policy matters in the media" (1998, pp. 205-206). Miller's concept of civic scientific literacy is interesting, but has several limitations. First, the vocabulary dimension seems to make knowing the meanings of the individual terms sufficient for reading. This is the simple view of reading. The view is flawed, because concatenating the meanings of individual words does not yield the meaning of a proposition (Anderson, 1985; Goodman, 1985; Smith, 1978). Reading is not additive in this way. Second, the vocabulary dimension appears to assume that only scientific constructs need to be known to understand scientific text. We shall argue in the fourth section that many literate constructs that are not specifically scientific are needed to understand scientific text. Third, the vocabulary dimension seems also to assume that words have one and only one meaning, and that their meanings can be isolated from the contexts in which the words are used. Even for strictly circumscribed scientific terms, their meanings are often sufficiently complex that they vary somewhat from context to context, requiring readers to make judgements in each case. Fourth, although the process of inquiry dimension could be interpreted to include the general reading competence necessary to interpret the argumentative structure of text, it really refers on Miller's conception to expository knowledge of the nature of science: the ability of individuals "to describe, in their own words, what it means to study something scientifically" (1998, p. 213). Although such knowledge and ability are important, they do not yield "competence to comprehend and follow arguments" in text, as the process of inquiry dimension is described by Miller. Fifth, and this is our primary problem with Miller's view, his two dimensions of scientific literacy separate reading science from scientific inquiry. Yet, the very interpretive tasks that he sees as central to inquiry, comprehending and following arguments, are central to good reading whatever the context, or so is our point.

The contrast, even tension, between reading and doing science appears in other contexts. For example, in a recent analysis of secondary school methods textbooks for mathematics, science, and social studies, Draper (2002) found that common advice to teachers trying to help their students read and comprehend content-area materials was to encourage them to read slowly, and to reread sections. The science textbooks sometimes described reading instrumentally and, curiously, as a teaching method, and set reading against the goals of doing science, which were described as engaging in discovery and inquiry.

Let us consider more closely this simple view of reading in order to grasp its appeal and seeming obviousness to many people as an account of reading. Here is an example that might be used to support the simple view. First, read the following passage:

SEAVIEW

During last night's severe winter storm the ten-story oil rig Seaview, operating on the Banks, capsized and sank. Rescue ships and aircraft are still in the area but no <u>survivors</u> have been found. The oil rig, with its 95 crew members, was operating for Petro Company in one of the richest fields in the world. It has been drilling since September 20. The pilot of one of the search and rescue aircraft reported that only debris could be seen on the surface. None of the rig's lifeboats were found. It is not known at this time what effect this disaster will have on Petro's future drilling program. There is as yet no official statement from the company.

Now, answer the five following questions:
1. The underlined word <u>survivors</u> means: A. people alive, B. people dead, C. lifeboats, D. life vests.
2. According to the article, the Seaview: A. turned over only, B. turned over and sank, C. was sinking, D. was nearly capsized.
3. According to the article, the future of Petro Company's drilling program: A. is bright and promising, B. is in serious difficulty, C. is unknown at this time, D. is certain at this time.
4. It can be inferred that the Seaview was drilling: A. on land, B. in the deepest part of the ocean, C. in a sheltered bay, D. in the ocean.

5. It can be concluded from the information in this passage that a probable cause of the sinking was: A. human error on the rig, B. the severe winter storm, C. poor rig construction, D. the huge size of the rig.

Check your answers against ours, which are: ABCDB. We are confident that your answers match ours because we have given this test to hundreds of people from elementary school to university and they usually get it all right. How should we understand this performance? One answer is that you and all of those people understood the Seaview passage. You were asked a word-meaning question, two literal comprehension questions, and two inferential comprehension questions, just like those that are found on many standardized tests of reading, and you got them all right. In order to accomplish this, you had to decode letters into sounds, recognize the words, and locate information in the text. The example supports the simple view of reading, because saying you could read the passage, according to that view of reading, is sufficient to account for these accomplishments. We do not notice anything strange because we also all believe we can read the Seaview passage, and understand what it means. The following example brings out the strangeness.

QUANTUM DAMPING

We assumed that the atomic energy levels were infinitely sharp whereas we know from experiment that the observed emission and absorption lines have a finite width. There are many interactions which may broaden an atomic line, but the most fundamental one is the reaction of the radiation field on the atom. That is, when an atom <u>decays</u> spontaneously from an excited state radiatively, it emits a quantum of energy into the radiation field. This radiation may be reabsorbed by the atom. The reaction of the field on the atom gives the atom a linewidth and causes the original level to be shifted. This is the source of the natural linewidth and the Lamb shift. (Louisell, 1973, p. 285)

The questions are as follows:
1. The underlined word <u>decays</u> means: A. splits apart, B. grows smaller, C. gives off energy, D. disappears.
2. According to the article, observed emission lines are: A. infinitely sharp, B. of different widths, C. of finite width, D. the same width as absorption lines.
3. According to the article, the most fundamental interaction that may broaden an atomic line is: A. the Lamb shift, B. the action of the atom on the radiation field, C. the emission of a quantum of energy, D. the reaction of the radiation field on the atom.
4. It can be inferred that when an atom decays it may: A. return only to a state more excited than the original one, B. not return to its original excited state, C. return to its original excited state, D. return to a state less excited than the original one.

5. It can be concluded from the information in this article that the assumption that atomic energy levels are infinitely sharp is: A. probably false, B. false, C. true, D. still under question.

Check your answers against ours: CCDCB. Again, we assume you got them all right. What is the point? If we constrain our conception of reading to the simple view, you must conclude that you and our other participants read both passages. This is because perfect performance on these items proves that one can do all that is required by the simple view of reading: decode letters into sounds, recognize the words, and locate information in the text. Yet, we face a problem. In the case of the Seaview passage, you understood it. However, except for a small number of people, including perhaps some of you, most readers do not have the faintest idea of what the Quantum Damping passage means. Even so, our participants answered all the questions correctly for that passage also. Given the parallel structure of the two examples, we can conclude that successful reading according to the simple view does not imply understanding, because, although the criteria for the simple view of reading were met in both cases, understanding occurred only in one. The simple view of reading can neither account for understanding nor differentiate cases of understanding from cases of failure to understand, except for cases of minimal understanding involving only word recognition or information location. Clearly, then, the simple view of reading does not address what we wish to achieve in science education. Just as clearly, the focus on word recognition and information location that defines the simple view of reading and that dominates reading in science instruction cannot resemble the reading that scientists do in their work. In teaching according to the simple view of reading, we are leading students to a false impression of their ability and to the perception that science does not have to make sense.

We can now circle back to the university students who reported finding the media reports easy to read but performed poorly on the interpretive tasks we set for them. Why did they interpret poorly? Our conjecture is that the interpretive tasks required them to go beyond decoding words and locating information in the text. They were asked to infer connections between statements that often were widely separated in the text; they were asked to infer pragmatic meanings that often were not displayed explicitly and literally; in short, they were asked to make interpretations that went beyond what was explicitly there in the text and they were unable to do it. However, why did they report finding the passages of suitable reading difficulty? Our hypothesis is that their theory of reading led them to this conclusion. They knew they could identify the words, and they knew they could locate information. Indeed, as we have reported, they were significantly more adept at locating information than they were at answering the interpretive questions. Therefore, all of their school and university experience told them that they had read successfully, even though they had not. Sadly, although according to the simple view of reading they had read, they did not understand. The simple view aims to reduce reading to rote recognition and location of information, but fails because it is easy to demonstrate how the satisfaction of these criteria can be achieved without understanding in any deeper sense than grasping surface meanings.

Finally, why did their science background appear not to help? Why did these university students perform no better than high school students who had much weaker science backgrounds? It is widely believed that more background knowledge is associated with improved reading comprehension. Phillips (1988) supplies a possible explanation of the lack of relationship between students' science backgrounds and their performance on the tasks we set for them. She found that sixth grade readers' background knowledge mattered only in the context of reading proficiency defined by the use of what she called "productive reading strategies", which include questioning your interpretations and considering alternative ones. Students in her study who used such productive reading strategies were able to compensate somewhat for their lack of background knowledge, although the best reading was found in the context of both productive strategies and background knowledge. The productive reading strategies used by the children corresponded in large measure to those identified by Collins, Brown and Larkin (1980) in their study of skilled adult readers, and strongly overlap with strategies often associated with critical thinking (Norris & Phillips, 1987). Therefore, it is reasonable to surmise that the university students in our study lacked productive strategies for reading the type of text found in the media reports of science. Consequently, they could not take advantage of their scientific background knowledge.

The point of this section is that science education needs to be guided by a different view of reading. The view of reading that currently underlies science teaching, and education generally, does not find its motivation in the creation of meaning and the search for understanding. Rather, it is motivated by a misguided reliance on facts, and on their transmission to and uptake by students. The evidence shows that students have mastered well the simple view of reading: they are adept at identifying words and locating information and at answering questions that expect no more than these tasks from them. However, when they are expected to interpret, to go beyond the literal meaning of what is written, which always is required in science, they perform far less well. In the final section, we turn to a view of reading for science education that is designed to address these concerns.

IV. READING AS INQUIRY

Schwab is credited with efforts in the 1960s to conceive of science as inquiry and to found science education upon that notion. Schwab was responding to two influences. The first influence was the radical increase in the rate of revision of scientific knowledge, "twenty to a hundred times what it was much less than a century ago" (Schwab & Brandwein, 1966, p. 18). The second influence was the teaching of science as dogma, which Schwab saw as inconsistent with the rapidity of scientific change. Science was taught "as a nearly unmitigated *rhetoric of conclusions*" (p. 24), "a structure of discourse which persuades men to accept the tentative as certain, the doubtful as the undoubted, by making no mention of reasons or evidence for what it asserts" (p. 24). It was this structure that was to be replaced by "the teaching of science as enquiry" (p. 65) in which both teaching and science were seen as processes of inquiry.

Schwab was the first director of the Biological Sciences Curriculum Study, so it is instructive to examine the view of reading adopted in those materials. In the Third Edition of the *BSCS Green Version* (BSCS, 1973), the Supervisor, Haven Kolb had the following to say in a preface for students:

There are two major aims in studying any natural science.

One aim is to become acquainted with scientific facts and with the general ideas that are built upon them . . .

The second aim in studying a natural science is even more important. It is to understand what science *is* – to feel its spirit, to appreciate its methods, and to recognize its limitations . . .

It might be possible, by using textbooks only, to achieve the first aim of science study. But to pursue the second aim – the more important aim – requires experience in scientific work. Such experience cannot be gained by reading; it cannot be gained by listening – even listening to the most accomplished scientists; it can be gained only by doing the kind of things scientists do in their laboratories. (pp. viii-ix)

Given that the BSCS curriculum is founded on science conceived as inquiry, the obvious inference is that reading is not inquiry, or, if it is, it certainly is not part of what defines science as inquiry. Furthermore, reading is not an important part of scientific work, because, according to this view, it is not one of the kinds of things that scientists do in their laboratories. Presumably, reading is not important even for scientists who are not experimentalists and who do not work in laboratories. Over several decades, the BSCS view on this matter seems not to have changed. For instance, they define science as inquiry on their 2004 website (http://www.bscs.org) as reflecting "the ways in which scientists conduct their work and develop their explanations . . . our materials provide students with opportunities to make observations, pose questions, study what scientists currently know, design and carry out investigations, and develop explanations based on evidence." Although this list of ways of conducting scientific work does not rule out reading, and indeed reading is certainly involved, reading itself is not portrayed as one of the ways of conducting scientific work.

The *National Science Education Standards* (National Research Council, 1996) offers a view of science as inquiry that creates more of a role for language. However, actively participating in scientific investigations, which includes experimenting, analyzing evidence and data, and proposing explanations, does not appear to make a central role for reading. Rather, reading, and writing and speaking, are named oddly as instructional strategies to be used by the teacher to gain information on students' current understandings. The Standards do say that school science programs should promote the abilities needed for effective communication, to write procedures, express concepts, summarize data, develop diagrams and charts, and construct reasoned arguments. All of these goals are important and we support them. Yet, nowhere is there a sense that reading is itself part of scientific inquiry, by casting reading as inquiry and reading as one of the activities that scientists perform a great deal.

The *Common Framework of Science Learning Outcomes* in Canada (Council of Ministers of Education Canada, 1997) includes in its definition of scientific literacy "the skills required ... for communicating scientific ideas and results" (p. 6). These skills involve "examining information and evidence, of processing and presenting data so that it can be interpreted, and of interpreting, evaluating, and applying the results" (p. 12). There seems to be a more central role for language, including reading and writing, in these objectives as indicated by a number of specific learning outcomes that are meant to ensue by the end of grade 12, such as:
- "use library and electronic research tools to collect information on a given topic" (p. 82);
- "compile and display evidence and information, by hand or computer, in a variety of formats, including diagrams, flow charts, tables, graphs, and scatter plots" (p. 83);
- "provide a statement that addresses the problem or answers the question investigated in light of the link between data and the conclusion" (p. 83);
- "communicate questions, ideas, and intentions, and receive, interpret, understand, support, and respond to the ideas of others" (p. 84)
- "select and use appropriate numeric, symbolic, graphical, and linguistic modes of representation to communicate ideas, plans, and results" (p. 84)
- "systhesize information from multiple sources or from complex and lengthy texts and make inferences based upon this information" (p. 84).

None of the above views, however, goes as far as it is clear we must go. In the remainder of this section, we shall defend a view of reading as inquiry, a view that makes reading intrinsic to science, as part of what science is. We shall begin with some needed concepts and distinctions, proceed to providing a conception of reading as inquiry, and end with an argument for a constitutive relationship between reading and science.

IV.1 Concepts and Distinctions

The central idea of our view of reading as inquiry is that reading is principled interpretation of text, which involves the reader in inferring meaning from text by integrating text information with the reader's relevant background knowledge. This central idea presupposes a set of concepts and distinctions: text information and the difference between literal and inferential interpretation, the distinction between comprehension and interpretation, and the relevance of readers' background knowledge to their interpretations. We have attempted to capture the central idea of reading as inquiry in a sketch provided in Figure 1. The meaning of the figure will be explained throughout the following sections.

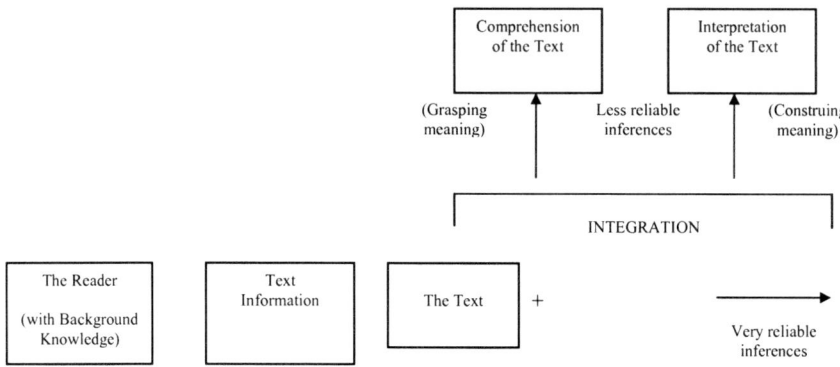

Figure 1. Reading as Inquiry

IV.1a Text Information. There is a potential circularity in conceiving of reading as integrating text information with readers' relevant background knowledge. To talk about text information implies that interpretation already has taken place. Otherwise, we should talk about ink marks on a page or pixel patterns on a screen, rather than information (Norris & Phillips, 1987). But it is not ink marks or pixel patterns that readers integrate with their knowledge. The same problem arises in trying to account for how we interpret the world. The logical positivist account is that we start with sense data, raw and uninterpreted, and then impose meaning on that data. However, it is widely accepted that data are theory-laden (Hanson, 1958). Therefore, interpretations of the world begin with other interpretations of the world. Uninterpreted foundations for knowledge simply do not exist.

A fallibilist account of scientific knowledge can begin from the stance that whenever humans reason about something they rely upon a set of assumed propositions (Harré, 1986; Norris, 1985; Shapere, 1982). For instance, a scientist might look at an instrument and observe an electrical current of 12 amperes. The observation is fallible. The particular instrument could be shown to be flawed, and even ammeters in general could be shown to be flawed, however improbable these outcomes might be. The observation involves interpretation. Claiming the existence of an electric current is to interpret nature, and using an ammeter to measure current is to interpret the behavior of the machine as a manifestation of electric current. The very construction of the instrument relied upon interpretations of nature. Although fallible, the observation is highly reliable, because it is based upon a body of justified scientific knowledge. Although interpretation is involved, it is not an interpretation that is made and justified on the spot by individual scientists every time a reading of current is taken. It is an interpretation that is built into the instrument, approved by consensus, and taken for granted by the community of scientists as the correct interpretation of the instrument's behavior. Thus, the interpretation is assumed implicitly and properly by scientists in the context of most scientific research. The required thinking has been done previously, both by the community of scientists and by the individual scientist in his or her training. It is through such contextualization of scientists' reported observations of electrical current magnitudes that we can see how reports of observations differ in an important sense from reports of inferences from observations. What scientists refer to as inferences do not have the same consensual and dependable status within the scientific community.

A parallel argument can be made for text information. Readers are justified in taking for granted certain propositions, such as, that the text is coherent, and that the writer is being cooperative in using language conventionally (Grice, 1989). These assumptions allow the reader to make interpretations of the meanings of many individual words, phrases, and sentences, quite automatically and reliably. Analogously to the example of scientific observation, the thinking has been done previously, both by the community of language users and by the individual reader in his or her language development. Therefore, inferring meaning from text involves different levels of inferential reliability. The reader's grasp of text information is dependent upon highly reliable and often automatically made inferences.

Imagine there to be two continua: obviousness of meaning and reliability of inferences. When the reader proceeds toward the nonobvious end of the meaning continuum, the inferences required by reading become increasingly unreliable; as the reader proceeds in the other direction on the meaning continuum, inferences become increasingly reliable, until they hardly seem like inferences at all (see Figure 2). For different readers, and for different occasions, different texts mark the extremities of the meaning continuum. The distinction between literal and inferential reading refers to the opposite ends of the function relating obviousness to inferential reliability. That is, inferences occur at every point along the function, but, near the obvious extremity, they are so reliable that they appear different in kind, so we mark them by a different term, "literal". When inferring meaning from text by integrating text information and knowledge, the reader is building upon a base of reliable inferences (interpretations of text whose meaning is obvious and that we have been calling "text information") to make less reliable inferences (interpretations of less reliability that integrate text information and knowledge).

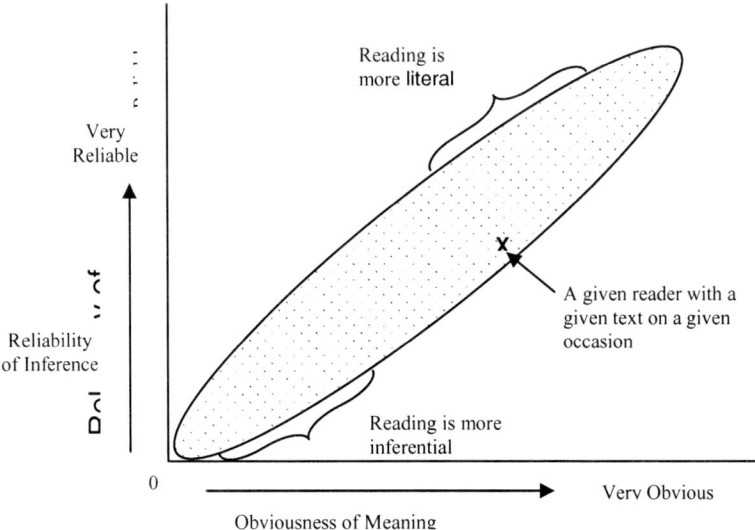

Figure 2. Relationship between Reliability of Inference and Obviousness of Meaning

IV.1b Comprehension and Interpretation. The *Webster's Third New International Dictionary* uses such phrases as the following in defining "comprehend" and "comprehension": to see the nature, significance, or meaning of; grasp mentally; attain to the knowledge of; the act or action of grasping with the intellect; apperceptive knowledge or knowing; the capacity or power of the mind for understanding fully. Understanding is offered as the synonym, and grasping names the act. In comprehension, the reader aims to grasp the meaning of something that has that meaning in it, as it were. Let us consider an example by reflecting on one of Aesop's Fables: *The Fox and the Crow*:

> A Crow was sitting on a branch of a tree with a piece of cheese in her beak when a Fox observed her and set his wits to work to discover some way of getting the cheese. Coming and standing under the tree he looked up and said, "What a noble bird I see above me! Her beauty is without equal, the hue of her plumage exquisite. If only her voice is as sweet as her looks are fair, she ought without doubt be Queen of the birds." The Crow was hugely flattered by this, and just to show the Fox that she could sing she gave a loud caw. Down came the cheese, of course, and the Fox, snatching it up, said, "You have a voice, madam, I see: what you want is wits." (Aesop's Fables, 1912)

The fable is designed to teach us lessons about life and people. But, what can we learn from this fable about reading? At one level, the crow correctly read the fox: She recognized the words he used and she grasped their literal meaning. But she missed the fox's intention to deceive her. This is why the ruse succeeded. So, the crow really did not comprehend the fox, because the fox was not complimenting her, he was attempting to deceive her.

If we focus on comprehension, we are forced to look beyond the simple view of reading to take into account factors other than word recognition and information location. Let us reflect upon a young child. Having read the fable, Suzie says, correctly, "It is a little story about a fox and a crow." Again correctly, she says, "The crow has a piece of cheese and the fox wants it." She continues, "The fox tells the crow that she is beautiful, that she could be a Queen, a bird-Queen, that is. The crow really liked what the fox said. She began to sing and dropped the cheese. The fox ran away with it." From Suzie's paraphrase and retelling, we know that there was a lot she understood beyond the literal. She grasped the chronology, the story-line, the characters, the events, and the resolution. She made appropriate inferences. Suzie surpassed what is expected by the simple view of reading.

However, we know there is more to the fable than Suzie related. She missed the lesson. She missed Aesop's intention to teach us something general about human behavior. She did not see the fox and the crow as abstractions, as representations of things beyond themselves. Although we are confident based upon her paraphrase that she understood much of the meaning of the fable, we are equally confident that she did not explore such issues as what the author intended, what might the author have intended, and what can be understood by the story that is not literally there in the words and sentences or available through simple inferences. Suzie needed to know the genre; she needed to know that *The Fox and the Crow* is a fable, and what being that genre implies. If she had known these things, she could have more fully grasped Aesop's meaning.

Understanding can go even further than what has been recounted so far. Suppose we were to consider the following questions regarding Aesop's fable: Which character did you like better? Why? How do you think the animals and their behavior in this fable compare to animal behaviors in other fables? Could the fox and the crow switch their character roles and the story still work? Do you believe there are cultural differences in judging which animals serve better in particular character roles? Do you know examples to support your case? Can you think of a person you have encountered who reminds you of the fox? The crow? Does the person's fox-like or crow-like character permeate his or her personality? Clearly, you can understand to a degree the *Fox and the Crow* without having thought of questions such as these and their answers. However, these sorts of questions point to a level of understanding that goes beyond even full comprehension, even a full grasp of the meaning. We have added *interpretation* to our answer to the question, "What is reading?"

We can imagine a parallel case with the Quantum Damping passage. It is a bit more difficult to deal with as an example because it is a selection from the middle of an extended text, rather than a complete text like *The Fox and the Crow*. However, a reader can interpret this passage very deeply also. The reader may contemplate why energy levels have a finite width, and speculate that it has something to do with the fact that the atom must absorb energy from a band of frequencies whose width must be finite, else the band would contain zero energy. On the other hand, it may be a relativistic effect. If we want to measure the energy of an atom, then the uncertainty principle tells us that a minimum period of time is required. The consequence of having to wait is that the energy can only be known to a certain level of accuracy, thereby giving a linewidth to the energy level. The reader might wonder whether in the passage the author is viewing the atom through the lens of quantum dynamics or through the lens of classical physics. How should this text be construed? What significance should be placed upon it?

Interpretation requires a personal judgement of significance based upon one's comprehension. Whereas comprehension implies success in understanding meanings that are present in the text, interpretation goes beyond this. Interpretation has a subtle yet importantly different sense. The Webster's uses such defining expressions as the following: to explain or tell the meaning of; translate into intelligible or familiar language or terms; to understand and appreciate in the light of individual belief, judgment, interest, or circumstances; explanation of what is not immediately plain or explicit; explanation of actions, events, or statements by pointing out or suggesting inner relationships or motives or by relating particulars to general principles. "Construe" is offered as one of the synonyms, and judging and explaining is the act.

In interpretation, the reader is making meaning that goes beyond comprehension of the text. Interpretation is the broader of the terms, since it includes comprehension. Whereas comprehension is about grasping the meaning of the text (if such exists), interpretation is about exploring meanings presupposed, implied, and reasonably justified by the text. Although appeal to linguistic intuitions rarely provides proof of a point, it is instructive to think of our intuitions in the use of these two terms. You will recall that "comprehend" means to grasp fully. Thus, the idea of re-comprehending seems a bit odd, because to comprehend implies to have reached the end of the road. "Interpret", on the other hand, means to make a judgement. Thus, the thought of re-interpreting works, because to interpret is to imply the road might continue. Since interpretation is more inclusive than comprehension, we shall use interpretation to name the goal of reading. What is reading? To read is to interpret.

IV.1c Relevance of Background Knowledge. Somehow, readers draw upon their background knowledge and integrate it with information from the text to create interpretations. Contrary to views often expressed, it is our view that reading does not involve selecting knowledge relevant to a reading task prior to reading and then using that knowledge to aid interpretation. We have called this "the god's eye view of reading" (Norris & Phillips, 1994a). Reading indeed does involve making relevant the knowledge one has to an interpretation of the text. The creation of relevance is part and parcel of the reading act. However, knowledge that is relevant to interpreting a particular text generally cannot be prespecified. So, for example, there is a growing recognition that the provision of prior definitions and formula, perceived from the god's eye view to be prerequisite to understanding subsequent material, does not serve the purpose, unless students can use the definitions and formula in constructing their interpretations (Shymansky, Yore & Good, 1991).

Having knowledge about a topic prior to reading about it is useless to a reader unless the reader can see the relevance of that knowledge by making inferential links between the knowledge and the text. That is, relevance as seen from the viewpoint of someone other than the reader – someone with a god's eye view that such-and-such knowledge is relevant to such-and-such topic – is worthless to readers who do not see the relevance themselves. Constructing inferential links that make knowledge relevant to text interpretations is, in the final analysis, what proficient reading is.

Inferential links between readers' knowledge and text must be forged (Collins, Brown & Larkin, 1980; Phillips, 1988, 1992). Without such links, knowledge remains inert, that is, irrelevant, because the reader has no effective way to use it. On the other hand, readers who lack the knowledge that, from some god's eye point of view, is judged relevant to a topic are able in principle to turn to their own knowledge as a resource. Proficient readers will be able to use this latter resource, and through this means compensate for the lack of knowledge that others have deemed prerequisite (Phillips, 1988). They do this by creating relevance for knowledge from sources that might not, from a god's eye perspective, be judged relevant.

IV.2 Reading as Principled Interpretations of Text

Under the concept of text, we include whatever is meant to be read, without distinction for the medium that displays the print: the panoply of literate objects including not only printed words and sentences, but also graphs, charts, tables, mathematical equations, diagrams, figures, maps, and so on. The possibility of more than one good interpretation exists for all text types, notwithstanding the fact that the leeway for proposing multiple interpretations varies from type to type. For example, a wave function may be meant to be read in one and only one way. However, no text can be written so as to avoid the need for interpretation by its readers, or so as to guarantee that they all will reach the same interpretation.

Reading is best understood as a constructive process. However, we reject the relativism associated with some versions of constructivism. Readers should adopt a critical stance toward text by engaging in interactive negotiation between the text and their background beliefs in an attempt to reach an interpretation that, as consistently and completely as possible, takes into account the text information and their background beliefs (Phillips & Norris, 1999). It is in the fashioning of interpretations that our position is constructivist, and in the fashioning under constraints of consistency and completeness that our position is not relativistic.

According to the simple view, reading is knowing all the words and locating information in the text. By contrast, we maintain that reading is not a simple concatenation of word meanings; is not characterized by a linear progression or accumulation of meaning as the text is traversed from beginning to end; and is not just the mere location of information. Rather, reading depends upon background knowledge of the reader, that is, on meanings from outside the text; it is dependent upon relevance decisions all the way down to the level of the individual word (Norris & Phillips, 1994a); and it requires the active construction of new meanings, contextualization, and the inferring of authorial intentions (Craig & Yore, 1996; Yore, Craig & Maguire, 1998). In short, *reading shares the features of all inquiry* (Norris & Phillips, 1987). First, *reading is iterative*. By this we mean that reading proceeds through a number of stages, each aimed at providing a more refined interpretation. A stage consists of steps, not necessarily followed in order: lack of understanding is recognized; alternative interpretations are created; judgement is suspended until sufficient evidence is available for choosing among the alternatives; available information is used as evidence; new information is sought as further evidence; judgements are made of the quality of interpretations, given the evidence; and interpretations are modified and discarded based upon these judgements and, possibly, alternative interpretations are proposed, sending the process back to the third step. Second, *reading is interactive*. Interaction takes place between information in and about the text, the reader's background knowledge, and interpretations of the text that the reader has created. When reading, people use the information that is available, including their background knowledge, in creative and imaginative ways. They make progress by judging whether what they know fits the current situation; by conjecturing what interpretation would or might fit the situation; and by suspending judgement on the conjectured interpretation until sufficient evidence is available for refuting or accepting it. The reader actively imagines, and negotiates between what is imagined and available textual information and background knowledge. Finally, in order to carry out such negotiation, *reading is principled*. The principles help determine how conjectured interpretations are to be weighed and balanced with respect to the available information. Completeness and consistency are the two main criteria for judging interpretations. Neither criterion by itself is sufficient; they must be used in tandem. In order to deal with situations where there are competing interpretations, the criteria must also be used comparatively. Readers must ask which interpretation is more complete, and more consistent, because often neither interpretation will be fully complete and fully consistent.

If reading is as expansive as we describe, then reading involves many of the same mental activities that are central to science (Gaskins, Guthrie, Satlow, Ostertag, Six, Byrne & Connor, 1994). Moreover, when the reading is of science text, it encompasses a very large part of what is considered doing science. It is not all of science, because it does not include the manipulative activities and working with the natural world that are so emblematic of science. However, the relationship between reading and science is intimate. If reading is as expansive as we describe, science educators need to be aware of the possibility that many students will bring to their science learning the simple view of reading. If science teachers continue to show little concern for text, see reading as merely a tool to get to science, or see reading as unimportant, then they are likely to reinforce the attraction that this simple view has, and unwittingly underestimate the complexity of reading in science.

IV.3 Reading as Intrinsic to Science

In contrast to the expansive view of reading and the multilevel engagements between readers and texts that we described in the previous section, there is the much simpler view we have reported. According to this simple view, knowing what the text *means* comes down to knowing what the text *says* (Olson, 1994), that is, of decoding the words. According to this view, texts that cannot be grasped immediately upon knowing what they say are problematic ones, because they introduce an obstacle between the reader and the world or theory they depict. Good texts are transparent because they can be seen through to their meaning. This is a perspective we hear frequently from those who are expert readers but who do not study reading.

Text is related to scientific thought and ideas in a more complex and circuitous way than implied by the simple view of reading. First, text is not speech written down, but rather is built upon a theory of speech (Olson, 1994). Writing, including scientific writing, is a form of idealization that leaves out much that is included in spoken communication (such as, intonation, repetition, stammers, incomplete and interrupted thoughts, facial expressions, and gesture); and is a form of construction that adds in much that is not included in speech (such as, sentence structure, punctuation, paragraphing, breaks between words, and long chains of expressions connected by tight logical links). Reading involves coping with both the expressed and unexpressed in the written word (Olson, 1994, p. 265). Readers lack the inferential cues provided to listeners by tone, gesture, facial expression, repetition, stammer. By way of compensation, they have at their disposal the characteristics of text not found in speech. In sum, speech and text are not the same, and do not impose the same interpretive demands. This latter point is crucial to science teaching. Readers must learn to cope with science text, must learn to use the resources of text to determine what they mean, or might mean. To accomplish these goals, readers require interpretive strategies that can be isolated and taught.

Second, the same text can express different thoughts, and the same thought can be expressed through different texts. This means that there is no direct connection between what a text says and what it means. What a text means always must be inferred from what it says plus other extratextual information. There is no alternative when reading but to bring to the text thoughts from outside of it. Readers must learn that neither they nor the text is supreme: each is a required source of information in the interpretive process; their relative weightings change from situation to situation depending upon, among other factors, familiarity with what is being read (Phillips & Norris, 1999).

Third, and as a corollary to the second point, although a scientific theory is independent of any given text (because the same theory can be expressed in many different ways), a scientific theory cannot exist outside of text altogether. Any scientific theory requires for its creation and its expression the use of text. Any attempt to provide a scientific theory without appeal to text runs quickly into insurmountable shortcomings of expressive power, memory, and attention. Even to express such a simple theory as that of an ideal gas requires appeal to mathematical equations, graphs, and diagrams, all of which are tools of literacy. There is no possible expression of the theory, except for some isolated aspect of it (and even such expressions are parasitic on text), outside of text of some sort. For instance, it is not simply a matter of convenience that the pressure-volume-temperature surface is used centrally in expressions of the theory of ideal gases. We just do not have another way to express the complex of thoughts that this surface, portrayed in text either graphically or in mathematical functions, represents. If somebody were to invent another mode of expression, it too would be within the realm of literacy and would impose the same interpretive demands.

The only contender for the literate tradition is the oral one, and oracy does not provide the tools needed or the forms of expression required to describe scientific theories. In saying this, we do not intend to denigrate oracy. Just as text provides expressive power not available in speech, speech marshals expressive power not found in text, and speech plays a crucial and irreplaceable role in the development, critique, and refinement of thoughts that go into theories. Halliday and Martin (1993) attempted to make the contrast metaphorically: "Written language is corpuscular and gains power by its density, whereas spoken language is wave-like and gains power by its intricacy" (p. 118). Operating from a psycholinguistic framework, Rivard and Straw (2000) found that among eighth-grade students "Talk or discussion appears to be important for sharing, clarifying, and distributing knowledge among peers" (p. 585); while "writing is an important discursive tool for organizing and consolidating rudimentary ideas into knowledge that is more coherent and well-structured" (p. 586).

Thus, scientific knowledge has an essential dependence upon texts, and the route to scientific knowledgeability is through gaining access to those texts. Although individuals can portray and learn much science within oracy, such access to scientific knowledge is parasitic upon access gained through text. This is so, because, without literacy, the knowledge would not have existed, been preserved, and inherited in the first place. Hence, science education must be premised upon an essential role of text in science.

Fourth, although a scientific theory is independent of any given text, an expression of a theory through text capitalizes upon three features of textual fixity: (a) fixity of physical presentation (allowing for variation in fonts, margins, spacings, media, and some orderings); (b) fixity of literal meaning; and (c) fixity of what is to be taken for granted. These fixities are what allow the *same* text to be transported across time and space to be traded, revisited, queried, and reinterpreted. "Writing puts language in chains; it freezes it, so that it becomes a *thing* to be reflected on" (Halliday & Martin, 1993, p. 118). "[W]riting fixes language, controls it in such a way that words do not scatter, do not vanish or substitute for one another. *The same words, time and again.* The mystery [that all readers need to see through in learning to read] lies to a large extent in this possibility of repetition, reiteration, re-presentation" (Ferreiro, 2000, p. 60). Yet, although the text is fixed, the interpretations of it are not. The interpretations are made against the backdrop of the relative fixity of physical presentation, literal meaning, and what is taken for granted. As Olson has said: "Consciousness of the fact or possibility that utterances can be taken literally − according to the very words − is at the heart of literate thinking" (1996, p. 149). Although literal meaning is only relatively stable, it affords a level of stability unprecedented in oracy. As argued by Tishman and Perkins (1997): "Written language, stabilized on paper, invites kinds of reflection not so natural to oral exchanges. The written statement is more easily examined, checked, contradicted, doubted, challenged, or affirmed" (p. 371). We would go further than "invitation", and argue that written language is what makes these activities possible in any depth. Readers must understand both the significance of the fixity of text and also that fixed texts invite and allow interpretation and reinterpretation.

Fifth, although the fixities of text make interpretation and reinterpretation possible, they also make some interpretations, if not impossible, then highly implausible. That is, the fixities impose constraints on interpretations that readers are obliged to take into account. Without the constraints provided by the text and the context, no interpretation would be possible, because without constraints the text could mean anything at all. So, the very fixities of text that make interpretation possible are the same features that ensure that not everything goes, by providing constraints on what can be meant. Therefore, the reader must take note of the very words, the very data, and other textual elements, and bears the burden of delivering interpretations under the constraints of those fixities. There is always room to maneuver, but the degrees of freedom are restricted. Students must learn that the very words and other textual elements matter as constraints on allowable interpretation, and learn how to discern the boundaries of those constraints.

Sixth, Halliday and Martin (1993) have expanded the notion of the fixity of text to the fixity of what is taken for granted:

Until information can be organized and packaged in [written language] ... knowledge cannot accumulate, since there is no way one discourse can start where other ones left off. When I can say

the random fluctuations in the spin components of one of the two particles

I am packaging the knowledge that has developed over a long series of preceding arguments and presenting it as 'to be taken for granted − now we can proceed to the next step'. If I cannot do this ... I will never get very far. (pp. 118-119)

Therefore, science is a result of cumulative discourse that trades on the fixities of text and on what is taken for granted by that text. This is not to imply that scientific knowledge accumulates linearly as in building a monument brick-by-brick. Yet, students need to learn that scientific discourse always attaches to and is dependent upon discourse that has gone before, even if it rejects the former, and serves as an attachment for discourse that is to come, even if it itself is rejected by that subsequent discourse.

The six points above are summarized in Figure 3. The implication for the relationship between scientific text and thought is that science would not be possible without text. Science is in part constituted by texts and by our means of dealing with them. Without the expressive power and relative fixity of text; and without the comprehension, interpretive, analytical, and critical capacities we have developed for dealing with texts; then Western science as we know it could never have come into being. The only alternative to literacy, oracy and the oral tradition, simply does not have at its disposal the tools for developing and sustaining science as we know it, although, as we have said, it plays an irreplaceable role in the development, critique, and refinement of scientific thought. Without text, the social practices that make science possible could not be engaged: (a) the recording and presentation and re-presentation of data; (b) the encoding and preservation of accepted science for other scientists; (c) the peer reviewing of ideas by scientists anywhere in the world; (d) the critical re-examination of ideas once published; (e) the future connecting of ideas that were developed previously; (f) the communication of scientific ideas between those who have never met, even between those who did not live contemporaneously; (g) the encoding of variant positions; and (h) the focussing of concerted attention on a fixed set of ideas for the purpose of interpretation, prediction, explanation, or test. The practices centrally involve texts, through their creation in writing and their interpretation, analysis, and critique through reading. The practices define partly what it is to engage in scientific inquiry and what it is to read. Thus, reading is properly seen as inquiry and as part of science.

- Coping with speech and coping with text are not the same
- Supremacy lies with neither the text nor the reader
- Text is an essential vehicle for the expression of scientific thought
- Although fixed, texts permit interpretation and reinterpretation
- The very words matter as constraints to interpretation
- Scientific knowledge relies upon the cumulative discourse made possible by text

---

Figure 3. Six Points about the Relationship between Scientific Text and Thought

V. IMPLICATIONS FOR PRACTICE AND FOR FUTURE RESEARCH

The view that underlies much that we have written in this paper is that science teachers must come to see themselves also as literacy teachers. For many science teachers, this shift would require a radical alteration in their self-conception. Our view is based upon the beliefs both that literacy is more fundamental than science epistemically, and that the educational goal of literacy attainment is more fundamental than the goal of learning science. In taking on the mantle of literacy teacher, science teachers would find themselves playing a role in an agenda more central to education than the teaching of science. Science teachers would look to the concepts, skills, understandings, and values that are generalizable to all reading and look for application within the reading of science. In this way, scientific inquiry would be seen as related to inquiry in other subjects with which it shares a common epistemology. Learning to read scientific text, according to our view, leads not only to the possibility of learning the substantive content of science, but also to learning epistemology, knowledge of which successful reading requires. In order to achieve this transformation in teacher outlook, much more needs to be known and much curriculum work needs to take place. We shall outline some of these needs in the remainder of this paper.

First, much more needs to be known about teaching and learning to read argumentative text, that is, text in which reasons are offered for conclusions. Although the domination by narrative of beginning reading programs is slowly coming to an end, its replacement hardly ever includes argumentative text (Phillips, Smith & Norris, 2005). Rather, what is found is an increased proportion of these programs devoted to informational and expository text of the type most frequently found in current science textbooks. However, it is domination by these latter genres that is precisely what many science educators are attempting to end. There is a need for research on teachable reading strategies for argumentative text and for persistent pressure on the educational system to take seriously very early in schooling explicit instruction on reading and writing argumentative text.

Second, it would be helpful if science educators could articulate a clear rationale for the scientific practices with text that ought to be brought into the science classroom. The rationale would be partly value driven by referring to and justifying the goals that would be achieved, but also empirically driven by drawing on available research on which translated practices work and which do not. There are many questions that need to be explored, including ones about the type of texts that might achieve the most desirable ends – genuine scientific research reports, genuine reports suitably translated for particular levels of schooling, or fictional texts purposely designed and created for the situation.

Third, and related to the second point, on the assumption that textbooks will be around for a while, models are needed for redesigned textbooks that incorporate more argumentative text, and these models require testing for their effectiveness and useability. Some possibilities include the more explicit and frequent treatment of reasons for conclusions, examples from frontier science where the scientific community has not reached consensus on an issue, and the use of media reports of science as texts to be interpreted and critically appraised.

Finally, and coming back to science teachers' views of themselves, reading in science needs to be an explicit part of science teacher education. Accomplishing this change will not be easy, and it is likely to be resisted. Science teachers in training, especially at the high school level, come with strong scientific backgrounds. During their scientific education, the importance of reading to science in the sense we have articulated in this paper is rarely articulated or even implied. Many are likely to come to their teacher education programs holding the simple view of reading. Research needs to be done to test this conjecture. If we are correct, then changing this view would be a good place to start, perhaps first by trying to change the views of scientists and science education instructors.

Stephen P. Norris
Department of Educational Policy Studies
University of Alberta

Linda M. Phillips
Centre for Research on Literacy
University of Alberta

PHILIP BELL

INQUIRY AS INSCRIPTIONAL WORK: A COMMENTARY ON NORRIS AND PHILLIPS

Norris and Phillips (this volume) advance a detailed argument that reading should be considered an important mode of scientific inquiry – that engagement with text is a complex set of intellectual activities inherently bound up in the practice of science and that students would benefit from being similarly engaged in the sophisticated reading of text as a constituent part of their inquiry activities. Norris and Phillips go on describe the default theories of reading held by students and outline a more sophisticated model of targeted reading as disciplinary interpretation. I am, overall, quite resonant to this perspective and in this commentary seek to elaborate upon it a bit and detail some areas where we need more work to further pursue such an approach.

Coming from a learning sciences orientation to this work, I work from two touchstones when considering these issues. The first touchstone concerns itself with the nature of scientific practice itself, especially in everyday and material terms as outlined by ethnographic research on scientific work. The second touchstone deeply considers the personal relevance of science to the students learning about it. It is from these two positions that I consider issues surrounding reading (and writing) in the science classroom.

CONSIDERING THE BREADTH OF SCIENTIFIC INQUIRY

In educational conversations, I have been repeatedly shocked how quickly deliberation about how best to support students in 'scientific inquiry' collapse to be discussions of how to scaffold them in doing systematic observation and hypothetico-deductive style experimentation (typically on tasks where the science was worked out centuries ago). More often than not, these two inquiry modes are presumed to be the only appropriate focus – as if they exhausted the range of desired intellectual activities we would want for science education, or that they present enough of an intellectual challenge that we would not want to muddy the waters further by educationally pursuing other modes of inquiry. Well, I believe we need to broaden our focus to other modes of inquiry, including reading – as laid out by Norris and Phillips.

So, what activities make up the intellectual life of natural scientists? What is the overall range of activities and how do they fit together in meaningful ways? How similar or different are these activities across subfields of science – or even across research groups in the same subfield? Then given the important distinction made by Duschl and Grandy (this volume) that we would not want to presume that we need to bring all of the complexity of science into education, the question

Richard A. Duschl and Richard E. Grandy (eds.), Teaching Scientific Inquiry: Recommendations for Research and Implementation, 263–267.
© 2008 *Sense Publishers. All rights reserved.*

becomes: which of these intellectual practices would we want to bring into the classroom, in some form, and for what purpose?

Personally, what I have found to be profoundly missing in our discussions of appropriate modes of scientific inquiry for the classroom are ways to include modes of inquiry that make the formal knowledge work of disciplines – or interdisciplinary efforts – visible to students and to engage them with it in educationally appropriate ways. Engaging students in reading the written literature surrounding a scientific problem or topic is a compelling way to engage them with this knowledge work. Schwab made a similar call for engaging more advanced science education students with products from the research literature (Schwab, 1978).

In reading about the research recounted by, and in some cases conducted by, Norris and Phillips, I was reminded of the early anthropological work at the Scripps Institute conducted by Latour and Woolgar (1979/1986). In that ethnographic research in a lab, Latour and Woolgar end up providing a detailed account of the material and social resources that are put into a complex orchestration in order to support the *inscriptional work* of the scientific team. The inscriptional work of the lab was identified to be the core intellectual activity of the team as defined by the focus of their activities. This inscriptional work included close engagement with the existing literature, the translation of data into appropriate data representations for use in publication, the authoring of research studies, interactions with editors, etc. The situation they describe involved significant individual specialization and collective activity, ultimately with the PI in position of pulling together the final articles. But the focus on inscriptional work – both in terms of reading the literature carefully and contributing to it through writing – is an image of scientific practice I believe is important to consider. It presents a relatively unique and generative image for education.

FRAMING STUDENTS' SCIENCE INQUIRY AS INSCRIPTIONAL WORK

"Science education must be premised upon an essential role of text in science" (Norris and Phillips, this volume). What would it mean if we were to systematically frame students' science inquiry as pedagogically-appropriate inscriptional work? Being able to do so by engaging them sensibly with the research literature and the popular science literature would have the important advantage of helping students see nascent lines of inquiry launched around a problem, accumulations of systematic data pieced together across studies, evolving knowledge of the topic over time, and policy implications of the work. These are dimensions of the nature of scientific work that are more difficult to make visible around decontextualized observation and experimentation.

Norris and Phillips frame students' interpretation of 'text' broadly to include whatever is meant to be read – graphs, charts, tables, diagrams, equations, text, and so on. In this broad sense, existing sustained inquiry project approaches can be interpreted as being quite congruent with a 'reading as inquiry' approach. There are many possible examples that could be mentioned. Palincsar and Magnusson (2001) engaged early elementary school students with notebooks written by a fictionalized scientist describing her experimental inquiry. Engaging with these notebooks allowed students to learn deeply about the specific inquiry and associated concepts and analytical techniques second-hand. They refer to this approach as "second hand investigations" for that reason. It offers a clear alternative to traditional textbook approaches while still fitting with that genre of instruction.

There are also sustained inquiry approaches that engage students with the research literature that surrounds an historical or contemporary research topic. In our research we backed into a focus on engaging students with accounts of the research literature. In 1993 we became engaged in studying the role of the Internet in the science classroom, and we had adopted a pedagogical stance of thinking of online resources as potential pieces of scientific evidence for students to explore (Bell, Davis & Linn, 1995). This approach evolved to focus on scaffolding students in explorations of contemporary scientific controversy (Bell, 2004b). I was intrigued to read about Norris and Phillips' account of a simple view of reading and 'certainty bias.' It seems to be epiphenomenal for personal epistemology patterns we've seen in students. As described by Norris and Phillips, students are inundated with texts best described as in keeping with a *rhetoric of conclusions* approach to knowledge claims (cf. Schwab, 1962). They originally approached our materials in a similar manner, and we found that we needed to cultivate a classroom culture of evidentiary inquiry over the course of weeks and months in order for students to engage in evidence interpretation and argument construction (Bell, 2004a). Given the inherent complexity of much of the scientific research literature, we found it was necessarily to craft intermediate versions of them – with appropriate simplifications and educational elaborations – for use with students (Bell, 2004b). Our approach of using a web-based learning environment as a learning partner allows us to scaffold students' process. Specifically, we prompt for interpretation of evidence, elaboration of thinking, and metacognitive reflection during the learning process. We have connected such instruction to positive shifts in student's epistemological understanding of science (Bell & Linn, 2002); they come to see scientific knowledge as something that changes over time, the enterprise being more evidentiary in nature and how the social deliberation process in the classroom can help explore a broader range of scientific ideas.

What do students gain by reading the existing research literature in science? It certainly provides students with more direct access to the knowledge work of science – in its uncertainty and in its full rhetorical envelope. If pedagogical issues have been dealt with, it can also allow students to learn about contemporary science topics and forms of inquiry. But perhaps more importantly, students get an opportunity to see that they are joining in on academic conversations about topics and issues that have spanned years and decades. All too often, that historical and sociological dimension of academic knowledge is not visible to non-specialists –

and yet it is a defining feature of the intellectual work (cf. Bruner, 1996). Students rarely get to see how scientific understanding refines over time, let alone come to understand that the communal level is a central feature of the enterprise.

REFLECTIONS ON "READING AS INQUIRY"

In order to pursue a "reading as inquiry" approach, I believe there are a few different issues that require additional clarity. First, it was not clear in Norris and Phillips how reading fits into the broader cloth of students' inquiry. The paper and the cited studies seem to frame "reading as inquiry" as an isolated instructional activity. The *inscriptional work* frame broadens things to include constructive acts associated with producing text (admittedly, a perspective telegraphed by Norris and Phillips in their piece). More broadly, we need to consider how inscriptional work fits with other modes of student inquiry so that we can design appropriate curricular sequences.

Second, in a related fashion we need to carefully consider the kinds of texts that students might be asked to read and clarify the epistemological, cognitive, and pedagogical demands associated with those texts. Engaging students with popular science articles is likely to be different than engaging them with data-based primary source research articles (or pedagogically-appropriate, isomorphic adaptations). They are likely to be textured differently in terms of the nature of the embedded truth claims. They likely serve different pedagogical roles in terms of content. Scientific policy papers can make the personal relevance of science more accessible.

Third, Norris and Phillips outline research and an argument that presumes an individual unit of analysis. At the end of the article they turn briefly to a discussion of social practices surrounding text in science. We need a parallel consideration of the social practices that would surround uses of text in the science classroom. Many of the participant structures developed through sustained inquiry projects would be relevant here, although taking a broader look at social processes parallel those in the sciences would also make sense. For example, it would be powerful for students to understand the central role of peer review in the shaping and accumulation of knowledge.

Norris and Phillips outline a complex set of argumentative forms and metalinguistic terms (e.g., cause, effect, observe, hypothesis, explanation, etc.) found in scientific texts. Inquiry-focused instructional efforts often spend significant time helping students develop an understanding of specific metalinguistic terms (Bell, 2004a; Herrenkohl, Palincsar, DeWater & Kawasaki, 1999). However, given that such efforts tend to be focused on specific subject matter topics, we tend to ask students to shoulder the burden of how such argument forms and metalinguistic terms show up in epistemologically distinct ways across different parts of the school day (e.g., the meaning of evidence in math vs. fields in science vs. history) (Stevens, Wineburg, Herrenkohl & Bell, in press).

I believe we need to broaden our view of scientific inquiry to include appropriate forms of inscriptional work associated with the scientific enterprise. Such a move will increase the epistemological fidelity between scientific practice and science instruction. It has the potential to help students understand the nature and substance of contemporary science.

Philip Bell
Cognitive Studies in Education
University of Washington

DREW GITOMER

PANEL B

I'm going to try to close this section by being a provocateur in some ways. I want to thank all the presenters and commentators today. They raised some huge issues, particularly policy issues. I want to focus on the policy issues because one of the key themes throughout this entire session is the fact that people are very committed to making educational change. And doing that is not just an intellectual exercise. It has significant policy implications. If we don't have policies that are consistent with the kinds of changes that are being advocated, those changes really won't make it into the schools. As soon as you're in a single school, you're in a political context. When you're in a district, it's even more so, or state or nationally. And so we didn't talk a lot about that, but I thought I would raise those issues and, as I said, try to be deliberately provocative, provocative to the extent that I will even talk about some of the virtues of science fairs, the scientific method in textbooks. I'll get to that.

I really want to talk about two issues. One is something that came up very significantly. That is: What are we doing here? Are we trying to create an educational environment to create scientists, or for all children? I want to suggest that that's a false choice; that, in fact, if we look at the current status of our education, we're not doing too hot a job on either end. Further, I don't think there are any really strong examples of us making a commitment to one or the other and being particularly successful. That's in any discipline.

I think our challenge is to really design programs for all that provide a breaking-off point for those who want to and can go more in depth into a science preparatory pursuit. What that means, this program for all can't trivialize or discourage; it needs to encourage people to be engaged in science. There are significant issues that we need to explore. When do you start making the break? What's the nature of that break? Also, I think there's also a research imperative to do ongoing data collection on how we're meeting those simultaneous goals. That also has implications for what we measure, rate progress with respect to progress on those particular goals.

I'm encouraged by our discussion of these immersion units, problem-based solving, that these, in fact, are contexts that can provide a science for all and yet encourage further pursuit for some subset of the population. What I want to focus my remarks on, though, is the implications for what I think is at least an emerging consensus, is this move towards an immersion problem-based kind of learning. I think that's a theme that really came forward today, that people are ready to embrace that as a consensus. I'm so dated, maybe we have that consensus.

Policy in general is largely about resource allocation. It's choosing from a range of options and possibilities. Danny Edelson talked about his journey, technology, curriculum and professional development. One of those issues that I

Richard A. Duschl and Richard E. Grandy (eds.), Teaching Scientific Inquiry: Recommendations for Research and Implementation, 268–283.
© 2008 *Sense Publishers. All rights reserved.*

thought about is: Where do we put our dollars and efforts predominately? A lot of the discussion seems to suggest that it's really in the professional development arena. But let me go a little further. I think there are two very different paths that policymakers have pursued and will pursue and really underlie our current debate. One is that we develop highly controlled curricula, which puts a low priority on professional development. It accepts the notion that we have teachers with limited resources, schools with limited resources. So let's have highly prescriptive curricula. That's certainly on the table.

But what is suggested in the discussion of the last couple of days is an endorsement of the very rich curriculum with significant and costly professional development that's required. Now, if we, in fact, endorse that, that has very significant policy implications. What I want to suggest and I'm not sure it's realizable, but I think it's provocative is that we need to target a small number of agreed problem context immersion units that really form the basis of a curriculum. I say that because there is a tremendous cost intellectually and economically to develop any one of these units. That cost is in the conceptual work to design the units, the challenges to understanding developmental progressions, the kinds of models that are developed, the misconceptions. We heard about all of those things today. Questioning strategies that need to be developed. What's the nature of science understandings that we want to support? What are the kinds of branching that might occur and what to do when those opportunities to branch out occur. When Rick presented the characteristics of these units, one of those is that professional development is very specific to the particular unit. [*Note: the immersion unit presentation is not currently in the volume – we might add the characteristic here as a table*] All of that suggests that in a world of constrained resources and constrained talent I wonder, why don't we focus on maximizing the potential of a strongly conceptualized set of units and reduce the overhead of curriculum development and really ultimately put the resources into the professional development that can lead to successful implementation of these units on a larger scale.

And so here's my sort of virtue of textbooks. They, in fact, are a model of economy of scale. Now, it may not be the content we want; but, in fact, we have lots of kids doing the same sort of thing. We want them doing different things. But there's a lesson there. So there's this economy of scale issue. But I want to argue that there's actually something else that's important about these immersion units and having a smaller number of them. And that is creating a set of common activity structures and provide the shared experience across students for what science is.

Here's my endorsement of the scientific method. That is a very familiar activity structure to kids across this country. That is what they think of as science. Okay, we don't want them to think of science in that way. What we do want them to think about it is in terms of some activity structures that they can recognize and that they can talk about even if they're in different schools. That they have common experience, and that common experience can't just be what we have now.

What I hope and what I've heard from descriptions of the success that people have had is that these activity structures also have the potential to create a motivation and identification with science in the same way that Joe Krajick talks about the young kids playing soccer. One of the things missing in the discussion – one of the things that was obviously present was whether we considered kids as doing science all throughout. There was never any discussion of kids thinking of themselves as scientists. Ultimately, we've got to grapple with that. That kid, that 7-year-old kid running around that soccer field – it's painful to watch, maybe, unless it's your kid – they think of themselves as athletes. That kid that goes to the piano recital and plays "Twinkle, Twinkle, Little Star," they think of themselves as musicians. We've got to get to a point where – it's not the full form, it's a minimal level in a lot of ways. But the key is that children think of themselves, they identify themselves with that culture at a very early age. That's why, as Danny Edelson pointed out, these kids will engage in reasoning about who's a better baseball player or a better athlete. It's because they have a psychosocial identification with something that starts at a pretty early age. We've got to start understanding how to cultivate those things. These units, I think, if they're common and shared, can start providing some of that possibility.

One of the things that didn't come up in the two days that's absolutely essential – I know folks are working in those environments, but we certainly are not doing a very good job of creating that sort of identity and motivation for large segments of our population. If we're going to have a successful society, that absolutely has to occur. One of the things I think is important about this soccer game or this little recital is that there are these core activities that are recognizable in the world of experts. The expert, they play their concerts; the professional soccer player plays their soccer game. I think immersion units have some of these core features that are what scientists do, albeit in a simplified, incomplete manner; but there's something identifiable that connects their practice to the practice of the professional.

The other thing I would say these commonstructures provide is an ability to make meaningful assessments of how people are doing. We talked a little bit about this yesterday. But in the current structure where everybody is doing different things, any sort of assessment is necessarily going to be impoverished because you can't really deal with the content issues. But if people are focused on similar kinds of core concepts, similar issues, we can find out how kids are doing around the country and make adjustments based on that information, and it's not such limited kinds of information that we have right now.

Then I want to suggest, finally, that if we could go to a model where we had a more constrained set of units that we all were focused on, we really would have research platforms where we could explore fundamental issues that have to be explored. Those are the things like sequencing, like interest, like curriculum design, professional development, questioning areas, all the things – there's lots of fields in science, in fact, where you do have common platforms upon which research from all sorts of universities and research labs actually study interesting questions. Right now we're in a place where the researchers largely do their own development and then do the research within that context. There's very little cross pollination and cross conversation. So you need to understand the person's environment and then the questions they're looking at. I think there comes a point where we actually may have a much more robust field if many of us are focused on these common platforms. I think it ultimately gives the research community ability to generalize about effective science education and to better understand the phenomena that underlie successful learning and instruction.

Now, there's a huge amount of work to get to a place where we would be comfortable with having some set of activities, and I would hope that the group in this room would play a major role in actually helping to shape that kind of agenda. But if we're going to be successful, I don't see any way but moving to something that has more of a comprising nature and a synthesis of the good work that's going on here.

So thank you.

Drew Gitomer
National Opinion Research Center
University of Chicago

CINDY E. HMELO-SILVER

PANEL B

I'm going to continue the bid for consensus. Part of the place where I come from is work on problem-based learning (PBL). As I was reading through many of the papers and listening to many of the presentations it seemed like problem-based learning to me.

The version of Problem-based learning I start with had its origins in medical schools (Barrows, 2000). It starts out with a problem scenario in which students try to figure out what's going on with a patient. They try to identify the facts of the problems and generate ideas about what they think is going on, what might be potential solutions. Their goal is to construct a causal explanation as they learn the basic biomedical sciences underlying the patient problem.

So problem scenarios sounds like something Danny Edelson was talking about and what Rick Duschl was talking about in immersion units. The expectation is that problem- or scenario- based learning approaches will generate the need for knowledge. You start getting at motivation as students begin to identify their knowledge deficiencies, and to help scaffold the process, there are different kinds of inscriptional devices that students and teachers use. In all these approaches, students then engage in some kind of inquiry. In the problem-based learning community, researchers talk about self-directed learning, which could involve using text-based resources, experimentation or modeling, to give a few examples. Next, learners apply this new knowledge to the problem they are addressing.

This sounds like what a lot of people have been talking about today. There may be some differences in some forms, but overall a lot of the functions seem very similar. I think problem-based learning doesn't say that you never give an explanation, (there are times for reading text) and the role of the teacher is pretty important. I had a recent paper on problem-based learning where I was trying to talk about problem-based learning and anchored instruction and project-based science (Hmelo-Silver, 2004). The journal editors were asking me to explain the differences. I don't think the differences are big. But they said tell us the differences and distinguish between the differences. And I don't know how much of that is splitting hairs.

In problem-based learning you try to have some kind of realistic problem that may have multiple solutions. In anchored instruction you have these complex narratives, often video-based. In project-based science you have driving questions. These represent examples of immersion units which provide opportunities for extended engagement with a scientific problem or question. In all of these the focus is on learning strategies and content in an integrated way. These approaches provide a shared context so that students can actually negotiate the knowledge as they work in groups. Often, they lead to producing an artifact.

There are also some differences. With the version of problem-based learning that I originally studied, we used a generic white board where students wrote down facts of the medical case, their ideas about solutions and the things they needed to learn. As I've moved this into work with pre-service teachers, I've begun to develop domain-specific scaffolding for instructional planning, in which we try to encourage teachers to think about what are the big ideas in a unit (Derry & Hmelo-Silver, 2005; Hmelo-Silver et al., 2005). We ask what would be evidence of that kind of understanding, and then what kind of activities they would use, using a backward design model (Wiggins & McTighe, 1998). So we've tried to scaffold that more specifically in the structure of the online whiteboards and personal notebooks. I think some of the work Nancy Nersessian is doing in biomedical engineering is similar as they're also trying to figure out how to do that domain-specific scaffolding.

One of the things we didn't talk a lot about today is scaffolding learning processes. We talked about having kids explain things, but we haven't talked a lot about how the teacher facilitates that. But implicit in lots of these models is the teacher as facilitator. Perhaps they're to provide some information when it's needed or asking open-ended questions, always working with the kid's ideas. That seems, also, again, pretty similar across what people have talked about. I think working with the kids' ideas connects to what David Hammer was talking about in terms of the kinds of resources people bring to the situation and working with those resources and using them as a starting point for both inquiry and further explanation and knowledge construction.

I think where things might also be different and they seem pretty subtle – is with different kinds of inscriptional devices. In "traditional" PBL, there is the generic whiteboard. There are computer-based tools used in other approaches. Danny Edelson talked about how he adapted tools to support the novice's use of professional kinds of tools. Joe Krajcik's work certainly makes use of computer-based tools to support learning – he talked a little bit about the modeling tools but also tools for planning and data collection. These tools serve both to scaffold the learners and provide a focus for their collaborative knowledge construction.

I suspect the tool differences are related to the developmental differences in the groups of kids that we work with and the size of the classes we're working with. As a teacher I can facilitate problem-based learning very well in a group of five without the cool technology. But if I've got seven groups working at the same time – (Joe's laughing) I need some help!

My take on today's presentations is that it looks like we have a decent amount of consensus here. But there are also differences. In one of the side conversations during a break today, I was talking to Danny Edelson and saying that I don't think the differences are that important. I think the similarities are more important than the differences. I've had arguments with people in the past who have talked about the pure versions of problem-based learning, whatever that means. To some degree I think we have to engage in some inquiry ourselves. We need to examine our terms for separating what we're calling these immersion environments or scenario-based environments or problem-based/project-based environments. We need to figure out what are the critical pieces, what do different components buy us and what are the affordances of having kids engage in different kinds of practices.

In my classes, when we start doing problems, most of our inquiry is text based. We spent a lot of time early on talking about how you know that that particular resources are any good at all and what are criteria for evaluating resources (e.g., being both useful and reliable, being aware of biases, etc) and engaging in discussions of just those kinds of issues. Pre-service teachers don't necessarily have a better understanding of this than the kids. I think we need to think about the pieces that are important. We were talking about the kinds of design decisions that were made. Which of those decisions are really critical? For example, consider extended units – is it a week, is it a month, or is it something that happens once a week over a whole year. How do we scaffold students going through different kinds of inquiry cycles?

My call would be trying to understand the critical similarities and the critical differences. I think it's partly a call for using some similar language instead of inventing a new term to describe each new environment that we come up with. Whichever word we all like, it may mean the same thing instead of different things. I think important issues for both research and practice are to understand what the crucial pieces are and how we can help teachers understand them and adapt them to their practice.

Cindy E. Hmelo-Silver
Graduate School of Education
Rutgers University

EUGENIA ETKINA

PANEL B

Our panel is called policy and practice for educational research. I thought I would talk about each of them a little bit but in a different order. I'll start with practice, then go to research and then to policy. I do so because I think policy should be based on research and research should be based on practice t because we need to study what we're doing.

I've identified three key issues in what we're talking about for two days. One is that, if we're going to do inquiry, it has to be supported by some curriculum materials. Secondly, if we want students to learn something, we need to assess them on that. So if we're going to do inquiry, then what is the assessment that we're going to implement. Third, if we want our students to be immersed in inquiry and want to assess them, then the teacher should definitely be prepared to do both. So there are three parts and then each part breaks into three others. So there will be nine units. Curriculum materials. The current situation is either read the text and tell me what it says, or here's the kit, FOSS (Full Option Science Study) kits. I know FOSS kits because I've trained many teachers to use many FOSS kits. FOSS kits, for example, don't have any textbooks with them. People who train teachers to use FOSS kits, typically don't train them in the content; they just tell them what part to take out first, what's second, then watch the video and follow the directions. So teachers don't understand the content in the kits. There is no inquiry, it's just going through the motions.

I was training middle school teachers to teach electronics in which they had capacitors, diodes and transistors; and none of those teachers ever had a physics course. So the reality is that inquiry is pretty difficult to organize. Are there ways to do it? We're developing a learning system that is based on sound scientific practices. For scientists to come up with ideas, they have to struggle with lots of information and data; some of it conflicting. And they come up with multiple explanations. If our students go through what scientists do, it will take long long time to develop conceptual understanding.

So how can we shorten it? We have an idea that one way to shorten it is to give learners something to observe that is very clear, very clean; it's not messy, as what scientists deal with. And then occasionally give them something that is not clean. Then they can come up with multiple explanations. But most often the data they collect or what they observe leads to something we want them to construct. They still need to test it. They still need to make predictions based on it. These few times when learners do come up with multiple explanations, they're all testable. But if we want them to construct the idea of two charges, we give them something to observe that is easier than lifting papers with the comb rubbed with your hair. Because this is a very complicated phenomenon. Instead we give learners two identical rods rubbed with two identical materials and then look at how the rods interact with the material that it's rubbed with. These we feel are clean observations and the idea is clean. But then there are some others where the idea is not clean. So this is one curriculum approach to take.

Assessment. In physics education research, the Force Concept Inventory (FCI) by David Hestenes and colleagues assesses student's conceptual understanding on force and motion. Since it was developed about twelve years ago it has been widely used to measure learning. The FCI showed that most college students in the introductory physics course didn't learn much. They could do complex problems, but they could not answer elementary questions such as if a mosquito hits the windshield of your car, then who is stronger, the mosquito or the car?

And then the students would ask in a mechanicals course, Professor, how do you want me to answer, how you taught us or how I really think?

So this assessment FORCE inventory showed the students do not understand basic concepts. I'm not sure we really want to focus on these concepts. Maybe we need a different assessment. Can we really think and approach things critically? This is what we were talking about yesterday, the whole day: What do we want our students to learn? Do we want them to be scientists or do we want them to be consumers? I don't like the word "consumers" because I don't like consumers, but I'd like them to be critical citizens so they can make decisions.

So if students develop an ability to say, well, this is my solution to the problem and here are the assumptions I've made. How do we assess what did they assume and what they think are the uncertainties? When we assess them on how well they understand whatever experiment they do with the data they collect and analyze, there are some assumptions. There is no instrument that does it. Are there scientific abilities we want them to develop? If we had an instrument that assessed it, then maybe, together with other conceptual tests, we could combine them and assess not only their knowledge but the abilities we want them to acquire. And so we're working on developing rubrics for assessing scientific abilities and using them in introductory labs. And we give them to the students for self-assessment so they can see how we want them to think.

The labs don't tell them what to do, but they tell them what to worry about. Did you draw a picture? Did you list your assumptions? Did you think about how your assumptions will affect the result of your experiment? Did you try to minimize uncertainties? Where do uncertainties come from?

So if we imagine, we develop these rubrics and we validate them, then they can become part of the national assessment. Instead of giving multiple choice tests, we can give students experiments to design, articles to read. Analyze the argument. If we have a rubric to assess their reasoning or their experimental ability, then it will be completely different teaching. So why don't we create a collection of these articles for different ages that students can read and rubrics that will assess their interpretation of these articles. This will be the assessment that goes into policy.

Now, the last thing is teacher preparation. Elementary teachers have one methods course. They usually have two science classes in college. You know from what I told you before how much science they learn in their science classes. I work with graduate students who are going to be physics teachers. They all have an undergraduate physics or chemistry or engineering degree.

They all went through the same classes, so they're still struggling with mosquitos and windshields. So their content knowledge not only is very fragile and superficial, they did not acquire it in the mode that we want them to teach. If they only have one methods course, what are the chances that they're going to relearn everything they learned and start teaching differently?

Maybe we need to create a sequence of courses, a much deeper approach, so our teachers can develop what's called PCK, pedagogical content knowledge. We need to go out and observe these teachers because there are claims being made that the number of content courses after six doesn't matter. (References??) And, that whether the teacher preparation program either is high quality or is alternate route doesn't matter on the quality of teaching. These are the kind of reports that are moving us to the alternate route. You don't need regular teacher preparation program; they can just start teaching and then take a couple of courses. I don't know where you live, but it's a serious thing in New Jersey. And it is getting more and more serious. So the question is how many courses of what quality are really sufficient? If we create a really superb teacher preparation program, can we follow into the classroom and see that the teachers really doing the things that we want them to do?

Are there instruments? In fact, there are instruments that you can take into the classroom, observe the teacher teaching and give the teacher some scores or some inquiry elements that we were talking about., The instrument called Reform Teaching Observation Protocol was developed in Arizona and validated How many people know about it? How many policymakers are going to use it to make decisions about the quality of teacher preparation programs?

So these are questions that connect practice, research and policy. I'm really passionate about it because our school of education is going through the process of accreditation. On top of that or before that, we need to have state approval for our teacher preparation programs. The matrices we need to fill out, with all the skills that our teachers should possess, are amazing. Furthermore, we need to demonstrate that they will do it when they graduate. We are responsible for what they do after they graduate. So because of this, I really want the policy to be informed by research. But research should be based on practice. I think what we heard here ties it all together. Thank you very much.

Eugenia Etkina
Graduate School of Education
Rutgers University

MARK WINDSCHITL

PANEL B

Part of the charge of our panel is to try to connect what we've been talking about over the last couple of days and move it into the practice and policy arena. I want to extend our conversation to a group of people that have really not been talked about much at all or just maybe in passing. These are the teachers. They are the fulcrum of any reform policies; they're the people who enact it. We haven't talked enough about their role in all of this.

I get to work with pre-service teachers in a two-quarter methods course at the University of Washington. I'm really interested in these people because they're in a transitional phase. They've just come off of forty classes in undergraduate science based on a quarter system, and they're coming to me and they're trying to make the transition into designing learning environments for young people. I always wonder, what does forty courses in science buy you? They spend a lot of time in science classes at the undergrad level, let me just say it that way.

I was wondering, what if by some miracle we did come to some consensus about inquiry and what it means? Would we then be undertaking a massive enlightenment project for teachers, illuminating their minds about the kinds of inquiry they ought to use in the classroom? I don't think people would disagree with me if I said the answer was "no" and "no." The reason is twofold. The first is because teachers already have durable beliefs and conceptions about inquiry, durable but fragmented. David Hammer and I were talking earlier about how this durable but fragmented idea about teachers' conceptions maps onto his manifold resources theory. I'll draw a better connection in a moment here. But the pre-service teachers already have existing conceptions, and we probably will have to facilitate a massive conceptual change program with teachers. It's not going to be a head-filling model. The other reason it's not an enlightenment project is because what inquiry "means" is not just in the heads of people; it's embedded in a self-reinforcing cultural system that supports methodologically and epistemologically simplistic versions of inquiry.

This diagram is an example of all the influences that impact what it means to do science.

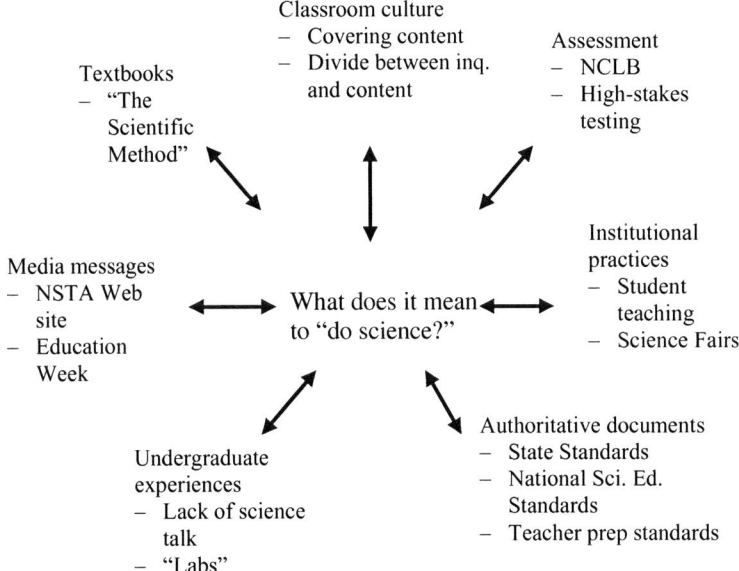

Figure 1.

I know that everybody in here can point to different places on this diagram and say,"Oh, I've got a story for you about how that really has shaped some teacher's thinking about inquiry that I know". You'll get a chance to look at it again. at the end of this talk.

I've written before about the possibility of a folk theory of doing science, and I draw this from cognitive anthropology. Folk theory is not the most rigorous analytical tool with which to look at individual conceptions about inquiry; however, if we use it as a lens to take a look at the layers and scope of the challenge that faces us, it's useful for that purpose.

Folk theory is a kind of culturally constructed common sense. It's not technically a cognitive organization, but it's a set of operating strategies for using knowledge in the world. It could be considered a set of shortcuts or idealizations, simplifying paradigms − about inquiry in this case − that work well together. But it differs from schemas, in-the-head schemas, because it's distributed across and includes social and institutional practices and material artifacts, authoritative documents, media messages and more.

Going back to the teachers that I work with, I'm very interested in how they understand what it means to do science and inquiry. My philosophy is, the only way to really find out in an authentic sense is to have them do it and study them intensively as they do it. Every year in my methods course I have my pre-service teachers do an independent project. To me, inquiry in this sense means they have to pose a question and develop a hypothesis and collect the firsthand data to answer that. Then they have to argue their case in front of their peers about their findings.

Like David Hammer, I have a good news/bad news scenario about how these pre-service teachers understand empirical hypothesis testing. The good news is they can do school science really well. They follow the scientific method. They really struggle with developing questions because they've never been asked to – only 20% of my students over eight years has reported that they've been able to ask their own question in a science class and test it out. Only one out of five! So they have trouble developing the question. But after that, they're very creative at designing a study, do a great job of collecting data, analyze it very well, represent it in graphical form. They do all of that without any trouble. But they have real epistemological problems, what I call "front-end" and "back-end" problems.

On the front end they pose questions like you see here: "Will my philodendron prefer Mozart of Pearl Jam?" What I mean by that, is they do not understand how questions are derived from tentative models. At the outset of an investigation, they pull questions out of thin air or just out of interest. This creates a problem because they're not proposing a model and they're not proposing to test a piece of a larger model. Another problem with this is it reinforces this artificial separation between understanding content and doing inquiry. The goal of *their* inquiries is to find empirical regularity. Those are the front end problems.

The back end problems, their goals are to find differences between groups or to find correlations in the data. That becomes the epistemological end in itself. They use relation-based reasoning rather than model-based reasoning. The data is never used as evidence for underlying mechanisms in a tentative model. So they've got front- and back-end problems. In the course of doing their inquiries, they move from the left-hand conception to the right-hand conception (in diagram below). The left hand is the traditional scientific method. In the journals they keep, they go through minor shifts in thinking. The right-hand diagram is actually not much different than the left. It just means there are iterations. You might get the data analysis and find out you've got to go back and collect different data. That's the only difference between the left- and the right-hand model.

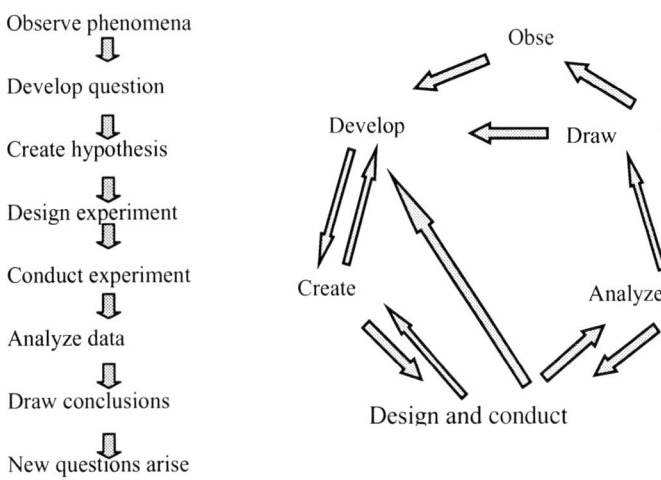

Observe phenomena
⇩
Develop question
⇩
Create hypothesis
⇩
Design experiment
⇩
Conduct experiment
⇩
Analyze data
⇩
Draw conclusions
⇩
New questions arise

Figure 2.

Most recently I've tried to force my students to integrate models into their inquiry and see how that works. They're now required to develop a study in which they have to specifically test a model. They have to go read four college-level research articles on the model of interest, the phenomena of interest, and they have to develop a tentative model. They do their study, and they're required at the end to take the empirical findings and argue for some underlying mechanism, for or against it. Well, what happened then? They generally thought of a model as a product that always trails scientific investigation. Consistent with the research, they do not see models as tools to generate ideas or questions. So this is a big conceptual change thing we would have to work on with teachers.

Only two out of twenty-one were able to develop a model, test it and argue for underlying mechanisms. It's interesting how the other 19 construed their arguments. At the end of their data analysis all, they would use what I call a "method-directed" argument. They would state the purpose of their argument was to justify their data analysis and the precision with which they collected the data. They were trying to justify their correlations or justify the T-value for differences between two groups. They didn't use any "theory-directed" argument, which means taking those correlations and then going and arguing for or against some theoretical or underlying mechanism in the original model. Only two out of twenty-one engaged in this theory-directed mode of argument.

Interestingly, they all staunchly defended the separation between the scientific method and modeling. They saw those two as really separate enterprises. The person who was most adamant about this was a young woman who had won several science fairs at the district, state and national level, I believe. This is a model one student came up with. This is one of the more impoverished models. This is not an unintelligent student; it's just that they're so unused to the discourse and inscriptional practices and argumentation practices involved with models. It's a different way of thinking to them.

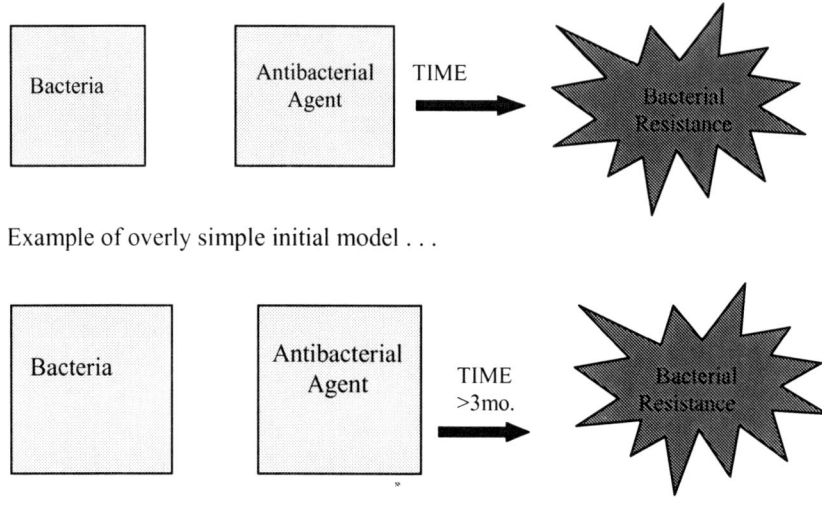

Example of overly simple initial model . . .

. . . and other superficial changes after investigation

Figure 3.

There were some bright spots. I scaffolded model-based discourse throughout the two quarters that we were together, in many different ways separate from the inquiry project the students did. It did have some good offshoots in that they internalized some of that model-based discourse, went into the classroom in their student teaching six months later and started using model-based discourse in their classrooms. I don't mean by that that they simply grafted the word "model" onto the word "concept," if they would have used that in the classroom. For example, one student teacher asked his students to take what they had learned about plate tectonics and the earth's structure as a model and then study another body in the solar system, either a planet or a moon, and create arguments either for or against why that body might have tectonic activity on it. So it's getting kids to understand that they can create models, that they are objects of intellectual critique, that models are created for a certain purpose, there may be multiple models. That, to me, is model-based discourse. So there was some success.

I'm ending with this slide so you can look at it.

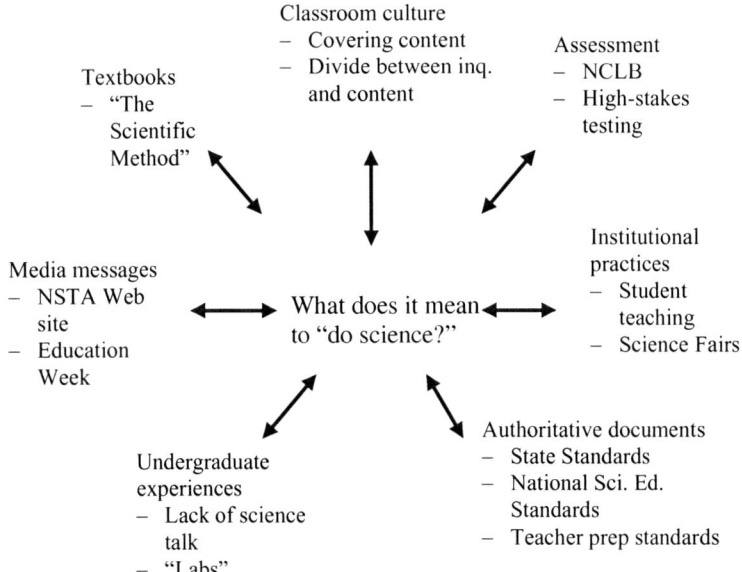

My closing statement is just to remind everybody of two points. Teachers already have existing conceptions about what it means to do science, even though these are fragmented ideas. So it's going to be difficult for us to work with those. Secondly, it's going to be a difficult project for us to help teachers form more robust forms of inquiry because of the folk theory that pervades institutional practices, authoritative documents, classroom culture, textbooks, media messages and so forth.

Mark Windschitl
College of Education
University of Washington

NANCY BRICKHOUSE

WHAT IS INQUIRY? TO WHOM SHOULD IT BE AUTHENTIC?

I suspect the dedication to science teaching/learning as inquiry of the scholars featured here can be traced to a core commitment that has been perhaps most frequently and forcefully articulated by Richard Duschl (with reference to Joseph Schwab) that science learners should understand science not as a rhetoric of conclusions, but as an "enquiry into enquiry." We believe it is insufficient to simply know the scientific canon. We want students to also know how it is we are able to reach agreement on such matters, to be able to recognize the significance and reasons for scientific disagreement, and to be able to use scientific knowledge in a complex and changing world.

What we really want is "authentic science inquiry." In these concluding remarks, I will describe some of what I learned about science from the scholars at this conference during our collective search for the Holy Grail of authentic science inquiry. I'll then conclude by problemetizing our understandings of authenticity and raise questions regarding where we should be looking for our models of science inquiry.

IN SEARCH OF AUTHENTIC SCIENCE INQUIRY

Typically we look to scholars in science studies to help us understand what authentic inquiry is so that we can model our instructional approaches after "authentic" science practices and feel confident that our students are developing important scientific competencies. This has lead to what I think are some real improvements in our understandings of inquiry. For example, Duschl and Grandy, in their introductory piece, explain how our understandings about science need to move away from the hypothetico-deductive views of science to views to include other essential elements, such as model-construction and revision. Norris and Phillips, in their chapter, show the significance of considering reading as a critical aspect of scientists' work – an aspect of inquiry that is too often down-played in its importance. Bell, in his commentary, makes a case for greater attention to the inscriptional work of scientists and how they both read the scientific literature carefully and contribute to it through writing. What is common in these arguments is a desire to expand our understandings of inquiry by examining closely the work of scientists.

One component of inquiry that we spent very little time at this conference talking about was generating good questions.[1] The ability to ask a good question requires substantive knowledge that is only acquired through sustained engagement. I am generally astounded at how often I see students asked to generate a question to research in a field of study they know nothing about.

Richard A. Duschl and Richard E. Grandy (eds.), Teaching Scientific Inquiry: Recommendations for Research and Implementation, 284–303.
© 2008 *Sense Publishers. All rights reserved.*

Most scientific inquiry carried out in schools does not require students to generate their own questions. Even some of the best models of inquiry articulated at this conference do not insist that students work at generating questions. Three studies that have focused on students' questions were all carried out with elementary-aged children in which their teachers were also researchers, and in the context of sustained scientific inquiry where questions were specifically elicited. Gallas (1995), for example, used children's questions as a way of framing the science talks that take place in her classroom. Van Zee, Iwasyk, Kurose, Simpson & Wild (2001) found they were most successful at generating thoughtful questions when children had conversations about familiar contexts in which they had made observations over a considerable period of time. Lucas, Broderick, Lehrer, and Bohanan (under review) structured their inquiry science instruction such that children first made observations and gained experience that would give them a good foundation for developing good questions. They then spent a substantial amount of classroom time helping the children generate their own criteria for evaluating the questions. The rubric they generated gave high priority to asking questions that could actually be answered, i.e., they were researchable and doable. In addition they valued questions that were specific, measurable, and logical. Questions that met these criteria could then be improved on by being generative and by contributing to the knowledge of the classroom as a whole.

What I find most striking in this descriptive study was the way in which the children developed criteria that was not only authentic to them, but that also overlapped in important ways with the practices of professional scientists. In particular, for children to recognize the value of asking questions that are not merely personally relevant, but also relevant to the ideas and practices of the classroom community, seems to me to be remarkable. [2] Mark Windschitl's university student whose research question was "Does my philodendron prefer Def Leppard or Mozart?" appears to not yet have reached the same point as these 6th graders.

IN SEARCH OF MODELS OF SCIENTIFIC INQUIRY

I heartily support these attempts to expand our understandings of inquiry to encompass a wider range of significant forms of scientific activity. However, I also want to push these points a bit further and ask: Should notions of authenticity rest on the closeness of classroom practices to the practices of professional scientists? Or to whom should inquiry be authentic?

In my commentary on Solomon's paper I argued that although this desire for authentic scientific inquiry is understandable, the purposes of science education are not the same as the purposes of science nor should they be the same. In many respects, the desire for science practice in classrooms to mimic professional science is both undesirable and impossible. But does that mean we give up on all notions of authenticity?

Conceptualizing authenticity is a matter that other science educators have taken up. These analyses are generally carried out by examining the practices of scientists. For example, Chinn & Malhotra (2001), taking a cognitive perspective, describe the ways in which scientific reasoning is at odds with the tasks children are given in science at school. Others, taking a socio-cultural perspective, criticize this notion of authenticity and argue for a conception of authenticity that is not located in the task, but rather is located in the interaction between an individual and a scientific community (Barab & Hay, 2001; Carlone & Bowen, 2003; Roth, 1995). In a recent articulation of authenticity Rahm, Miller, Hartley & Moore (2003) draw upon systems theory to argue that "authenticity is no longer taken as being located in the scientists' science, the learner, the task, or the environment, but instead perceived as an emergent property of these components as they interact in a complex manner" (Rahm et al., 2003, p. 738). They argue for two characteristics of an "emergent notion of authenticity: (a) the need for sustained involvement and experiences over time, and (b) the need for ownership of such experiences by the participants (teachers, students, and scientists)."

My query, however, is if we accept that our primary obligation is to educate students to eventually become adults who understand and appreciate science, and are able to critically engage with scientific communities, then why do we so often rely solely on professional scientists as models of scientific practice? Or take enculturation into communities of professional science as our one and only desired goal? Aren't there other communities whose purposes may not necessarily include generating new knowledge, but from whom we might learn more about a possible variety of models of scientific inquiry?

These other forms of inquiry have been studied, but not nearly as intensively as studies of professional science. For example, Layton, Jenkins, MacGill and Davies (1993) describe "workshop science" as places where ordinary people are engaged in solving personally significant problems for which science is potentially useful. They found that formal scientific knowledge needed to be transformed in order to be used in practical settings. Similarly, Aikenhead (2005) found that nurses made frequent judgments regarding the quality of the evidence they collected, but made little use of formal science concepts typically taught in school. Eisenhart and Finkel (1998) have examined the way in which science is employed in a variety of settings that they describe as marginal to mainstream scientific practices. Pamela Lottero-Perdue (2005) has examined the ways in which women in a nursing mothers support group engaged with and critiqued scientific texts about breastfeeding. These women examined funding sources to judge the reliability of studies reported in scientific texts. They also used their scientific knowledge of how breastfeeding works and their own experiences to assess scientific texts. In cognitive anthropology, Kempton, Boster & Hartley (1995) studied cultural models of the environment and how these models shape both reasoning and decisions regarding actions toward the environment. They found widely held models of pollution, the ozone hole, and photosynthesis were misapplied in an attempt to understand global warming. Environmental decisions were also strongly affected by other kinds of values such as religious values and commitments to their children.

How might these studies and others like them inform our understandings of inquiry? Perhaps we need to think in terms of multiple models of inquiry that serve different purposes. In models of inquiry that are focused on generating new knowledge, then the generation of first-hand data may be of considerable importance. In models of inquiry that are more focused on applications, generating first-hand evidence may not be nearly as important as critical reading or understanding how scientific institutions operate.

Much of my work has focused on issues of equity and access to science. As researchers, I believe it is a moral obligation for us to better understand how to shape school science in ways that provides this kind of access to children who must depend on public schools to gain the kind of competence they need to be seen by others as scientifically competent. At the same time, I worry that we will not succeed in authentically engaging all students in science if we focus only on narrow forms of inquiry (Calabrese Barton & Brickhouse, 2006). Furthermore, I worry that we will not support our students in acquiring the full range of scientific practices that they may need to act intelligently in a future dominated by science and technology.

NOTES

[1] Helen Longino was the first to point this out during the conference.
[2] Thanks to Leona Schauble for bringing my attention to this manuscript.

Nancy Brickhouse
School of Education
University of Delaware

GREGORY J. KELLY

CONTINUING CONVERSATIONS REGARDING INQUIRY

There were many issues discussed at the Inquiry conference that invite comment and continued conversation. I shall limit myself to comments on two issues, clarification regarding the social in social epistemology and the views of the nature of science implied in views of inquiry.

First, the argument I made in my contributing paper focused on the social dimensions of epistemology particularly as related to disciplinary knowledge. This focus requires some clarification, specifically in terms of how such a view intersects with individual learners' cognitive frameworks. In my contributing paper I advocate a focus on the epistemic practices associated with producing, communicating, and evaluating knowledge in science and school science. These epistemic practices refer to the specific ways members of a community propose, justify, evaluate, and legitimize knowledge claims within a disciplinary framework. This view may be (mis)interpreted to suggest that social factors trump cognitive considerations and that a social view competes with studies of individual learners' frameworks.

Cognition refers to the mental processes connected with understanding, formulation of beliefs, and acquisition of knowledge (Flew, 1979). As Toulmin (1979) eloquently points out, the physiological processes of our brains are necessarily interior, but the mental processes become part of our inner lives through internalization of processes that are first performed publicly. This suggests a focus on the ways knowledge is constructed socially − processes that are particularly salient for science and science learning. For example, claims in science must pass through rigorous social processes (e.g., the product of adherence to research protocols, articulation in common specialized language, peer review, and so forth) to become certified as knowledge. The justification of a claim is thus not an individual matter, as knowledge is entered into a community through social processes. Nevertheless, scientists maintain their individual judgment and agency (Giere, 1999), while operating in a broader intellectual ecology (Toulmin, 1972). Similarly, in school science, what counts as a contribution to the collective knowledge is decided in a social way − preferably through marshalling and considering evidence, and not arbitrarily by an authority. The view of social in social epistemology thus does not suggest that non-epistemic factors overwhelm the merits of ideas. Rather, the processes of reasoning are themselves social – for example, the articulation of an argument, persuasion with evidence by a peer, evaluation of a research design, and so forth. Furthermore, the normative considerations of Habermas (1990) and Longino (2002), described in my contributing paper, are explicitly aimed at creating communities of knowers that are open to debate, value evidence, and make decisions based on the respective merits of various arguments. Thus, the reasoning involved in acquiring knowledge (i.e., cognition) is (at least partially) social.

I noted in my contributing paper that there is a prevalence of studies concerning epistemology on the ways that individual learners conceptualize knowledge and how these conceptualizations influence learning (Hofer, 2001; Schraw, 2001). I offered a contrasting view that considers how issues about *what counts* as knowledge are considered in social spaces. These two views, while differing, are potentially complementary. Indeed, the ways that individuals conceptualize knowledge may very well be influenced by the experiences they have in social settings where issues of knowledge are considered. There is potential in developing studies that examine how epistemic criteria are established within groups and how these criteria come to influence the learning of students.

The second issue I wish to consider here surfaced at the conference in the form of discussions about inquiry and the nature of science. There was some discussion about whether specific aspects of the nature of science need to be taught explicitly, mirroring a discussion that has been in the literature for a number of years (Akerson, Abd-El-Khalick & Lederman, 2000; Bell, Lederman & Abd-El-Khalick, 1998; Bianchini & Colburn, 2000; Cunningham & Helms, 1998; Palmquist & Finley, 1997; Sandoval & Reiser, 2004). My purpose is not to settle this issue. Rather, I shall discuss the relevance of the explicit nature of the nature of science as related to the educational context used to illustrate my theoretical arguments in my contributing paper (Kelly, this volume). While I do not oppose the teaching of certain aspects of the nature of science in an explicit/reflective way, I will illustrate that such explicit teaching is beside the point for the goals of the university oceanography course described in my contribution to this conference. This is true even though some aspects of the "consensus" position on the nature of science are presumably relevant to the course goals.

The central point of my contributing paper was to identify research directions that take advantage of advances in sociocultural learning theory and are cognizant of the social nature of scientific knowledge and practices. Thus, the paper illustrates how views of epistemology and learning can be mutually informing regarding learning scientific inquiry. As an illustrative example, I described briefly some aspects of a university oceanography course that has been part of my research program for about a decade. The goals of this course are to increase science literacy among the general student population by improving their ability to analyze the basis for scientific claims. They learn how to assess the merits of scientific arguments through a variety of educational experiences including writing their own arguments supported by large-scale geological data sets and critiquing (informally and formally through peer review) arguments of other students. The processes of writing arguments with scientific evidence (drawing on theory, making choices about uses of data, making inferences, and so forth) and engaging in peer review constitute practices that intersect with aspects of what has been characterized in the literature as a "consensus" position regarding the nature of science. For example, Schwartz, Lederman & Crawford (2004, p. 613) describe eight aspects of the nature of science: tentativeness, empirical basis, subjectivity, creativity, sociocultural embeddedness, observation and inference, laws and theories, interdependence of the previous seven aspects. This consensus position in similar forms appears elsewhere (e.g., Abd-El-Khalick & Akerson, 2004; Akerson et al., 2000; Bell, Lederman & Abd-El-Khalick, 2000).

The oceanography students' use and critique of evidence in argument intersects with each of these consensus position aspects of the nature of science. For illustrative purposes, I shall focus on just one, but the argument would hold for any of the first seven. Consider the description of observation and inference:

> Science is based on both observation and inference. Observations are gathered through human senses or extensions of those senses. Inferences are interpretations of those observations. Perspectives of current science and scientists guide both observations and inferences. Multiple perspectives contribute to valid multiple interpretations of observations. (Schwartz et al. 2004, p. 613)

I find this to be sensible definition for a reasonable idea. While a Wittgensteinian analysis may prove in any instance that an observation is an inference of a particular sort requiring a certain degree of acculturation to a social norm, I shall take the definition at face value for now, and ask Wittgenstein to wait for a moment. The question I raise is: What counts as knowing this definition? Schwartz et al. (2004) found that the subjects of their study, preservice secondary science teachers, needed both a context for reflection and active reflection experiences that included "explicit and guided attention" (p. 634) to improve their views of the nature of science as measured primarily by a survey and interviews. This has been shown in many studies (e.g., Abd-El-Khalick & Akerson, 2004; Akerson et al., 2000).

The purpose of the oceanography course is not that the students know that scientists make observations and inferences (*propositional knowledge* about science) but rather than they are able to put their knowledge of theory, observation, and inference to use for the purpose of creating an argument (*procedural knowledge* predicated on knowledge of geological theory, rhetorical and argumentation practices, and mathematical/graphical information). What makes the procedural knowledge reflective is entirely different than what makes statements *about* science reflective. Therefore, the reflection in the oceanography course is tied to a set of activities in the course aimed at engaging students in the processes of learning and applying knowledge – knowledge that may be properly called scientific and epistemological. Thus, the nature of science issues are not separated out into an independent structure, partially because the point is for the students to assess issues of evidence in the context of relevant geological content. The explicit and reflective processes are centered on the creation and assessment of evidence, including discussions about observations and inferences in geology/oceanography, but knowing the lexical definition of each is not the goal. We are not concerned with changing their "views" about science. We are concerned about developing epistemic practices for using evidence to assess the merits of claims in science contexts. This is an activity view rather than a "conceptions" view of knowledge about the nature of science.

Teaching the "observation and inference" aspect of the nature of science independent of the process of making observations is not consistent with the theory of language informing our studies. Now is the time to enter Wittgenstein (1958). My view, and that of my co-researchers, is that one learns meaning through multiple experiences with using words in context. While one context might be explicit instruction (within a context for reflection), stating the lexical definition – as assessed through interviews or surveys or other reactive research methods – does not constitute understanding. Our view is that meaning is made through engagement over time in multiple instances of use of a lexicon in relation to a particular social language (Gee, 1999). Our oceanography course aims at meanings as evidenced by use of language in context. For example, understanding "observation" means being able to include appropriate observations in the formulation of a scientific argument. So, while explicit instruction about the seven aspects of the consensus position of the nature of science would probably do no harm, this is simply not the point of the educational experience given the course goals. The point is for students to be able apply their knowledge of science through application to complex socioscientific issues.

My conclusion is that issues of the Nature of Science (for me this ought to be called the natures of the sciences, given the variations across disciplines) neither should be limited to any set of assertions about science nor should pedagogy about science be limited to explicit instruction narrowly construed. Rather, we need to recognize that there are times where explicit, reflective instruction about the propositional content of a view of the nature of science might make sense (as is the case for Schwartz et al., 2004, where the relevant population is prospective teachers), but others times when the pedagogical goals differ, that steering the conversation into a framework of explicit instruction regarding the tenets of the nature of science can be limiting.

Gregory J. Kelly
College of Education
Pennsylvania State University

MARK WINDSCHITL

OUR CHALLENGE IN DISRUPTING POPULAR FOLK THEORIES OF "DOING SCIENCE"

> I was less than a novice at completing full inquiries and have always been given the question to start with...my previous investigations have never included assumptions, predictions, argumentation, or initial theories, and none required me to support my claims, present to peers, or look up other studies. I have now changed my mental model of inquiry which was step-by-step, very orderly, where one step has to be finished before the next step starts...
>
> – quote from the inquiry project journal of a pre-service secondary science teacher who had just completed over 30 undergraduate science courses and a bachelor's degree in earth and space sciences.

Recent scholarship confirms that the disciplinary pursuits of scientists are methodologically and epistemologically more complex than what popular views of school inquiry suggest – views currently dominated by the idea of "The Scientific Method" as an unproblematic and universally applicable protocol (Darden, 1991; Giere, 1991; Kitcher, 1993; Latour, 1999; Rudolph, 2003). Among the fundamental characteristics of discipline-based inquiry rarely incorporated into school science, at any level, is the use of models to generate meaningful questions for investigation and the use of argument to link evidence with the revision of these tentative models. Indeed, these represent entire discourses that are absent from the school science experience.

As new visions are developed around what school science *should* become with regard to links to the discipline and practices that foster deep learning for students, it would be shortsighted if not reckless for us to ignore the impending challenges of helping classroom teachers align their thinking and curriculum with radically different ways of approaching classroom inquiry. We risk much if we consider the moves from research to practice as an "enlightenment project" in which we disseminate new ideas to teachers, expecting that they will find these ideas inherently sensible, potentially fruitful in the classroom, and meritorious simply because they represent practices that are closer to those of real science.

To be honest, no experienced researcher or professional developer is naive enough to think such fundamental change of teacher practice on a national scale can be so easily accomplished. So it will be difficult, but I believe that it is important at this point to understand *why* we face an uphill battle – important because if we seek changes in school science we must appreciate the mechanisms and the contexts that sustain the current status quo. There are two interrelated conditions that constrain our efforts to re-shape school science.

The first condition is that many teachers already operate on closely-held conceptions of what it means to "do science." These models, based principally on the Scientific Method, are defined by aims and activities that work remarkably well in the classroom environment, and if we have learned anything from the conceptual change literature, we know that such utilitarian views are quite resistant to alteration let alone wholesale replacement. The second condition, related to the first, is that teachers are part of a larger cultural system of activity that perpetuates publicly shared conceptions of science. This system includes media portrayals of science, institutional practices such as science fairs, the use of textbooks as de facto curricula and a host of other mutually-reinforcing influences that shape how we all think about science inquiry.

These two conditions can be understood through the lens of "folk theory" – a construct drawn from cognitive anthropology which can help clarify the nature and scope of our task. I employ this perspective because it effectively integrates the intellectual, social, institutional, and material challenges of the move from research to practice. It is a useful theoretical tool to capture the current "meaning of inquiry" from both individual and cultural perspectives.

USING THE IDEA OF "FOLK THEORIES" TO UNDERSTAND SCHOOL SCIENCE PRACTICES

Folk theories (also known as cultural models) are pre-supposed, taken-for-granted theories about the world that are widely shared by most members of a society (although not to the exclusion of alternative models) and that play an enormous role in individuals' understandings of the world and their behavior in it[1] (Boudon, 1986; D'Andrade, 1995; D'Andrade & Strauss, 1992; Holland & Quinn, 1987; Shore, 1996). In everyday life, people form, transform, and operate on "theories" of common constructs such as "bachelor-hood", "child-rearing", or "patriotism." Such personal theories, models, or everyday explanations are largely subconscious, or at least not easily articulated in full detail, and are often incomplete (Gee, 1999). Some aspects of these folk theories reside in individuals' heads while others are shared across people, texts, and other media, and are embodied in various social and educational practices (Hutchins, 1995; Shore, 1996). From a pragmatic standpoint, folk theories help people to "make sense of actions, fathom the goals of others, set goals for action, direct actualization of these goals, and produce verbalizations that may play parts in all the aforementioned projects as well as in interpretation of what has happened" (Holland & Quinn, 1987, p. 373). Furthermore, institutions create forces (authoritative documents, apprenticeships, sanctions, rewards) that ensure repetition, ritualization, and reproduction of many folk theories.

"Doing science" is the subject of folk theory shared by teachers and the entire science education community. The folk theory that goes with inquiry (as practiced by scientists) roughly includes the idea that people conduct investigations in order to "find something out," but that there are different forms of scientific inquiry that are more or less scripted, more or less social, directed to different ends and enacted in different situations. Different "theories" of inquiry then, encapsulate viewpoints on who conducts inquiry, how it unfolds, and for what purposes. Just as with other folk theories, the meaning of science inquiry does not reside in any dictionary and cannot be reduced to symbolic representations in people's heads. Rather, it is situated in specific cultural and disciplinary practices and is continually transformed through these practices. Every day, thousands of science teachers enact their favored models of scientific investigations, and in doing so, reinforce various dimensions of a folk theory of inquiry (or "doing science") as they plan classroom lessons, interact with colleagues, adopt textbooks, talk about their work at conferences, host science fairs, draft local standards for learning, write about their practice for publication, and supervise beginning teachers.

HOW DO TEACHERS THINK ABOUT AND ENACT SCIENTIFIC INVESTIGATIONS?

To first explore the folk theory of inquiry at the individual level, I refer to a series of studies with pre-service secondary science teachers in a methods class I teach (Windschitl, 2001, 2003, 2004). Over the past seven years I have examined the thinking and activity of dozens of aspiring middle and high school science teachers in an attempt to find out how well prepared they are to participate in the discourse of theory, models, evidence, and argument. The studies all revolve around one central feature: The pre-service participants were asked to design and carry out an independent scientific inquiry. My methodological conviction has always been that one can only understand how others understand science by engaging them in authentic activity. Because these studies took place in a methods course, the projects served the dual purpose of immersing participants in a learning experience and providing a context in which to study how these individuals thought about scientific activity. The major data sources for each of the studies were journals that the participants kept of their experiences, participant interviews, course artifacts, and videotapes of final arguments to peers.

The participants in these studies all entered our teacher education program with at least a bachelors degree in an area of science and they graduated with a Masters in Teaching degree. On the average, these teacher candidates had completed over 30 college courses in science and some had research experience either as undergraduates or in careers prior to teaching (although only 20% of students in this course over eight years have reported opportunities to design and conduct independent investigations at any point during their pre-college and college careers!).

Although "inquiry" can have many meanings in science and in classrooms, I require my students to engage in *empirical hypothesis testing* as a broad but fair emulation of what scientists do. They were asked to develop a question, do background study on the phenomena of interest, change the question if necessary, design a study, execute it, and present the findings to their peers. Participants have studied a wide range of phenomena. Some examples include: bird-feeding patterns, how cigarette smoke disperses in a home, how trees at different latitudes change color in the fall, how orange juice oxidizes under different conditions, and the absorption capability of various fabrics based on their fiber structure.

So...what *do* new teacher learn from four years of undergraduate work in science? Generally speaking, participants took inquiry to be more than posing and finding the answer to a question; there were implicit rules and operating strategies that defined their conceptualizations. Their putative folk theory, though influenced by the idea of the Scientific Method, was not entirely synonymous with it. Some facets of this folk theory were congruent with authentic science inquiry (e.g. empirical investigations involve developing questions, designing studies, and collecting and analyzing data; set-backs are to be expected; etc.). Others seemed to represent a limited view of scientific inquiry (e.g. there is a scientific method, although it is not linear; the ultimate goal of inquiry is to determine whether a relationship exists between two variables; etc.). More problematically, several facets were misrepresentations of some of the most fundamental aspects of scientific inquiry (e.g. hypotheses function as guesses about outcomes but are not necessarily part of a larger explanatory framework; background knowledge may provide ideas about what to study, but this knowledge is not in the form of a theory, explanation, or other model; empirically testing relationships and drawing conclusions about these relationships are epistemological "ends-in-themselves"; and, models or theories are optional tools you might use at the end of a study to help explain results).

Almost entirely absent from participants' journals and interviews were references to the epistemological bases of inquiry – talk of claims and arguments, alternative explanations, the development of models of natural phenomena, etc. Most of the participants, for example, based their inquiry questions not on a hypothesized model, but on what seemed interesting, do-able, and novel. One participant wrote in his inquiry journal:

> I am thinking about how noise pollution changes the environment. The effects of loud noise on plant growth/photosynthesis? What about setting up two plants each in the same window, playing music for a length of time each day and measuring changing heights, weights?

Another participant wrote:

> ...we were thinking things up half the time that were measurable and then the other half of the time we were completely shifting our focus to what interested us about pollution in places near our house. Then we would filter these ideas back through the measurability factor and usually we'd have to start again....We went toward having the plants in containers and exposing

them to different types of air pollution. What about cigarette smoke? Fire smoke? And carbon monoxide even!

A third participant who had an extensive background in chemistry wrote: "We're going to bubble [car] exhaust through water and see how acidic it gets over time." The types of thinking reflected by these findings are consistent with what Driver et al. (1996) refer to as *relation-based reasoning*.[2] This is characterized by a focus on correlating variables or finding a linear causal sequence. This typically requires an intervention to seek cause or predict an outcome. The nature of explanation for relation-based reasoning is inductivist and refers to relations between features of a phenomenon which are observable/taken for granted. Alternative variables may be entertained, however, only one relationship is assumed to be true. This form of reasoning was pervasive in the pre-service teachers' inquiries. They employed interventions or planned observations of behavior of phenomena needed to find explanations. These involved identification of influential variables, controlled interventions of phenomena such as fair-testing, and discovering outcomes as related to conditions.

Inquiries framed by relation-based reasoning can short-circuit the process of argumentation about claims. If the goal of the inquiry is merely to support a claim for a relationship between observable variables, the argument could well focus only on the definitions, quality and accuracy of the data (methodological soundness), as well as analysis and data representation. Without considering an underlying theoretical model, the findings could not be used to support an argument for unseen mechanisms.

None of the participants employed the most authentic form of thinking: model-based reasoning. This form of reasoning takes inquiry to be an empirical investigation involving the collection of evidence in order to check, test, or develop a model or theory, or to compare theories. The nature of explanation that follows is that models and theories are conjectural, that explanation involves coherent stories that posit *theoretical* entities, that explanation involves discontinuity between observation and theoretical entities, and that multiple possible models are entertained. Using model-based reasoning, argument not only includes querying data and its analysis, but more importantly questioning aspects of the model that underlie the study such as assumptions made about the model, challenging the coherence of the model, and appeals to alternative explanations. Discourse about the inquiry is framed, in part, by the theorized entities, properties and relationships posited in the model.

Because participants used relation-based rather than model-based reasoning, once a question was formulated their thinking was focused on connections between the different stages of the study and what the final results would show. The procedure was important for its own sake – to test hypotheses about the relationship between observable variables – but it was not conceived as a way to describe, explain, or predict unseen entities.

CAN FOLK THEORY BE CHALLENGED?

Because of the absence of model-based reasoning in the three previous studies, we conducted a fourth in which participants were immersed in modeling discourse throughout the 2-quarter course (Windschitl & Thompson, 2006). Participants were involved in creating and refining models and looking at common curriculum in terms of the models they contained and the types of questions one could use to interrogate models as objects of intellectual critique. For their independent inquiry projects participants were required to develop an initial model of the phenomenon they were studying by synthesizing at least four "college-level" papers on the topic. They were asked to identify which aspect of the model they were going to test, and at the end of the study they were to construct an argument that supported or refuted some theoretical aspect of their original model.

What we found was that these pre-service teachers were able to talk cogently about scientific models that were presented to them (for example, describing the contexts within which they would be valid, the completeness of the models, what would enhance or limit the predictive power of the models, etc.). However, they had tremendous difficulty in developing their *own* models, in particular for grounding their own inquiries. This task appeared to demand skills beyond a generic understanding of the nature and function of models. For example, we found that participants developed one of five different kinds of models to base their inquiries on. Seven participants portrayed their models as *causal systems or event chains,* six participants portrayed their models as *descriptive systems,* four participants expressed their models as *simple hypotheses,* two participants portrayed their models as *a series of statements related to a central topic but the statements themselves were not related to each other,* and two participants drew models not of the phenomena to be investigated, but of *the experimental procedure itself.* Many participants who had sparse "theoretical" models (three or four inter-related concepts) or simple descriptive models included additional relevant information in their final presentations or in their journals about the central phenomena but labeled it separately from the model as "background." This information could have been integrated into the stated models to create richer networks that involved theoretical mechanisms. For example, one participant studied the effects of grapefruit seed extract on bacterial growth. Her model was a flowchart of experimental procedure rather than a set of interrelationships among conceptual entities. In her final presentation, however, she presented as "background information" the results of previous studies that showed how the extract broke down the cell membranes of the bacteria. Clearly she understood something of the underlying mechanism of the anti-bacterial agent, but she did not recognize it as a key component of her model.

Only a fraction of the participants were able to coordinate their initial models with their actual inquiry question. Only two participants both represented unobservable mechanisms in their model and then tested for their effects.

Consequently they were the only ones who could engage in a theory-directed argument about what the observable outcomes suggested of underlying theoretical processes (i.e. employing model-based reasoning).

Perhaps most problematically the "scientific method" remained the dominant procedural framework for their thinking about inquiry – to the exclusion of considering models as the grounding for investigateable questions and object of inquiry's pursuits. Almost all participants wrote in their journals that they were following the "Scientific Method" even though we, as instructors, repeatedly discussed with participants the oversimplification and inauthentic nature of this framework. Indeed several participants explained how the SM and the process of modeling were separate enterprises; even one of the two participants that had used model-based reasoning in her inquiry made such a statement.

The bottom line from all four of our studies is that most pre-service teachers, even those with degrees in science, have robust frameworks for thinking about inquiry that are incongruent with authentic science in some fundamental ways. Figure 1 summarizes this framework.

Facets of doing science described by participants that were *congruent with authentic science inquiry*:

- empirical inquiries involve developing testable questions, designing studies, collecting and analyzing data;
- many data points are necessary to make claims about the differences in two or more comparison groups;
- authentic inquiry often involves changing the question or the design of the study after the study begins;
- procedural and material set-backs are to be expected;
- the lack of conclusive results or results that run counter to original hypothesis cannot be considered a failure.

Facets that seemed to represent a *limited view of scientific inquiry*:

- there is a "scientific method," although it is not linear;
- inquiries are generally synonymous with controlled experiments;
- the ultimate goal of inquiry is to determine whether a relationship exists between two variables;
- comparisons between experimental conditions determine the answer to your inquiry question, however, statistical tests of significance are not part of inquiry.

Facets that were misrepresentations of some of the most fundamental aspects of scientific inquiry:

- hypotheses function as guesses about outcomes, but aren't necessarily part of a larger explanatory framework;
- background knowledge may be used to give you ideas about what to study, but this knowledge is not in the form of a theory or scientific model;
- theory is an optional tool you might use at the end of a study to help explain results;
- one does not use empirical studies to theorize about unseen entities or processes;
- models are primarily to help one understand established scientific ideas;
- models trail empirical inquiry, they are not used to generate ideas that would drive investigations;
- the purpose of "argument" is to support the empirical findings of differences or correlation by citing the appropriateness of study design, the completeness and precision of data collection, and the accuracy of data analysis;
- the scientific method is in most ways distinct from the use of models in its aims and processes.

Figure 1. Elements of pre-service teachers' conceptual frameworks for doing inquiry.

FOLK THEORY AT THE CULTURAL LEVEL

As described earlier, folk theory is not just an "in-the-head" phenomenon, it is embodied in authoritative documents (such as state standards), institutional practices (such as science fairs), academic forums (such as NSTA conference presentations), and media messages (such as radio and TV depictions of science and scientists) that reinforce epistemologically and methodologically simplistic forms of inquiry (see Figure 2).

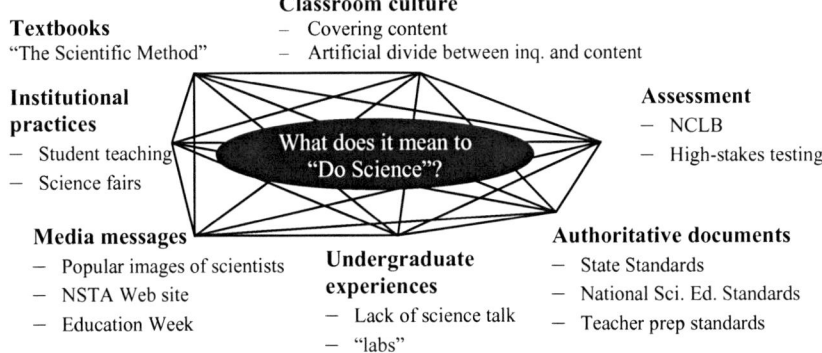

Figure 2. Folk theory as a matrix of mutually reinforcing institutional and social practices, media messages, authoritative documents.

A major influence in the public discourse around inquiry is the science textbook. The empirical investigations described in textbooks bear little resemblance to authentic science. Chinn and Malhotra (2002) examined 468 inquiry tasks in nine textbooks written for upper elementary and middle schools; none of these activities required students to develop their own questions, in only 2% of these activities were students asked to select their own variables, and there were few opportunities to think about controlling variables. They conclude:

> ...the goal of simple inquiry tasks is only to uncover easily observable regularities (e.g. plants grow faster in the light than in the dark), or the salient structures of objects, not to generate theories about underlying mechanisms. In short, the ultimate goal of most authoritative research is the development and revision of theoretical models. The goal of most simple inquiry is a Baconian gathering of facts about the world (p. 187).

These results were similar to an analysis of 90 high school texts by Germann, Haskins, and Auls (1996). In addition to textbooks, official curricula can have a widespread effect on reinforcing particular images of inquiry. The National Curriculum for England and Wales, for example, portrays inquiry as multiple variable problems where students identify which independent variable affects a given dependent variable in a range of physical and biological contexts (Driver, Leach, Millar & Scott, 1996). This view of scientific investigation ignores theoretical models and fails to portray the much wider forms of inquiry scientists undertake.

There are still other kinds of public discourse which influence how people make sense of, enact, and reproduce certain forms of pedagogical inquiry. A recent broadcast by National Public Radio (2002) told the story of a high school biology class that had adopted "an inquiry approach" to learning. The segment opened with sound bytes of students' conversations as they roamed the halls of their school testing different surfaces for bacteria. One of the students remarked that they were "doing a different activity every day" (suggesting that at least some inquiries can take place in the course of one class period). The radio story then cut to the teacher who asserted that inquiry was "allowing students to ask their own questions and giving them the tools to find their own answers." A few moments later, however, a second biology teacher in the school is asked for her opinion; she describes her own more traditional instruction and cites key contrasts with the featured inquiry approach of her colleague. As the story unfolds, inquiry (as opposed to didactic methods) is characterized by the first teacher as a "thinking rather than a memorizing" approach. The second, more traditional colleague, questions the whole notion of "slipping content into inquiry" and claims that content in an inquiry classroom is only addressed when it is "disguised as essays" during assessments. The story seems to be about irreconcilable differences between inquiry instruction and direct teaching methods – inquiry is left characterized as activity-rich but lacking in content.

These media messages, texts, and institutional practices reinforce and are reinforced by teachers' classroom practices. For most science teachers, much of what they have learned about inquiry and models comes from their experiences as undergraduates, however this coursework often consists of confirmatory laboratory activities (Trumbull & Kerr, 1993) and lectures where instructors rarely discuss science as a discipline or how models and theory help generate new knowledge (Bowen & Roth, 1998; King, 1994; Reinvention Center at Stonybrook, 2001; Wenk & Smith, 2004). Many of our nation's science teachers come out of their undergraduate years as virtual strangers to authentic forms of inquiry.

In the case of the pre-service teachers described previously, they connect these limited views of inquiry with what they see in K-12 classrooms. One of my methods students wrote this in her inquiry journal about her observations of the sixth grade classroom in which she would be teaching:

> Today at school the teacher had students start inquiry experiments...They first submitted three questions they could study using plants. The teacher then pointed them towards the ideas that were the most workable. It was really cool to see some of the ideas they came up with. Some students are feeding their plants Coke and others are testing the difference between real and fluorescent light. They are also using dyes and various types of soil – even gravel and sand. One group is testing the effects of music on plants. I was really surprised by the sophistication of their ideas – A lot of questions similar to the ones in our methods class. There was even a plant hanging upside down. It was very cool to see the inquiry method implemented.

To summarize, the examples in this section are not meant to be part of an in-depth analysis of cultural discourses around inquiry teaching; rather, they have been included to provide evidence that different ideas about inquiry exist not only "in the heads" of science teachers, but are codified in authoritative documents, reinforced by textbooks, broadcast in the media, and embodied in the practices of educators. It is easy to see how the folk theory of inquiry comes full circle in the classroom – new teachers are ready to reproduce simple forms of inquiry, reinforced by a lifetime of exposure to culturally pervasive messages, and young students under their guidance begin their own initiation into the rituals of school science.

FINAL THOUGHTS

What puts the "folk" in folk theory is that 1) much of the scientific method is intuitively sensible and 2) it works in the classroom. These alternative conceptions of doing science, however, should not be considered entirely counter-productive. For many teachers, the scientific method is a safe "entry point" into hands-on or even data-based work in the classroom. If not for this scripted recipe, educators especially at the elementary and middle school levels would likely be reluctant to do such work with students. As Keesing (1987) suggests, folk theories are:

> ... a set of operating strategies for using cultural knowledge in the world. They comprise sets of shortcuts, idealizations, and simplifying paradigms that work well together but do not have to all fit together without contradiction into global systems of coherent knowledge (p. 379).

From a cultural reproduction perspective, D'Andrade (1987) adds that models such as the Scientific Method persist because their formulaic and linguistically economic construction signals a kind of wisdom handed down from authoritative figures to practitioners (in this case from scientists to educators). Considered this way, the text of my pre-service teachers' journals describe what could be considered by practicing teachers perfectly acceptable accounts of "doing science." The idea of a self-contained procedure (disconnected from theories or scientific models) with orderly steps and much of the epistemological complexity stripped away is, unfortunately, a useful framework for activity. Directing students in lockstep through a series of well-defined procedures may satisfy the narrow intellectual aims for what passes for inquiry in many classrooms, and this highly prescribed investigative style may be the only form of inquiry seen as manageable in overcrowded classrooms. The scientific method, however, is a procedure, but not a way of thinking. Indeed one can complete the technical aspects of many types of classroom inquiries without thinking beyond the procedures. I say this with some confidence because for years, this is how *I* taught inquiry in *my* middle school classroom.

The folk theories described in this report, although mostly tacit and somewhat fragmented, may have more influence within the science education community than authoritative documents offering guiding principles or prototypical examples of authentic forms of inquiry. As thousands of new science teachers enter classrooms each year, they will likely be influenced by and, in turn, contribute to a folk theory of inquiry. The ideas I have presented here are not meant as an extended polemic about the state of school science. Rather, it is a purposeful articulation of the nature of the challenges we face in advancing the work of educational researchers, learning theorists, and developmentalists from theory to practice. Knowing where we want to go is important but knowing the terrain we must traverse to get there is a well-advised approach to realizing meaningful change.

NOTES

[1] Folk theories share important features with schemas or scripts. These are representations that like folk theories play a dynamic role in guiding expectations and actions and are a shared possession by bearers of a culture. But FT are not tied to physical settings which serve as cues for expectations and behavior. Also, ideas of schemas or scripts have not been attributed to culture but to some pan-human experience. Scripts assume, without cultural background that individuals' understanding of the world are accumulated through generalizations of knowledge from one first hand experience to

another. Finally, fold theories not only include knowledge structures of individuals but also reside in social practices, institutional regularities, media messages, and cultural artifacts.

[2] From their study of 180 middle and high school aged students Driver et al. (1996) constructed a framework for characterizing features of students' epistemological reasoning about inquiry in science. At the least sophisticated level was phenomenon-based reasoning, in which students characterized empirical investigation as an unproblematic process of finding out "what happens" or simply making phenomena happen so that subsequent behavior can be observed. No pre-service teachers in the current study demonstrated that type of reasoning. The next level, which most older students in Driver et al's study subscribed to was "relation-based reasoning." The most sophisticated level they referred to was referred to as "model-based reasoning."

Mark Windschitl
College of Education
University of Washington

RICHARD E. GRANDY AND RICHARD A. DUSCHL

CONSENSUS: EXPANDING THE SCIENTIFIC METHOD AND SCHOOL SCIENCE

A commitment to inquiry is a hallmark of USA science education. At the same period (1955 to 1970) when scientists were leading the revamping of science education to embrace inquiry approaches, historians and philosophers of science were revamping ideas about the nature of scientific inquiry and cognitive psychologists were revamping ideas about learning. A reconsideration of the role of inquiry in school science, it can be argued, began approximately 50 years ago, but the widespread reconsideration has led to a proliferation of meanings associated with "inquiry". The purpose of our conference was to provide a structure for discussion of science education with the goal of summarizing and synthesizing developments in three domains:

(1) Science studies, e.g., history, philosophy and sociology of science
(2) The learning sciences, e.g., cognitive science, philosophy of mind, educational psychology, social psychology, computer sciences, linguistics, and
(3) Educational research focusing on the design of learning environments that promote inquiry and that facilitate dynamic assessments.

The goal was to use learning sciences and science studies perspectives to establish consensus positions and thus reduce the variety of inquiry images found in school science. The list of scientific inquiry practices that emerged at the conference includes:

posing questions
refining questions
evaluating questions
designing experiments
refining experiments
interpreting experiments
making observations
collecting data
representing data
analyzing data
relating data to hypotheses/models/theories
formulating hypotheses

Richard A. Duschl and Richard E. Grandy (eds.), Teaching Scientific Inquiry: Recommendations for Research and Implementation, 304–325.
© *2008 Sense Publishers. All rights reserved.*

CONSENSUS FINDINGS

    learning theories
    learning models
    refining theories
    refining models
    comparing alternative theories/models with data
    providing explanations
    giving arguments for/against models and theories
    comparing alternative models
    making predictions
    recording data
    organizing data
    discussing data
    discussing theories/models
    explaining theories/models
    reading about data
    reading about theories/models
    writing about data
    writing about theories/models

We note that any of the practices of inquiry on this list involve evaluating and communicating concepts, methods, and arguments. Scientific inquiry, at its core, is about acquiring data and transforming that data first into evidence and then into explanations. In the aftermath of the conference, we believe there are eight points of consensus and one important point of dissensus that broaden our understanding of scientific inquiry.

The eight consensus points are that:

1. Although children and their learning situations share some characteristics with scientists, there are many important characteristics they do not share. Children are like scientists in that they notice at least some regularity and pose hypotheses. However, children are (at least initially) unlike scientists because:

    Children have no social structure to support inquiry
    Scientists have strong motivations for inquiry
    Scientists actively look for evidence
    Scientists read about data and theories/models
    Scientists write about data and theories/models
    Scientists debate the merits of theories/models
    Scientific theories/models typically invoke hidden or not directly observable variables, entities and processes
    Scientific theories/models are constrained by related theories and models
    Scientific theories often rely on mathematics to represent data
    Scientific theories often rely on mathematics to represent models/theories
    Scientists evaluate theories/models against evidence

Many of these activities require epistemic and social abilities – including the abilities to appropriately discuss, read and write science content – and these abilities can only develop within an appropriate social environment (Brewer; Schauble; Nersessian; Abd-El-Khalick; Longino; Stich; Siegel; this volume).

2. The traditional concept of The Scientific Method greatly oversimplifies the nature of science. It oversimplifies the cognitive elements by omitting essential epistemic and social elements that are part of the dialogical exchanges around the evaluation (epistemic) (Chinn & Samarapungava; Krajcik; this volume) and communication (social) of evidence and explanations. Moreover, The Scientific Method focuses narrowly on the experimental context. There are good arguments for ensuring that school science includes a much wider range of activities, including the epistemic and social inquiry practices (Solomon; Brickhouse; Kelly; Rudolph; this volume).

3. "Authentic science" includes all of the elements in 2, but "school science" cannot incorporate all of these into every 50-minute class or even into a weeklong instructional sequence. New curriculum frameworks and sequences for science learning are needed (Hmelo-Silver; Krajick; this volume).

4. Many of the activities on the expanded list, such as writing arguments for or against a model involve cognitive processes that inextricably include epistemic elements (what counts as evidence) and social elements (what to present and who is the audience). (Norris & Phillips; Bell; this volume). Together these elements are most effectively incorporated as part of extended sequences of instruction or coordinated progressions of learning (e.g., full-inquiry or immersion units for shorter 4-6 week time scales; learning progressions across multiple grade levels) that bring together the cognitive, epistemic and social (Hammer; Sandoval; Edelson; Bordeaux; this volume).

5. While we have some general knowledge about students' cognitive capacities at various ages, we know very little about their abilities under conditions of sustained good instruction. Moreover, we know almost nothing about the relevant social skills and epistemic abilities that also need to develop (Schauble; Stich; Helmo-Silver; this volume).

6. Science teaching and learning occur in complex cultural situations that involve participatory practices. We know very little about how the socio-cultural setting enhances or inhibits learning these practices In particular, we know very little about the dynamics of epistemic authority in the student-teacher relationship and its effect on learning outcomes. It is important to recognize that the relation is both bidirectional and asymmetric (Nersessian; Kelly; Stich; this volume).

7. Better science education for teachers is essential because most university science courses reinforce the inadequate conception of scientific method and largely omit the epistemic and social elements. The capacity of science teachers to deliver effective inquiry instruction is limited by their lack of experience with effective inquiry that incorporates the social and epistemic elements (Etkina; Windschitl; this volume).

8. Even when we have better answers to the questions inherent in the consensus points above and achieve a better understanding of what is *theoretically* possible in the classroom and school, we will be faced with practical and political limitations of what can and should be implemented (Gitomer; this volume).

Dissensus Point: Some participants felt strongly that science teaching should include explicit teaching of what is distinctive about scientific inquiry. Some suggestions were based on claims that science offers "causes", "mechanisms" or "causal explanations". Other participants were very skeptical about the feasibility of generally demarcating science from non-science by general principles. We will return to this topic at the end of our discussion of expanding the conception of scientific method.

Looking across the 8 consensus points two broad themes emerge. The first is the 'Scientific Method Theme'. The second is the 'School Science Theme'. For each theme, the 8 consensus points suggest revisions of prevailing images of both scientific method and school science.

The two themes, of course, are not mutually exclusive. A view of scientific method that embraces dialogic processes has implications both for what should be included and excluded in school science.

SCIENTIFIC METHOD

In matters of K-12 science education, the guiding conception for the scope and sequence of curriculum materials is that of the 'scientific method' (SM). Grounded in narrow interpretations of positivistic views about scientific inquiry, the accepted SM is based on a view that doing science is doing experiments. The bulk of investigations conducted in schools are designed to reinforce concept objectives and not promote inquiry. When processes of science *are* the goals of investigations they are frequently separated from conceptual frameworks.

New research from the learning sciences (Sawyer, 2006; NRC, 2007) stresses the importance of connectedness for sequences of curriculum around (1) fundamental ideas in science (e.g., particulate nature of matter; evolution; energy transfer; etc.), and, (2) important scientific practices (e.g., building theories and models; constructing arguments; using specialized ways of talking, writing and representing phenomena). An approach to school science that over-emphasizes hypothesis testing as the scientific method constrains what takes place in science classrooms.

Twentieth century philosophy of science can be partitioned into 3 major developments: an experiment-driven enterprise, a theory-driven enterprise, and a model-driven enterprise. The experiment-driven enterprise gave birth to the movements called logical positivism or logical empiricism, shaped the development of analytic philosophy and gave rise to the hypothetico-deductive conception of science. The image of scientific inquiry was that experiment led to new knowledge that accrued to established knowledge. How knowledge was discovered and refined was not on the philosophical agenda, only the final justification of knowledge was deemed important. This early 20th century perspective is referred to as the 'received view' of philosophy of science.

This 'received view' conception of science is closely related to traditional explanations of 'the scientific method'. The steps in the method are:

1. Make observations
2. Formulate a hypothesis
3. Deduce consequences from the hypothesis
4. Make observations to test the consequences
5. Accept or reject the hypothesis based on the observations.

The 'Scientific Method' conference theme suggests that current images about the nature of scientific inquiry are dated and narrowly defined. If we contrast the list of science inquiry practices in Consensus Point 2 with the traditional Scientific Method, we can see that although all 5 steps involve cognitive tasks, only the last involves an epistemic 'what counts' task. In contrast, many of the activities on the Consensus Point 2 list include social or epistemic elements. In fact, many of the items on the list involve all three. Recall that Consensus Points 1 and 2 from the conference participants was science as a practice has many social and epistemological characteristics that are critical to engaging in the discourse and dialogical strategies at the core of what it means to being doing scientific inquiry.

We must question the usefulness of scientific method frameworks that do not attend to the theory refinement dialogic practices embedded in science as a way of knowing. We must also question the usefulness of scientific method frameworks that do not attend to the epistemic practices inherent in constructing and evaluating models. Theory-building and model-building practices provide the contexts where epistemic abilities, social skills and cognitive capacities are forged. Missing from the traditional 5 step hypothetico-deductive laboratory-based view of scientific method are the important dialogic and epistemic processes involved in model and theory building.

Consider the example of writing about a theory – this is obviously a cognitive task. As Norris and Phillips (this volume) point out though, writing also requires social judgment since the writer is writing for an audience. Writing successfully for an audience requires the writer to have a nuanced conception of the belief and motivation structures of the reader. If the writer does not engage the motivational structure of the reader, the reader will read superficially if at all. If the writer does not engage the readers' belief structure in a relevant way, the readers' beliefs will not change and the reader may not even pay attention to the arguments.

The writing-about-a-theory task is also epistemic because the presumptive point of the writing is to adduce evidence that will encourage belief in or doubt about the theory and so it is essential that one make epistemic judgments about the relations between evidence and theory. Although one can formulate ideas in solitude, if you are part of a scientific community in which there is an ongoing discussion of a question against a background of shared theoretical assumptions, then what counts as a relevant conjecture, inference or hypothesis is constrained by social and epistemic considerations.

Expanding the Scientific Method

In the introductory chapter of the book, we framed one aspect of the nature of science inquiry debate in terms of 7 tenets of logical positivism. It is important not to simply reject logical positivism without understanding it. If we do so we risk both losing some of the insights provided and perspectives on some of the oversimplifications that were involved. Similarly, we do not reject the idea of scientific method, but want to radically supplement it. Let us return to the 7 tenets and consider them as framing elements for expanding the scientific method. We place them in a new order that reflects the improvement and refinement practices of scientific inquiry. We also organize the 7 Tenets Table to contrast "The Received View" of the nature of science with "The New View." The *Nature of Science 7 Tenets* Table 21.1 is placed at the end of the chapter.

The organization of the table's rows and columns are meant to highlight the reactions, objections and insights the 7 tenets raised. Consideration for both the insights and limitations of logical positivism and early 'Kuhnian' responses to logical positivism have expanded our perspectives about the nature of science, the growth of scientific knowledge, and the goals/limitations of science. For each of the 7 tenets we contrast the "Traditional" and the "Revised" views for the nature of science (NOS). Reading down the traditional Received NOS Views column paints a picture of the commitments held by the logical positivists. Reading down the Revised NOS Views column reveals commitments from those adhering to naturalized philosophical views. The Revised NOS View stresses the dialogic and dialectical processes/practices of science and does so with respect to conceptual as well as methodological changes in scientific inquiry. The main points from the 7 Tenets Table that provide a basis for an expanded scientific method include:

- The bulk of scientific effort is not theory discovery or theory acceptance but theory improvement and refinement.
- Research groups or disciplinary communities are the units of practice for scientific discourse.
- Scientific inquiry involves a complex set of discourse processes.

- The discourse practices of science are organized within a disciplinary matrix of shared exemplars for decisions regarding the a) values, b) instruments, c) methods, d) models, and e) evidence to adopt.
- Scientific inquiry has epistemic and social dimensions, as well as conceptual.
- Changes in scientific knowledge are not just in conceptual understandings alone; important advancements in science are also often the result of technological and methodological changes for conducting observations and measurements.
- What comes to count as an observation in science evolves with the introduction of new tools, technologies and theories.
- Theories can be understood as clusters of models where the models stand between empirical/conceptual evidence and theoretical explanations.
- Theory and model choices serve as guiding conceptions for deciding 'what counts' and are an important dynamic in scientific inquiry.
- Rubrics for a rational degree of confirmation are hopeless, dialogue over merits of alternative models and theories are essential for refining, accepting or rejecting them and are not reducible to an algorithm.

The expanded scientific method, then, would be inclusive, not exclusive, of the 3 sequential 20th century images of the nature of science: Hypothetico-deductive experiment driven science; Conceptual Change theory driven science; Model-based driven science. The expanded scientific method recognizes the role of experiment and hypothesis testing in scientific inquiry, but emphasizes that the results of experiments are used to advance models and build theories. Thus, the expanded scientific method makes a further recognition that the practices of scientific inquiry involve important dialogic and dialectical practices that function across conceptual, epistemic and social dimensions.

The Dissensus

Any characterization of scientific method - enhanced, expanded or otherwise - raises the question of whether science as a way of knowing is distinctive from other ways of knowing that also involve conceptual, epistemic and social discourses. Philosophers of science refer to this issue as the demarcation between science and other forms of inquiry. There are two related but somewhat distinct demarcation questions. First, some individuals see a distinction between legitimate science and activities that purport to be scientific but are not, e.g., astrology, creation science, etc. Second, there are some who see a distinction between scientific inquiry and other forms of legitimate but non-scientific activities such as historical research or electrical engineering. Some conference participants felt strongly that it not only is possible to make a sharp general demarcation, but that it is an important part of teaching science to teach that demarcation. Other participants were skeptical that such a demarcation is possible.

Several participants suggested that scientific inquiry involves mechanistic explanations. This is clearly too narrow as magnetism and gravitation are not mechanical. Another suggestion was that scientific explanations are causal. This suggestion has two problems; one is that it seems to rule out statistical explanations that are not necessarily causal. The second is that three centuries of debate over the nature of causation in philosophy have produced no consensus on what constitutes causation.

Another suggestion was that scientific explanations/hypotheses must be testable. While this seems right in spirit, decades of attempts by philosophers to make this concept precise have also consistently failed. Another group of participants argued that the distinction between scientific and non-scientific hypotheses is real, but is not a matter for which we can formulate explicit rules. For them, the only individuals able to appropriately make the distinction between testable and non-testable hypotheses are those who are deeply embedded in the practices of the specific science and have sophisticated knowledge. Today there are domains of science that do not begin inquiry with stating hypotheses but rather are guided by patterns of discovery from huge data sets (e.g., human genome project).

We are not suggesting that it is impossible to distinguish scientific inquiry from pseudo-science. But we believe that to the extent that the distinction can be made, it has to be made locally, from the perspective of the particular field at a specific time. A naturalized approach to understanding science means that researchers observe what scientists do, not just what scientists say about what they do. The naturalized approach to the history of science strongly suggests that the nature of scientific activities has changed over time and we expect change to continue. Knowledge of the relevant scientific principles and criteria for what counts as an observation are important elements in distinguishing science claims and developing demarcation capacities; e.g., distinguishing science from pseudoscience. However, we are skeptical that a general demarcation criterion can be abstracted from the concrete historically situated judgments.

Moreover, history suggests that at times new hypotheses or theories may provoke disagreement about testability even among the most expert scientists. Mach and others rejected atomism as unscientific around the turn of the century (Brush, 1968). The majority of earth scientists rejected continental draft as unscientific during the 1920s and later (van Waterschoot van der Gracht, 1928; LeGrand, 1988). In the present scene, many physicists regard string theory as the most important break through since quantum mechanics. Others disagree – one recent book title states that string theory is "not even wrong" because it makes no testable predictions (Woit, 2006; Smolin, 2006).

SCHOOL SCIENCE

Scientific ideas and information are rooted in evidence and by the scientific models and theories that frame investigations and inquiries. All elements of scientific inquiry − questions, methods, evidence and explanations − are open to scrutiny, examination, justification and verification. In parallel with our expanded scientific method, we expand the list of essential features of inquiry. The five essential features of inquiry presented in *Inquiry and the National Science Education Standards* (National Research Council, 2000) are:

- Learners are engaged by scientifically oriented questions.
- Learners give priority to **evidence**, which allows them to develop and evaluate explanations that address scientifically oriented questions.
- Learners formulate **explanations** from evidence to address scientifically oriented questions.
- Learners evaluate their explanations in light of alternative explanations, particularly those reflecting scientific understanding.
- Learners communicate and justify their proposed explanations.

Based on our list formulated at the conference after long discussion, we argue for adding:

- Learners respond to criticisms from others.
- Learners formulate appropriate criticisms of others.
- Learners seek criticism of own explanations.
- Learners reflect on alternative explanations and not have a unique resolution.

The importance of these additional elements follows from emphasis on dialogic processes and attention to epistemic and social learning goals. Together the two expansions of scientific method and features of inquiry have major implications for school science. Two important areas impacted by these expansions that were discussed at the conference were (1) the design and format of instruction sequences, and (2) the science learning experiences of preservice teacher education.

Instructional Sequences

Consensus point # 4 noted that many cognitive processes used in scientific inquiry inextricably include epistemic elements (e.g., deciding what counts as evidence) and social elements (e.g., determining what to present and who is the audience). Coordinating instruction around all of these components requires alternative conceptions of how to situate learning and teaching. Such coordination requires attention to the content and contexts that promote acquisition and use of the complex inquiry practices listed above.

Research (NRC, 2007) shows that prevailing models of science teaching are lesson based rather than unit based, emphasize concept learning rather than knowledge system learning, and focus inquiry lessons on completing experiments rather than on testing and revising explanatory models. Science learning is partitioned into discrete units that contribute neither to the development of scientific knowledge systems nor to the development of cognitive, epistemic and social practices that under gird the evaluation of such systems.

The consensus position is to focus on core knowledge and inquiry practices of school science as well as the themes and components that will serve as contexts that facilitate the cognitive activities and dialogical processes embedded in doing science. The commitment is to curriculum frameworks that promote extended instructional sequences around 'big ideas' and unifying themes and principles of science

Coordination around 'big ideas' and inquiry practices provides opportunities for learners to put knowledge to use and to engage in important dialogical discourse processes that support learning and reasoning. Knowledge-in-use discourse strategies and literacy practices enable learning and reasoning to progress beyond conceptual learning goals alone and facilitate development of important epistemic and social dimensions of scientific inquiry; e.g., arguing, modeling, explaining, questioning, valuating, representing, among other (Norris & Phillips, this volume; Bell, this volume). The move from week long modules to extended month long sequences shifts away from a curriculum position that asks, "what do we want students to know and what do they need to do to know it", to a curriculum position that asks, "what do we want students to be able to do and what do they need to know to do it".

There was a consensus, following from the discussions above that we want learners to initiate and take responsibility for as many of the dialogical practices in Table 21.1 as possible. As noted in Consensus Point 1, beginning learners and their environments lack many of the characteristics of mature scientists and their contexts. For learners who are not yet capable of taking full responsibility, the teacher (or perhaps peers) must take more of the initiative. Designing an inquiry curriculum for the long term means thinking about how to shift the classroom environment from the teacher directed right hand side of the table toward the student directed left hand side. We emphasize that the rate at which this can be done will vary from learner to learner, as well as from row to row, and that some of the rows will depend on others. In addition, the shifts may take years. We will discuss in a later section the very crucial fact that the ability to carry out the activities listed on the right are often beyond the current capabilities of many science teachers, although not beyond their potential capabilities.

When developing dialogic practices we should build on prior student abilities whenever possible. For example, with reference to the first row, we know that at an early age students spontaneously generate questions, but those questions are not necessarily scientific. Unfortunately, what happens in many classroom environments is that instead of learning to only ask the scientific questions,

Table 21.1. The Learner/Teacher Continua – Dialogic Practices

Learner can:	Teacher Models & Monitors:
Engage in asking scientifically oriented questions	*The asking of scientifically oriented questions*
Give priority to using evidence in responding to questions	*Giving priority to use of evidence in responding to questions*
Formulate explanations from evidence	*Formulation of explanations from evidence*
Connect explanations to scientific knowledge	*Connecting explanations to scientific knowledge*
Communicate and justify explanations	*Communicating and justifying explanations*
Respond appropriately to criticism of explanations	*Responding appropriately to criticism of explanations*
Formulate appropriate criticism of alternative explanations	*Formulating appropriate criticism of alternative explanations*
Criticize their own explanations	*Criticisms of explanations*
Construct tests to discriminate between explanations	*Constructing tests to discriminate between explanations*

Reflect on the fact that sometimes there are multiple explanations and no currently definitive answer	*The fact that sometimes there are multiple explanations and no currently definitive answer.*

environments is that instead of learning to only ask the scientific questions, students stop asking questions at all. There was also consensus (#6) that although we have some idea what learners are capable of we lack systematic extensive research of what is possible given a consistent and thorough full inquiry curriculum starting in kindergarten. In particular, our consensus list highlights activities involving models, none of which are even mentioned in the traditional "Scientific Method".

Preparation for engaging in scientific inquiry is linked to students' opportunities to engage with (1) the acquisition of measurements and observations that become data and (2) the unfolding or transformations of data into evidence that is used to construct and evaluate models and explanations. One framework for coordinating instruction that addresses the activities in Table 21.2 is the evidence-explanation (E-E) continuum (Duschl, 2003). The dialogical strategy within the E-E continuum expects students to make and report judgments, reasons and decisions during three critical transformations:

1. Selecting data to become evidence.
2. Analyzing evidence to generate models and/or locate patterns of evidence, and
3. Aligning patterns of evidence with scientific explanations and theories.

Emphasizing these transformation provides discourse opportunities for teachers and students to use and practice the dialogic practices in Table 21.2 that coordinate the cognitive, epistemic and social practices of science. Part of consensus point #1, also much discussed at the conference, was that many students lack motivation to study science. Research shows that students in dialogically oriented classrooms are more highly motivated (NRC, 2007).

The curriculum changes suggested demand much more time is spent on any one given school science topic with the obvious consequence that fewer objectives and topics will be addressed. This implies that fewer topics are covered, and this has led to resistance the dialogical approach from teachers, administrators, and parents. We will address this topic below.

The consensus #4 among conference participants is to coordinate science curriculum around immersion units and learning progressions. Immersion units, which can be project-based science, problem-based science, or full-inquiry science modules, are designed for 4-8 weeklong time scales. Learning progressions are longer to accommodate the difficulties of acquiring understanding and competencies of certain conceptual understandings and inquiry practices, They would be years long, spanning multiple grade levels (e.g. grades 3-6) The 2007 NRC report *Taking Science to School: Learning and Teaching Science in Grades K-8* presents a comprehensive review of learning progressions.

The unfolding of data and evidence takes time. By pausing instruction to allow students to discuss and debate what they know, what they believe and what evidence they have to support their ideas, their thinking is made visible thus enabling teachers and peers opportunities for monitoring and assessing communication of information and thinking. The extended curriculum time on thematic topics makes room for student conversations that promote representations of reasoning as specified in the expanded scientific method and features of inquiry list. The longer immersion units and learning progressions typically contain complex tasks, performances, and investigations. By design, the instructional sequences embed activities that facilitate feedback on the conceptual, cognitive, epistemic, and social goals of the unit:

- Developing scientific reasoning
- Communicating scientific ideas,
- Assessing the epistemic status of scientific claims.

The recommendations for longer instructional sequences are accompanied by recommendations for further research on the design and effectiveness of these sequences. According to Schauble (this volume), there are three important questions concerning development of scientific thinking:
1. What are the origins of the forms of scientific thinking?
2. Where does it go?
3. What are the supports that produce change in thinking?

While we have good research on Question 1, we know very little about Questions 2 or 3.

In summary, we must assist learners with the development of dialogical processes for both the construction **and** the evaluation of knowledge claims. Thus, students should be given extended opportunities to explore and critique the relationships between evidence and explanation. To this end, a consensus recommendation is to situate inquiry into longer thematic instructional sequences, where the theme is defined not by the conceptual structures of scientific content alone. Rather, the instructional sequence is designed to support the processes of acquiring and evaluating evidence, as well as other discourse, cognitive and reasoning skills.

Another implication for contemporary science education practice concerns how best to develop learners' understanding of the nature of science. Our review of the *7 tenets* and endorsement of immersion unit approaches to the teaching of scientific inquiry suggest that an alternative approach to using generic lists of views about the nature of science is to engage learners in extended learning sequences that by design have embedded elements that engage learners in the practices that embrace the debates surrounding science as a way of knowing.

The 'School Science Theme' suggests that more nuanced approaches informed by the learning sciences and science studies need to be adopted. A recommendation is that new frameworks are needed for both the selection of content and context and the design of instructional sequences. Clearly, considerations for expanding the Scientific Method (SMe) have implications for School Science. For example, the SMe suggests K-12 students have opportunities to examine (1) the development of the data and (2) the transformations of data. The general dialogical strategy is to allow students to make and report judgments, reasons and decisions during 4 critical epistemic decisions. The first is determining which feasible observations or measurements will provide the appropriate data for the inquiry. The second is selecting which data ought to become evidence. The third is analyzing evidence to generate models and/or locate patterns of evidence. The fourth is evaluating which scientific explanations best account for the models and patterns of evidence. This four-step approach would have the advantage of providing opportunities for students to engage in, and importantly for teachers to monitor, the cognitive, epistemic and social practices of doing science.

Next consider how each of the 7 Tenets in Table 21.1 has implications for School Science. Each Tenet is elaborated on with an example of how it might be implemented in a classroom.

1. Contexts of Theory Development – provide opportunities for students to engage in activities involving the growth of scientific knowledge, e.g., respond to new data, to new theories that interpret data, or to both. Employ a dialectical orientation to school science to model how science and scientists are being *responsive* to anomalous data and worthy alternative conceptual frameworks when refining and modifying theory. The *responsive* label is preferred over saying that scientific claims are at best only *tentative*, thus avoiding anti science connotations that scientific claims being tentative are unsupported by evidence or scientific reasoning.

2. Individual/Group Inquiry – use distributed learning models during investigations – avoid when possible all students doing the same investigation, rather have individuals/small groups complete components of an inquiry and then report back to large group the data and the evidence obtain. Whole class discussions can then focus on a larger corpus of data and evidence in the development of explanations.

3. Observation/Theory – employ dialectical processes regarding which theoretical frameworks are being used as guiding conceptions when critiquing or making decisions about what data to collect, what questions to ask, what data to use as evidence, among others.

4. Theory Evaluation – engage learners in the development of criteria for theory/explanation evaluation. This includes making decisions regarding when is it plausible to pursue the development and interrogation of theory. Furthermore, there is the need to have learners consider alternative explanations and participate in dialogical activities that debate and argue the merits/demerits of the alternative models and theories.

5. Theory Language – scientific explanations are not strictly linguistic statements. Scientists develop and use physical models, computational models, mathematical models and iconic models, among others, to explain scientific investigations. So, too, should students be encouraged to develop and to use model-based approaches in scientific investigations and inquiry.

6. Shared Exemplars – provide attention to the theoretical or model-based frameworks that guide scientific inquiry. Establish shared goals by arranging inquiry around big ideas, driving questions, immersion units and focusing meaning making on use of critical evidence, pivotal cases, and knowledge integration. Develop in students the idea that science is about engagement in theory change and model building and revision.

7. Scientific Progress – provide learners opportunities to engage in the examination of alternative explanations and guiding conceptions when developing accounts of phenomena and mechanisms. Allocate room in the curriculum for learners to engage in serious discussions about the criteria that are used to assess and make judgments about knowledge claims.

Consensus Point 5 states we have very little knowledge about the classroom learning conditions that promote and sustain the kinds of learning and teaching described above. A question for the field then is how general or how specific we need to be in the design of science learning experiences. The 'School Science Theme' suggests that we need to focus on the domain-specific features of knowing, reasoning, and learning across longer periods of time, e.g., years and months as opposed to weeks and days. Hence, the recent petitions from the NAEP 2009 Science Framework and the NRC (2007) report "Taking Science to School: Learning and Teaching Science in Grades K-8" is to adopt 'learning progressions' perspectives for science learning that focuses on 'big ideas' and 'scientific practices'.

One consequence of adopting learning progressions and immersion units is that fewer conceptual domains will be part of overall curriculum frameworks. Breadth of conceptual knowledge must yield some to depth of conceptual knowledge in order to facilitate acquisition of epistemic and social inquiry practices. The School Science implication is that reasoning, representing and reporting need to be at the core of science learning and teaching. Consensus Point 6 states that while we know a considerable amount about students' cognitive capacities at various ages, we know very little about the relevant social skills and epistemic abilities that also need to develop. Many of these skills and abilities are forged from experiences external to formal schooling. Ignoring them constrains the chances of connecting school science to life experiences.

Teacher Professional Development

In addition to inquiry learning goals, classroom dialogic practices, and constraints of student cognitive development outlined in the previous section, there are practical constraints that bear on suggestions for incorporating an enhanced scientific method or learning progression frameworks in school science. Consensus Point 7 speaks to negative impact university science courses can have on teachers' images of science. Postsecondary undergraduate science courses tend to reinforce the traditional of scientific method, the received views of the nature of science and thus largely omit the dialogic, epistemic and social elements of science inquiry practices. The capacity of science teachers to deliver effective instruction, inquiry or otherwise, is limited by their experiences with science and science instruction that incorporate the dialogic, epistemic, and social elements. The constraints on implementing the enhanced Scientific Method and the School Science recommendations are multiple but we highlight two:
- The cognitive demands placed on teachers during engagement with our elaborated sense of inquiry activities embedded in immersion units and learning progressions; and
- The implications for preservice and inservice science teacher development programs.

Teacher knowledge and beliefs impact the dynamics of science classroom learning environments. Two approaches with regard to teacher knowledge and school science prevail:
1. *The Deficit Model* – develop curricula to safeguard against lacunae of teacher knowledge;
2. *The Professional Model* – establish curricula that require teachers' to develop expertise with implementing dialogical approaches to teaching.

Research results on precollege students' images of science by Driver, Leach, Millar & Scott, (1994) have been shown by Windschitl (this volume) to be consistent with those of science majors entering preservice teacher education programs. Driver, et al (1994) developed a three level framework for evaluating epistemological reasoning in science. The least sophisticated level is *phenomenon based reasoning*, the next level is *relation-based reasoning* and the most sophisticated level is *model-based reasoning*. They found that students have a strong bias for experiment-driven notions of doing science grounded in a strong reliance on sense perception as the most reliable evidence. Students did not value how experiments are situated within larger explanatory models that can be used to guide the design and the evaluation of experiments, measurements and observations.

Science education goals have been framed as content vs. process, understanding vs. doing, knowing vs. reasoning, and discovery vs. inquiry, among others. The field of science education has witnessed rapid evolution in the last 50 years with respect to both theoretical models of learning and views about the nature of science. New tools, technologies and theories are constantly shaping the scientific landscape in a variety of ways, thus the emergence of dialogic and model-based practices in the revised nature of science views. Similar considerations for the role of modeling practices in science classrooms and of model-based reasoning (Lehrer and Schauble, 2006; NRC, 2007; Kelly, this volume; Chinn & Samarapungava, this volume) has contributed to discussions that the important school science inquiry practice perspectives are:
1. Building theories and models,
2. Constructing arguments.
3. Using specialized ways of talking, writing and representing phenomena.

Researching preservice science teachers' design and implementation of science investigations, Windschitl found that the college students in teacher preparation programs lack the above inquiry perspectives. They, like the Driver et al precollege sample, are also functioning at or below the lowest epistemological reasoning level. What makes this finding for college students so surprising is that they had completed 20+ college level science courses! The preservice teachers' folk theory or image of doing science is strongly grounded in the traditional the scientific method and devoid of any references to the epistemic frameworks of science: e.g., claims, arguments, alternative explanations, models, etc. When asked to design an investigation, the teacher candidates based their questions on topics of personal interest and things that were doable or novel rather than on extant scientific models or theoretical frameworks.

The implication for School Science is clearly we need to consider reforming more than just the K-12 science education programs. The enhanced scientific method should be part of the college science courses taken by preservice and inservice teachers. Teaching the dialogic processes of science requires teachers capable of engaging learners in the activities listed in Table 21.2 and so opportunities to engage in 'immersion unit' type science learning should be a part of science teacher education programs One of the points noted by many conference participants is that in order to truly engage students in dialogue, teachers must relinquish the complete authoritarian stance that is common in science classrooms. Relinquishing this role is usually not easy for teachers. Making these changes is not entirely, perhaps not mainly, a conceptual task, but involves exactly the social and epistemic skills we have been discussing in the enhanced scientific method. Consensus point #6 is that the teacher-student relationships in such classrooms are complex and bi-directional, though certainly not symmetric. That point also emphasizes that we need much more research on the processes that occur in these classrooms.

New models of school science and of children's science learning require new models of teacher professional development. There is a great deal of research to be done, and there are significant challenges as well in the face of alternative certification programs that allow school districts to hire science teachers unfamiliar with how students' engage in science learning. The implications of such research for the design of science education programs and for the professional development of teachers who implement the programs are significant. Answers to these questions will fundamentally change our images of school science. And, herein, as well, lies a significant tension for the design of science education programs of study.

Teachers, administrators and parents who are still immersed in the older scientific method and the older conception of inquiry often resist dialogically oriented curricula. There is evidence, though more would be strongly desirable, that dialogically oriented curricula which incorporate epistemic and social dimensions produce deeper conceptual knowledge, which is better retained (NRC, 2007). Such curricula also consistently produce higher levels of motivation. These facts must be brought into more prominence, in order to provide motivation for change in teachers, administrators and parents.

ESTABLISHING CONSENSUS ON SCHOOL SCIENCE INQUIRY

This edited volume has set out arguments and positions for shifting the focus of scientific inquiry in school science. We finish with some reflections on practical issues and constraints that face those who would transform schools and classrooms. On the one hand, the institutional culture of public education is severely constrained by economical, ideological and pedagogical conditions. As stated above and in Consensus Point 8, there are numerous practical and political limitations and policy issues that need to be confronted if changes and reforms to School Science are going to prevail. Such constraints have the effect of promoting certain forms of curriculum, instruction, and assessment practices while denying others on the basis of cost effectiveness; e.g., professional development for K-12 teachers. On the other hand, research on learning and on science learning environments contribute a richer understanding of the classroom contexts and conditions that promote scientific reasoning and understanding. In the end we must ask, do we fit the research on learning into the instructional culture of schools or do we change the culture of schools to accommodate the learning research. There are significant policy and practice issues that come to the table.

Table 21.2 Nature of Science 7 Tenets

Traditional Tenets from Logical Positivism	Received NOS Views	Reasons for Revised Tenets	Revised NOS Views
1. There is an important dichotomy between contexts of discovery and contexts of justification.	*Logical positivism's focus was on the final products or outcomes of science. Of the two end points, justification of knowledge claims was the only relevant issue. How ideas, hypotheses and intuitions are initially considered or discovered was not relevant.*	*Theory change advocates value understanding the growth of knowledge begins. Perhaps the most important element Kuhn and others added is the recognition that most of the theory change is not final theory acceptance, but improvement and refinement.*	*The bulk of scientific inquiry is neither the context of discovery nor the context of justification. The dominant context is theory development and conceptual modification. The dialogical processes of theory development and of dealing with anomalous data occupy a great deal of scientists' time and energy.*
2. The individual scientist is the basic unit of analysis for understanding science.	*Logical positivists believed scientific rationality can be entirely understood in terms of choices by individual scientists.*	*Kuhn's inclusion of the scientific community as part of the scientific process introduced the idea of research groups or communities of practice as being the unit of scientific discourse. This shift from individual to group produced negative reactions from many philosophers. Including a social dimension was seen as threatening the objectivity and rationality of scientific development. Teams*	*Scientific rationality can be understood in terms of dialogic processes taking place as knowledge claims and beliefs are posited and justified. Scientific discourse is organized within a disciplinary matrix of shared exemplars; e.g., values, instruments, methods, models, evidence.*

CONSENSUS FINDINGS

		of scientists engage in investigations.	
3. There is an epistemologically significant distinction between observational and theoretical (O/T) languages based on grammar.	Logical Positivism focused on the application of logic and on the philosophy of language to analyze scientific claims. Analysis void of contextual and contingent information produces a grammar that fixes criteria for observations.	The O/T distinction debate showed that our ordinary perceptual language is theory laden, what we see is influenced by what we believe. New theories leading to new tools and technologies greatly influenced the nature of observation in science and the representation of information and data.	What counts as observational shifts historically as science acquires new tools, technologies and theories. Science from the 1700s to the present has made a transition from a sense perception dominated study of nature to a tool, technology and theory-driven study of nature.
4. Some form of inductive logic would be found that would provide a formal criterion for theory evaluation.	There exists an algorithm for theory evaluation. Given a formal logical representation of the theory and data, the algorithm would provide the rational degree of confirmation the data confer on the theory.	Seeking an algorithm for a rational degree of confirmation is hopeless. Scientists working with the same data can rationally come to differing conclusions about which theory is best supported by given evidence. There is ongoing debate about how much variation is rational and how much is influenced by other factors.	Dialogue over the merits of competing data, models and theories is essential to the process of refining models and theories as well as accepting or rejecting them.
5. Scientific theories can most usefully be thought of as sets of sentences in a	Logical positivists advocated the position that	Model-based views about the nature of science embrace, where hypothetical-	Modern developments in science, mathematics, cognitive sciences, and computer

323

formal language.	theories are linguistic in character and could be described with deductive-nomological procedures.	deductive science does not, the dialogic complexities inherent in naturalized accounts of science. Scientific representations and explanations take many different forms: mathematical models, physical models, diagrams, computation models, etc.	sciences have extended the forms of representation in science well beyond strictly linguistic and logical formats. One widespread view is that theories should be thought of as families of models, and the models stand between empirical/conceptual evidence and theoretical explanations.
6. Different scientific frameworks within the same domain are commensurable.	Logical positivists sought to establish criteria that supported the claim that there are normative dimensions to scientific inquiry. The growth of scientific knowledge is a cumulative process.	Science communities are organized within disciplinary matrices. Shared exemplars help to define science communities. Scientific frameworks on different sides of a revolutionary change are incommensurable. Hypothesis testing takes place within more complex frameworks requiring more nuanced strategies for representing and reasoning with evidence.	Different scientific frameworks within the same domain share some common ground. But they can disagree significantly on methodology, models and/or relevant data. The issue is the extent to which knowledge, beliefs, reasoning, representations, methods, and goals from one research domain map to another research domain. The social and epistemic contexts are complex indeed.
7. Scientific development is cumulatively progressive.	Logical positivists held that the growth of scientific knowledge is cumulative and	Theory choice is an important dynamic of doing science and it influences how investigations are designed and	The Kuhnian view that 'revolutions' involve the abandonment of established guiding conceptions and methods challenges the

	continually progressive. Scientists work with common theory choices.	*conducted. On what grounds (e.g., rational vs. irrational) scientists make such choices is a matter for further research and debate.*	*belief scientific development is always cumulatively progressive. New guiding conceptions inform what counts as an observation or a theory. Such changes reinforce beliefs that all scientific claims are revisable in principle. Thus, we embrace the notions of the 'tentativeness' of knowledge claims and the 'responsiveness' of scientific practices.*

BIBLIOGRAPHY

Abd-El-Khalick, F. (2002). The development of conceptions of the nature of scientific knowledge and knowing in the middle and high school years: A cross-sectional study. Paper presented at the annual meeting of the National Association for Research in Science Teaching, New Orleans, LA.

Abd-El-Khalick, F. (this volume). Modeling science classrooms after scientific laboratories: Sketching some affordances and constraints drawn from examining underlying assumptions.

Abd-El-Khalick, F. & Akerson, V. L. (2004). Learning as conceptual change: Factors mediating the development of preservice elementary teachers' views of nature of science. *Science Education, 88*(5), 785-810.

Abd-El-Khalick, F., BouJaoude, S., Duschl, R. A., Hofstein, A., Lederman, N. G., Mamlok, R., Niaz, M., Treagust, D. & Tuan, H. (2004). Inquiry in science education: International perspectives. *Science Education, 88*(3), 397-419.

Abd-El-Khalick, F. & Lederman, N. G. (2000). The influence of history of science courses on students' views of nature of science. *Journal of Research in Science Teaching, 37,* 1057-1095.

Abraham, M. R. (1998). The learning cycle approach as a strategy for instruction in science. In B. J. Fraser & K. G. Tobin (Eds.), *International Handbook of Science Education* (pp. 513-524). Dordrecht, the Netherlands: Kluwer.

Ackerman, R. J. (1985). *Data, instruments, and theory: A dialectical approach to understanding science.* Princeton, NJ: Princeton University Press.

Aesop's Fables. (1912). (V.S. Vernon, trans., A. Rackham, illus.) New York: Avenel Books.

Aikenhead, G. S. (2005). Science-based occupations and the science curriculum: Concept of evidence. *Science Education, 89,* 242-275.

Akerson, V. L., Abd-El-Khalick, F. & Lederman, N. G. (2000). Influence of a reflective explicit activity-based approach on elementary teachers' conceptions of nature of science. *Journal of Research in Science Teaching, 37,* 295-317.

Allchin, D. (2003). Scientific myth-conceptions. *Science Education, 87,* 329-351.

Allchin, D. (2004). Should the sociology of science be rated X? *Science Education, 88,* 934-946.

Alsop, S. & Watts, M. (1997). Sources from a Somerset village: A model for informal learning about radiation and radioactivity. *Science Education, 81,* 633-650.

Amann, K. & Knorr-Cetina, K. (1990). The fixation of (visual) evidence. In M. Lynch & S. Woolgar (Eds.), *Representation in science practice* (pp. 85 – 121). Cambridge: MIT.

American Association for the Advancement of Science. (1990). *Science for all Americans.* New York: Oxford University Press.

American Association for the Advancement of Science. (1993). *Benchmarks for scientific literacy – Project 2061.* New York: Oxford University Press.

American Association for the Advancement of Science. (1998). *Blueprints for reform: Science, mathematics, and technology education.* New York: Oxford University Press.

American Psychological Association. (n.d.). Learner-Centered Psychological Principles. Retrieved November 9, 2004, from the American Psychological Association Web site: http://www.apa.org/ed/lcp/lcp14.html

Anderson, J. R. (1983). *The architecture of cognition.* Cambridge, MA: Harvard University Press.

Anderson, R. C. (1985). Role of the reader's schema in comprehension, learning and memory. In H. Singer & R. B. Ruddell (Eds.), *Theoretical models and processes of reading* (pp. 372-384). Newark, DE: International Reading Association.

BIBLIOGRAPHY

Antony, L. (1993). Quine as Feminist. In L. Antony & C. Witt (Eds.), *A mind of one's own: Feminist essays on reason and objectivity*. Boulder: Westview Press.

Atkins, L. J. (2004). *Analogies as categorization phenomena: Studies from scientific discourse.* Unpublished Doctoral dissertation, University of Maryland, College Park.

Atkinson, D. (1999). *Scientific discourse in sociohistorical context: The philosophical transactions of the Royal Society of London 1675-1975*. Mahwah, NJ: Lawrence Erlbaum.

Bachelard, G. (1940). *The Philosophy of no*. Paris: Paris University Press.

Baker, L. & Brown, A. L. (1984). Metacognitive skills and reading. In P. D. Pearson, R. Barr, M. L. Kamil & P. Mosenthal (Eds.), *Handbook of reading research* (pp. 353-394). New York: Longman.

Baillargeon, R. (1987). Object permanence in 3.5- and 4.5-month old infants. *Developmental Psychology, 23*, 655-664.

Barab, S. & Hay, K. E. (2001). Doing science at the elbows of experts: Issues related to the science apprenticeship camp. *Journal of Research in Scinece Teaching, 38*, 70-102.

Baron, J. (1990). Performance assessment: blurring the edges among assessment, curriculum, and instruction. In A. Champagne, B. Lovitts & B. Calinger (Eds.), *Assessment in the service of instruction* (127-148). Washington, DC: American Association for the Advancement of Science.

Barsalou, L. W. (1999). Perceptual symbol systems. *Behavioral and Brain Sciences, 22*, 577-609.

Bazerman, C. (1985). Physicists reading physics. *Written Communication, 2*, 3-23

Bazerman, C. (1988). *Shaping written knowledge: The genre and activity of the experimental article in science*. Madison, WI: University of Wisconsin Press.

Beamish, J. & Herman, T. (2003). Adsorption and desorption of helium in aerogels. *Physica B, 329-333*, 340-341.

Bechtel, W. & Richardson, R.C. (1993). *Discovering complexity: Decomposition and localization as strategies in scientific research*. Princeton: Princeton University Press.

Belenky, M. F., Clinchy, B. M., Goldberger, N. R. & Tarule, J. M. (1986). *Women's ways of knowing: The development of self, voice, and mind*. US: Basic Books.

Bell, P. (2004a). Promoting students' argument construction and collaborative debate in the science classroom. In M. C. Linn, E. A. Davis & P. Bell (Eds.), *Internet environments for science education*. Mahwah, NJ: Erlbaum.

Bell, P. (2004b). The educational opportunities of contemporary controversies in science. In M. C. Linn & E. A. Davis & P. Bell (Eds.), *Internet environments for science education*. Mahwah, NJ: Erlbaum.

Bell, P. (this volume). Inquiry as inscriptional work: A commentary on Norris and Phillips.

Bell, P. & Linn, M. (2000). Scientific arguments as learning artifacts: Designing for learning from the web with KIE. *International Journal of Science Education, 22*(8), 797-817.

Bell, P. & Linn, M. C. (2002). Beliefs about science: How does science instruction contribute? In B. K. Hofer & P. R. Pintrich (Eds.), *Personal epistemology: The psychology of beliefs about knowledge and knowing* (pp. 321-346). Mahwah, NJ: Lawrence Erlbaum.

Bell, P., Davis, E. A. & Linn, M. C. (1995). The knowledge integration environment: Theory and design. In J. L. Schnase & E. L. Cunnius (Eds.), *Proceedings of Computer Support for Collaborative Learning '95* (pp. 14-21). Mahwah, NJ: Lawrence Erlbaum Associates.

Bell, R. L., Blair, L. M., et al. (2003). Just do it? Impact of a science apprenticeship program on high school students' Understandings of the Nature of Scientific Inquiry. *Journal of Research in Science Teaching, 40*(5), 487-509.

Bell, R. L., Lederman, N. G. & Abd-El-Khalick, F. (1998). Implicit versus explicit nature of science instruction: An explicit response to Palmquist and Finley. *Journal of Research in Science Teaching, 35*, 1057-1061.

Bell, R. L., Lederman, N. G. & Abd-El-Khalick, F. (2000). Developing and acting upon one's conception of the nature of science: A follow-up study. *Journal of Research in Science Teaching, 37*, 563-581.

Bell-Basca, B. S., Grotzer, T. A., Donis, K. & Shaw, S. (2000). *Using domino and relational causality to analyze ecosystems: Realizing what goes around comes around*. Paper presented at the annual meeting of the National Association of Research in Science Teaching, New Orleans.

Bianchini, J. A. & Colburn, A. (2000). Teaching the nature of science through inquiry to prospective elementary teachers: A tale of two researchers. *Journal of Research in Science Teaching, 37*, 177-210.

BIBLIOGRAPHY

Bianchini, J. A. (1997). Where knowledge construction, equity, and context intersect: Student learning of science in small groups. *Journal of Research in Science Teaching, 34*, 1039-1065.

Biological Sciences Curriculum Study. (1973). *BSCS Green Version* (3rd ed.). Chicago: Rand McNally.

Black, M. (1962). Models and archetypes. In M. Black (Ed.), *Models and metaphors* (pp. 219-243). Ithaca, NY: Cornell University Press.

Bloor, D. (1976.). *Knowledge and social imagery*. London: Routledge and Kegan Paul.

Blumenfeld, P., Soloway, E., Marx, R., Krajcik, J. S., Guzdial, M. & Palincsar, A. (1991). Motivating project-based learning. *Educational Psychologist, 26*(3&4), 369-398.

Boltzmann, L. (1899/1974). On the development of the methods of theoretical physics in recent times. In B. McGuinness (Ed.), *Ludwig Boltzmann: Theoretical physics and philosophical problems: Selected writings* (pp. 77-100). Dordrecht, Holland: D. Reidel.

Bordeaux, J. (this volume). A Commentary on "Engineering Pedagogical Reform" by Dan Edelson.

Boudon, R. (1986). *The analysis of ideology*. M. Slater (Trans.). Chicago: University of Chicago Press.

Bowen, G. M. & Roth, W. M. (1998) *Isolation of variables and enculturation to a reductionist epistemology during ecology lectures*. Paper presented at the Annual Conference of the American Educational Research Association, San Diego, CA.

Bowen, G. M. & Roth, W. M. (2002). Why students may not learn to interpret scientific inscriptions. *Research in Science Education, 32*, 303-327.

Boyd, R. (1992). Constructivism, realism, and philosophical method. In J. Earman (Ed.) *Inference, explanation, and other frustrations: Essays in the philosophy of science* (pp. 131-198). Berkeley: University of California Press.

Boyd, R., Gasper, P. & Trout, J. D., Eds. (1991). *The philosophy of science*. Cambridge, MA: MIT Press.

Bransford, J., Brown, A. & Cocking, R., Eds. (1999). *How People Learn.* Washington, DC: National Academy Press.

Brewer, W. F. (1999a). Perceptual symbols: The power and limitations of a theory of dynamic imagery and structured frames. *Behavioral and Brain Sciences, 22*, 611-612.

Brewer, W. F. (1999b). Schemata. In R. A. Wilson & F. C. Keil (Eds.), *The MIT encyclopedia of the cognitive sciences* (pp. 729-730). Cambridge, MA: MIT Press.

Brewer, W. F. (1999c). Scientific theories and naive theories as forms of mental representation: Psychologism revived. *Science & Education, 8*, 489-505.

Brewer, W. F. (2001). Models in Science and Mental Models in Scientists and Nonscientists. *Mind & Society, 2*, 33-48.

Brewer, W. F. (this volume). In what sense can the child be considered to be a "little scientist"?

Brewer, W. F. & Chinn, C. A. (1994). Scientists' responses to anomalous data: Evidence from psychology, history, and philosophy of science. *Philosophy of Science Association, Volume 1*, 304-313.

Brewer, W. F., Chinn, C. A. & Samarapungavan, A. (2000). Explanation in scientists and children. In F. C. Keil & R. A. Wilson (Eds.), *Explanation and cognition* (pp. 279-298). Cambridge, MA: MIT Press.

Brewer, W. F. & Lambert, B. L. (2001). The theory-ladenness of observation and the theory-ladenness of the rest of the scientific process. *Philosophy of Science, 68*, S176-S186.

Brewer, W. F. & Mishra, P. (1998). Science. In W. Bechtel & G. Graham (Eds.), *A companion to cognitive science* (pp. 744-749). Oxford: Blackwell.

Brewer, W. F. & Samarapungavan, A. (1991). Children's theories vs. scientific theories: Differences in reasoning or differences in knowledge? In R. R. Hoffman & D. S. Palermo (Eds.), *Cognition and the symbolic processes: Applied and ecological perspectives* (pp. 209-232). Hillsdale, NJ: Erlbaum.

Brickhouse, N. (2001). Embodying science: A feminist perspective on learning. *Journal of Research in Science Teaching, 38*, 282-295.

Brickhouse, N. (this volume). Should social epistemology be rated X?

Brickhouse, N. (this volume). What is inquiry? To whom should it be authentic?

Brown, A. L. (1992). Design Experiments: Theoretical and methodological challenges in creating complex interventions in classroom settings. *Journal of the Learning Sciences, 2*(2), 141-178.

BIBLIOGRAPHY

Brown, A. L. & Campione, J. (1994). Guided discovery in a community of learners. In K. McGilly (Ed.), *Classroom Lessons: Integrating Cognitive Theory and Classroom Practice* (pp. 229-270). London: MIT Press.

Brown, J. S., Collins, A. & Duguid, P. (1989). Situated cognition and the culture of learning. *Educational Researcher, 18*, 32-42.

Brown, A. L. & DeLoache, J. S. (1978). Skills, plans, and self-regulation. In R. Siegler (Ed.), *Children's thinking: What develops?* (pp. 3-35). Hillsdale, NJ: Erlbaum.

Bruner, J. (1961). The Act of Discovery. *Harvard Educational Review, 31*, 21-32.

Bruer, J. (1993). *Schools for Thought: A Science of Learning in the Classroom*. Cambridge, MA: MIT Press.

Bruner, J. (1996). *The culture of education*. Cambridge, MA: Harvard University Press.

Burr, J. E. & Hofer, B. K. (2002). Personal epistemology and theory of mind: deciphering young children's beliefs about knowledge and knowing. *New Ideas in Psychology, 20*(2-3), 199-224.

Cacioppo, J. T., Petty, R. E., Kao, C. F. & Rodriguez, R. (1986). Central and peripheral routes to persuasion: An individual difference perspective. *Journal of Personality and Social Psychology, 51*, 1032-1043.

Calabrese Barton, A. & Brickhouse, N. W. (2006). Engaging girls in science. In L. Smulyan (Ed.) *Handbook of Gender and Education* (pp. 221-235). Thousand Oaks, CA: Sage.

Campbell, N. R. (1957). *Foundations of science*. New York: Dover. (Originally published as *Physics: The elements*, Cambridge: Cambridge University Press, 1920).

Carey, S. (1985a). Are children fundamentally different kinds of thinkers and learners than adults? In S. F. Chipman, J. W. Segal & R. Glaser (Eds.), *Thinking and learning skills: Vol. 2: Research and open questions* (pp. 485-517). Hillsdale, NJ: Erlbaum.

Carey, S. (1985b). *Conceptual change in childhood*. Cambridge, MA: MIT Press.

Carey, S. (1986). Cognitive science and science education. *American Psychologist, 41*, 1123-1130.

Carey, S. (1991). Knowledge acquisition: Enrichment or conceptual change? In S. Carey & R. Gelman (Eds.), *The epigenesis of mind: Essays on biology and cognition*. Hillsdale, NJ: Erlbaum.

Carey, S. & Smith, C. (1993). On understanding the nature of scientific knowledge. *Educational Psychologist, 28*(3), 235-251.

Carey, S. & Spelke, E. (1996). Science and core knowledge. *Philosophy of Science, 63*, 515-533.

Carlone, H. & Bowen, G. M. (2003). The fallacy of "Authentic" Science Classrooms: Missing Aspects of Practicing Science Communities. In the *Proceedings from the 7th International History, Philosophy of Science and Science Teaching Conference*, Winnipeg, pp. 187-198.

Carlsen, W. S. (1991). Questioning in classrooms: A sociolinguistic perspective. *Review of Educational Research, 61*, 157-178.

Carlsen, W. S. (1997). Never ask a question if you don't know the answer: The tension in teaching between modeling scientific argument and maintaining law and order. *Journal of Classroom Interaction, 32*, 14-23.

Carnap, R. (1956). The methodological character of theoretical concepts. In H. Feigel & M. Scriven (Eds.), *Minnesota studies in the philosophy of science : Vol. 1. Foundations of science and the concepts of psychology and psychoanalysis* (pp. 38-76). Minneapolis: University of Minnesota Press.

Cartwright, N. (1983). *How the laws of physics lie*. Oxford: Oxford University Press.

Cartwright, N. (1999). *The dappled world: A study of the boundaries of science*. Cambridge: Cambridge University Press.

Catley, K., Lehrer, R. & Reiser, B. (2005). *Tracing a prospective learning progression for developing understanding of evolution*, commissioned by the National Academies Committee on Test Design for K-12 Science Achievement, National Academies, Washington, D.C., http://www7.nationalacademies.org/bota/Test_Design_K-12_Science.html.

Champagne, A. & Newell, S. (1994), Directions for Research and Development: Alternative Methods of Assessing Scientific Literacy. *Journal of Research in Science Teaching, 29*, 841-860.

Chi, M. T. H. (1992). Conceptual change within an across ontological categories: Implications for learning and discovery in science. In R. N. Giere (Ed.), *Cognitive models of science* (Vol. 15, pp. 129-186). Minneapolis: University of Minnesota Press.

BIBLIOGRAPHY

Chi, M. T. H. (2005). Commonsense conceptions of emergent processes: Why some misconceptions are robust. *The Journal of the Learning Sciences, 14*, 161-199.

Chi, M. T. H., de Leeuw, N., Chiu, M. & LaVancher, C. (1994). Eliciting self-explanations improves understanding. *Cognitive Science, 18*, 439-477.

Chi, M. T. H., Peltovich, P. J. & Glaser, R. (1981). Categorization and representation of physics problems by experts and novices. *Cognitive Science, 5*, 121-152.

Chinn, C.A.& Samarapungavan, A. (this volume). Learning to use scientific models: Multiple dimensions of conceptual change.

Chinn, C. A. (2006). Learning to argue. In A. M. O'Donnell & C. E. Hmelo-Silver (Eds.), *Argumentation and technology* (pp. 355-383). Mahwah, NJ: Erlbaum.

Chinn, C. A. & Brewer, W. F. (1993). The role of anomalous data in knowledge acquisition: A theoretical framework and implications for science instruction. *Review of Educational Research, 63*, 1-49.

Chinn, C. A. & Brewer, W. F. (2000). Knowledge change in response to data in science, religion, and magic. In K. S. Rosengren (Ed.), *Scientific, and Religious Thinking in Children*.

Chinn, C. A. & Brewer, W. F. (2001). Models of data: A theory of how people evaluate data. *Cognition and Instruction, 19*, 323-393.

Chinn, C. A. & Malhotra, B. A. (2002a). Children's responses to anomalous scientific data: How is conceptual change impeded? *Journal of Educational Psychology, 19*, 327-343.

Chinn, C. A. & Malhotra, B. A. (2002b). Epistemologically authentic reasoning in schools: A theoretical framework for evaluating inquiry tasks. *Science Education, 86*, 175-218.

Chinn, C. A., O'Donnell, A. M. & Jinks, T. S. (2000). The structure of discourse in collaborative learning. *Journal of Experimental Education, 69*, 77-97.

Chinn, C. A. & Samarapungavan, A. (2001). Distinguishing between understanding and belief. *Theory Into Practice, 40*, 235-241.

Chinn, C. A. & Samarapungavan, A. (2005). *Learning to use scientific models: Multiple dimensions of conceptual change.* Paper presented at the Inquiry Conference on Developing a Consensus Research Agenda, Rutgers University, NJ.

Clancey, W. J. (1997). *Situated cognition: On human knowledge and computer representations.* Cambridge: Cambridge University Press.

Clement, J. (1982). Students preconceptions in introductory mechanics. *American Journal of Physics, 50*(1), 66-71.

Clough, E. E. & Driver, R. (1985). Secondary students' conceptions of the conduction of heat: Bringing together scientific and personal views. *Physics Education, 20*, 176-182.

Cobb, P. (1994). Constructivism in mathematics and science education. *Educational Researcher, 23*(4).

Cobern, W. W. (1996). Worldview theory and conceptual change in science education. *Science Education, 80*(5), 579-610.

Cognition & Technology Group at Vanderbilt. (1997). *The Jasper Project: Lessons in curriculum, instruction, assessment, and professional development.* Mahwah, NJ: Lawrence Erlbaum Associates.

Cohen, D. (1995). Argument is war and war is hell: Philosophy, education, and metaphors for argumentation. *Informal Logic, 17*(2).

Cole, M. & Engestrom, Y. (1993). A cultural-historical approach to distributed cognition. In G. Salomon (Ed.), *Distributed cognitions: Psychological and educational considerations* (pp. 1-46). Cambridge: Cambridge University Press.

Collins, A. (1992). Toward a design science of education. In E. Scanlon & T. O'Shea (Eds.), *New directions in educational technology* (pp. 15-22). Berlin: Springer.

Collins, A., Brown, J. S. & Duguid, P. (1989). Cognitive apprenticeship: Teaching the crafts of reading, writing, and mathematics. In L. B. Resnick (Ed.), *Knowing, Learning, and Instruction*. Hillsdale, NJ: Lawrence Erlbaum.

Collins, A., Brown, J. S. & Larkin, K. M. (1980). Inference in text understanding. In R. J. Spiro, B. C. Bruce & W. F. Brewer (Eds.) *Theoretical issues in reading comprehension* (pp. 385-407). Hillsdale, NJ: Erlbaum.

Collins, A. & Ferguson, W. (1993). Epistemic forms and epistemic games: Structures and strategies to guide inquiry. *Educational Psychologist, 28*(1), 25-42.

Collins Block, C. & Pressley, M. (2002). *Comprehension instruction: Research-based best practices*. New York: Guilford.
Confrey, J. (1990). A review of the research on student conceptions in mathematics, science, and programming. *Review of Research in Education, 16*, 3-56.
Connexions. (n.d.). Retrieved February 14, 2005, from the Connexions Project Web site: http://cnx.rice.edu/
Costa, V. B. (1995). When science is "another world": Relationships between worlds of family, friends, school, and science. *Science Education, 79*(3), 313-333.
Council of Ministers of Education, Canada. (1997). *Common framework of science learning outcomes K to 12*. Toronto: Author.
Covington, M. V. (2000). Goal theory, motivation, and school achievement: An integrative review. *Annual Review Pschology, 51*, 171-200.
Craig, M. T. & Yore, L. D. (1996). Middle school students' awareness of strategies for resolving reading comprehension difficulties in science reading. *Journal of Research in Development in Education, 29*, 226-238.
Crane, D. (1972). *Invisible colleges: Diffusion of knowledge in scientific communities*. Chicago: University of Chicago Press.
Crawford, T., Kelly, G. J. & Brown, C. (2000). Ways of knowing beyond facts and laws of science: An ethnographic investigation of student engagement in scientific practices. *Journal of Research in Science Teaching, 37*, 237-258.
Crawford, V. & Toyama, Y. (2002). *World Watcher / Looking at the environment curriculum: Final external evaluation report* (Project Report). Menlo Park, CA: Center for Technology in Learning, SRI International.
Cunningham, C. M. & Helms, J. V. (1998). Sociology of science as a means to a more authentic, inclusive science education. *Journal of Research in Science Teaching, 35*, 483-499.
Cushing, J. T. (1994). *Quantum mechanics: historical contingency and the copenhagen hegemony*. Chicago: University of Chicago Press.
D'Andrade, R. (1987). A folk model of the mind. In D. Holland & N. Quinn (Eds.), *Folk theories in language and thought*. Cambridge: Cambridge University Press.
D'Andrade, R. (1995). *The development of cognitive anthropology*. Cambridge: Cambridge University Press.
D'Andrade, R. & Strauss, C. (1992). *Human motives and folk theories*. Cambridge: Cambridge University Press.
Darden, L. (1991). *Theory Change in Science: Strategies from Mendelian Genetics*. New York: Oxford University Press.
de Vries, E., Lund, K. & Baker, M. (2002). Computer-mediated epistemic dialogue: explanation and argumentation as vehicles for understanding scientific notions. *The Journal of the Learning Sciences, 11*(1), 63-103.
DeBoer, G. E. (1991). *A history of ideas in science education*. New York: Teachers College Press.
Design-Based Research Collective. (2003). Design-based research: An emerging paradigm for educational inquiry. *Educational Researcher, 32*(1), 5-8.
Diakidoy, I. A. N. & Ioannides, C. (2004). Elementary school children's ability to distinguish hypothetical beliefs from statements of preference. *Journal of Educational Psychology, 96*(3), 536-544.
Dillon, J. T. (1994). *Using discussion in classrooms*. Buckingham: Open University Press.
diSessa, A. A. (1988). Knowledge in pieces. In G. Forman & P. B. Pufall (Eds.), *Constructivism in the computer age* (pp. 49-70). Hillsdale, NJ: Lawrence Erlbaum Assoc.
diSessa, A. A. (1993). Toward an epistemology of physics. *Cognition and Instruction, 10*, 105-225.
diSessa, A. A. (2000). *Changing minds: Computers, learning, and literacy*. London: MIT Press.
Dole, J. A. & Sinatra, G. M. (1998). Reconceptualizing change in the cognitive construction of knowledge. *Educational Psychologist, 33*, 109-128.
Donald, M. (1991). *Origins of the modern mind: Three stages in the evolution of culture and cognition*. Cambridge, MA: Harvard University Press.
Doran, R. & Tamir, P. (1992). An international assessment of science practical skills. *Studies in Educational Evaluation, 18*, 263-406.

BIBLIOGRAPHY

Doran, R., Boorman, J., Chan, F. & Hejaily, N. (1993). Alternative assessment of high school laboratory skills. *Journal of Research in Science Teaching, 30*, 1121-1131.

Doran, R., Lawrenz, F. & Helgeson, S. (1994). Research on assessment in science. In D. Gabel (Ed.), *Handbook of research on science teaching and learning* (pp. 388-442). New York: Macmillan.

Draper, R. J. (2002). Every teacher a literacy teacher? An analysis of the literacy-related messages in secondary methods testbooks. *Journal of Literacy Research, 34*, 357-384.

Driver, R. A., Asoko, H., Leach, J., Mortimer, E. & Scott, P. (1994). Constructing scientific knowledge in the classroom. *Educational Researcher, 23*(4), 5-12.

Driver, R. A. & Easley, J. (1978). Pupils and paradigms: A review of literature related to concept development in adolescent science students. *Studies in Science Education, 5*, 61-84.

Driver, R. A., Leach, J., Millar, R. & Scott. P. (1996). *Young people's images of science.* Philadelphia: Open University Press.

Driver, R. A., Newton, P. & Osborne, J. (2000). Establishing the norms of scientific argumentation in classrooms. *Science Education, 84*(3), 287-312.

Duell, O. K. & Schommer-Atkins, M. (2001). Measures of people's beliefs about knowledge and learning. *Educational Psychology Review, 13*, 419-449.

Dunbar, K. (1995). How scientists really reason: Scientific reasoning in real-world laboratories. In R. J. Sternberg & J. E. Davidson (Eds.) *The Nature of Insight* (pp. 365-395). Cambridge, MA: MIT Press.

Dunbar, K. (1999). How scientists build models: In Vivo Science as a window on the scientific mind. In L. Magnani, N. J. Nersessian & P. Thagard (Eds.), *Model-based reasoning in scientific discovery.* New York: Kluwer Academic/Plenum Publishers.

Duschl R. A. (1990). *Restructuring science education: The importance of theories and their development.* New York: Teachers' College Press.

Duschl, R. A. (2000). Making the nature of science explicit. In R. Millar, J. Leach & J. Osborne (Eds.), *Improving science education: The contribution of research* (pp. 187-206). Philadelphia: Open University Press.

Duschl, R.(2004a). Assessment of inquiry. In J. M. Atkin & J. C. Coffey (Eds.), *Everyday assessment in the science classroom.* Arlington, VA: NSTA Press.

Duschl, R. A. (2004b). Understanding dialogic argumentation. Paper presented at the Annual Meeting of the American Educational Research Association, San Diego.

Duschl, R. A., Ellengoben, K. & Erduran, S. (1999). Understanding dialogic argumentation among middle school science students. Invited paper at the annual meeting of the American Educational Research Association (AERA), Montreal, April 1999.

Duschl, R. A. & Gitomer, D. (1997). Strategies and challenges to changing the focus of assessment and instruction in science classrooms. *Educational Assessment, 4*(1), 337-73.

Duschl, R. A. & Grandy, R. E. (this volume). Reconsidering the character and role of inquiry in school science: Framing the debates.

Duschl, R. A. & Osborne, J. (2002). Supporting and promoting argumentation discourse in science education. *Studies in Science Education, 38*, 39-72.

Dym, C. L. & Little, P. (2000) *Engineering design: A project-based Introduction.* New York, NY: John Wiley & Sons.

Ebbers, M. & Rowell, P. (2002). Description is not enough: Scaffolding children's explanations. *Primary Science Review, 74* (Sep-Oct), 10-13.

Edelson, D. C. (1998). Realising authentic science learning through the adaptation of scientific practice. In B. J. Fraser & K. G. Tobin (Eds.), *International handbook of science education* (pp. 317-332). The Netherlands: Kluwer Academic Publishers.

Edelson, D. C. (2001). Learning-For-Use: A framework for the design of technology-supported inquiry activities. *Journal of Research in Science Teaching, 38*(3), 355-385.

Edelson, D. C. (2002). Design research: What we learn when we engage in design. *Journal of the Learning Sciences, 11*(1), 105-121.

Edelson, D. C. (this volume). Engineering pedagogical reform: A case study of technology-supported inquiry.

Edelson, D. C. & Gordin, D. N. (1998). Visualization for learners: A framework for adapting scientists' tools. *Computers and Geosciences, 24*(7), 607-616.

Edelson, D. C., Gordin, D. N., Clark, B. A., Brown, M. & Griffin, D. (1997). WorldWatcher [Computer Software]. Evanston, IL: Northwestern University.
Edelson, D. C., Gordin, D. N. & Pea, R. D. (1999). Addressing the challenges of inquiry-based learning through technology and curriculum design. *Journal of the Learning Sciences, 8*(3&4), 391-450.
Eflin, J., Glennan, S. & Reisch, G. (1999). The nature of science: A perspective from the philosophy of science. *Journal of Research in Science Teaching ,36*(1), 107-116.
Eisenhart, M. & Finkel, E. (1998). *Women's science: Learning and succeeding from the margins.* Chicago: University of Chicago Press.
Engestrom, Y. (1999). Activity theory and individual and social transformation. In Y. Engestrom, R. Miettinen & R. L. Punamaki (Eds.), *Perspectives on activity theory* (pp. 19-38). Cambridge: Cambridge University Press.
Engestrom, Y. & Miettinen, R. (1999). Introduction. In Y. Engestrom, R. Miettinen & R. L. Punamaki (Eds.), *Perspectives on activity theory* (pp. 1-16). Cambridge: Cambridge University Press.
Erduran, S. (1999). *Merging curriculum design with chemical epistemology: A case of teaching and learning chemistry through modeling.* Unpublished Ph.D. dissertation, Vanderbilt University, Nashville, TN, USA.
Etikna, E. (this volume). Panel B.
Etkina, E., Matilsky, T. & Lawrence, M. (2003). Pushing to the edge: Rutgers Astrophysics Institute motivates talented high school students. *Journal of Research in Science Teaching, 10*, 958-985.
Faraday, M. (1831). *Experimental researches in electricity.* 3 vols. New York: Dover.
Faucher, L., Mallon, R., Nazer, D., Nichols, S., Ruby, A., Stich, S. & Weinberg, J. (2002). The baby in the lab-coat: Why child development is not an adequate model for understanding the development of science. In P. Carruthers, S. Stich & M. Siegal (Eds.), *The cognitive basis of science* (pp. 335-362). Cambridge, UK: Cambridge University Press.
Feldman, D. H. (1994). *Beyond universals in cognitive development.* Norwood, NJ: Ablex.
Ferreiro, E. (2000). Reading and writing in a changing world. *Publishing Research Quarterly, Fall*, 53-61.
Feyerabend, P. (1993). *Against method* (3rd ed.). New York: Verso.
Flew, A. (1979). *A dictionary of philosophy* (2nd ed.). New York: St. Martins Press.
Forman, E. A. & Ansell. E. (2002). Orchestrating the multiple voices and inscriptions of a mathematics classroom. *The Journal of the Learning Sciences, 11*, 251-274.
Forrester, J. W. (1971). Counterintuitive behavior of social systems. *Technology Review, 73*, 52-67.
Friedman, M. (1999). *Reconsidering logical positivism.* Cambridge: Cambridge University Press.
Fuller, S. (1988). *Social epistemology.* Bloomington: Indiana University Press.
Galison, P. (1987). *How experiments end.* Chicago: University of Chicago Press.
Galison, P. (1997). *Image and Logic: A material culture of microphysics.* Chicago: University of Chicago Press.
Gallas, K. (1995). *Talking their way into science: Hearing children's question and theories, responding with curricula.* New York: Teacher College Press.
Gaskins, I. W., Guthrie, J. T., Satlow, E., Ostertag, J., Six, L., Byrne, J. & Connor, B. (1994). Integrating instruction of science, reading, and writing: Goals, teacher development, and assessment. *Journal of Research in Science Teaching, 31*, 1039-1056.
Gee, J. (1994). Science talk: How do you start to do what you don't know how to do? Paper presented at the annual meeting of the American Educational Research Association, New Orleans, April 1994.
Gee. J. P. (1999). *An introduction to discourse analysis: Theory and method.* New York: Routledge.
Gee, J. P. & Green, J. L. (1998). Discourse analysis, learning, and social practice: A methodological study. *Review of Research in Education, 23*, 119-169.
Gellatly, A. (1997). Why the young child has neither a theory of mind nor a theory of anything else. *Human Development, 40*, 32-50.
Gentner, D. (1989). The mechanisms of analogical learning. In S. Vosniadou & A. Ortony (Eds.), *Similarity and analogical reasoning* (pp. 199-241). Cambridge: Cambridge University Press.
Gentner, D., Brem, S., Ferguson, R. W., Markman, A. B., Levidow, B. B., Wolff, P. & Forbus, K. D. (1997). Analogical reasoning and conceptual change: A case study of Johannes Kepler. *The Journal of the Learning Sciences, 6*(1), 3-40.

BIBLIOGRAPHY

Germann, P. J., Haskins, S. & Auls, S. (1996). Analysis of nine high school biology laboratory manuals: Promoting scientific inquiry. *Journal of Research in Science Teaching, 33*, 475-499.

Gesell, A. (1926). Maturation and infant behavior pattern. *Psychological Review, 36,* 307-319.

Giedd, J. N. (2004). Structural magnetic resonance imaging of the adolescent brain. Ann. N. Y. Acad. Sci. 1021, 77-85.

Giere, R. N. (1988). *Explaining science: A cognitive approach.* Chicago: University of Chicago Press.

Giere, R. N. (1991). *Understanding scientific reasoning* (3rd ed.). New York: Harcourt Brace Jovanovich College Publishers.

Giere, R. N. (1992). *Cognitive models of science. Minnesota Studies in the Philosophy of Science, 15.* Minneapolis: University of Minnesota Press.

Giere, R. N. (1994). The cognitive structure of scientific theories. *Philosophy of Science, 61,* 276-296.

Giere, R. N. (1996). The scientist as adult. *Philosophy of Science, 63,* 538-541.

Giere, R. N. (1999). *Science without laws.* Chicago: University of Chicago Press.

Giere, R. N. (2002). Models as part of distributed cognitive systems. In Magnani et al. (Eds.).

Gilbert, D. T. (1991). How mental systems believe. *American Psychologist, 46,* 107-119.

Gilbert, D. T., Tafarodi, R. W. & Malone, P. S. (1993). You can't not believe everything you read. *Journal of Personality and Social Psychology, 65,* 221-233.

Gitomer, G. (this volume). Panel B.

Gitomer, G. & Duschl, R. (1998). Emerging isues and practices in science assessment. In B. Fraser & K. Tobin, (Eds.), *International Handbook of Science Education* (pp. 791-810). Dordrecht: Kluwer Academic Publishers.

Glaser, R. (1992). Expert knowledge and process of thinking. In D. F. Halpern (Ed.), *Enhancing Thinking Skills in the Sciences and Mathematics.* Hillsdale, NJ: Erlbaum.

Glaser, R. (1995). Application and theory: Learning theory and the design of learning environments. Paper presented at the 23rd International Congress of Applied Psychology, July 17-22, 1994, Madrid, Spain.

Glenberg, A. M. & Langston, W. E. (1992). Comprehension of illustrated text: Pictures help to build mental models. *Journal of Memory and Language, 31,* 129-151.

Godfrey-Smith, P. (2003). *Theory and reality.* Chicago: The University of Chicago Press.

Goldman, A. (1999). *Knowledge in a social world.* New York: Oxford University Press.

Goldman, A. (2001). Social epistemology. *Stanford Encyclopedia of Philosophy.* Retrieved from http://plato.stanford.edu.

Goldman, S., Duschl, R., Ellenbogen, K., Williams, S. & Tzou, C. (2002). Science inquiry in a digital age: Possibilities for making thinking visible. In H. van Oostendorp (Ed.), *Cognition in a digital age.* Mahwah, NJ: Erlbaum Press.

Gooding, D. (1990). *Experiment and the making of meaning: Human agency in scientific observation and experiment.* Dordrecht: Kluwer.

Goodman, K. S. (1985). Reading: A psycholinguistic guessing game. In H. Singer & R. B. Ruddell (Eds.), *Theoretical models and processes of reading* (pp. 259-272). Newark, DE: International Reading Association.

Goodman, N. (1983) *Fact, fiction and forecast* (4th ed.). Cambridge, MA: Harvard University Press.

Goodwin (1994). Professional vision. *American Anthropologist, 96*(3), 606-663.

Gopnik, A. (1996a). Reply to commentators. *Philosophy of Science, 63,* 552-561.

Gopnik, A. (1996b). The scientist as child. *Philosophy of Science, 63,* 485-514.

Gopnik, A. & Meltzoff, A. N. (1997). *Words, thoughts, and theories.* Cambridge: MIT Press.

Gopnik, A., Meltzoff, A. N. & Kuhl, P. K. (1999). *The scientist in the crib: Minds, brains, and how children learn.* New York: William Morrow.

Gopnik, A. & Sobel, D. M. (2000). Detecting blickets: How young children use information about novel causal powers in categorization and induction. *Child Development, 71*(5), 1205-1222.

Gopnik, A., Sobel, D. M., Schulz, L. E. & Glymour, C. (2001). Causal learning mechanisms in very young children: Two-, three-, and four-year-olds infer causal relations from patterns of variation and covariation. *Developmental Psychology, 37,* 620-629.

Gopnik, A. & Wellman, H. M. (1992). Why the child's theory of mind really *is* a theory. *Mind & Language, 7,* 145-171.

Gopnik, A. & Wellman, H. M. (1994). The theory theory. In L. A. Hirschfeld & S. A. Gelman (Eds.), *Mapping the mind: Domain specificity in cognition and culture* (pp. 257-293). Cambridge: Cambridge University Press.
Gordin, D. N. & Pea, R. D. (1995). Prospects for scientific visualization as an educational technology. *Journal of the Learning Sciences, 4*(3), 249-279.
Gorman, M. (1997). Mind in the world: Cognition and practice in the invention of the telephone. *Social Studies of Science, 27*, 583-624.
Gorman, M. E. & Carlson, W. B. (1990). Interpreting invention as a cognitive process: The case of Alexander Graham Bell, Thomas Edison, and the telephone. *Science, Technology, and Human Values, 15*, 131-164.
Grandy, R. E. (1997). Constructivisms and objectivity: Disentangling metaphysics from pedagogy. *Science and Education, 6*, 43-53.
Grandy R. E. (2003a). Kuhn's world changes. In T. Nickles (Ed.), *Thomas Kuhn*. Cambridge: Cambridge University Press.
Grandy, R.E. (2003b). What are models and why do we need them? *Science & Education, 12*, 773-777.
Grandy, R.E. & Duschl, R. A. (this volume). Consensus: Expanding the scientific method and school science.
Greeno, J. G. (1989). A perspective on thinking. *American Psychologist, 44*, 134-141.
Greeno, J. G. (1998). The situativity of knowing, learning, and research. *American Psychologist, 53*, 5-24.
Grice, H. P. (1989). *Studies in the way of words*. Cambridge, MA: Harvard University Press.
Griesemer, J. R. (1991a). Material models in biology. In *PSA 1990*. East Lansing, MI: PSA.
Griesemer, J. R. (1991b). Must scientific diagrams be eliminable? The case of path analysis. *Biology and Philosophy, 6*, 177-202.
Gross, A. (1990). *The rhetoric of science*. Cambridge; Harvard University Press.
Grosslight, L., Unger, C., Jay. E. & Smith, C. (1991). Understanding models and their use in science: Conceptions of middle and high school students and experts. *Journal of Research in Science Teaching, 28*, 799-822.
Grotzer, T. A. & Perkins, D. N. (2000). *A taxonomy of causal models: The conceptual leaps between models and students' reflections on them*. Paper presented at the National Association of Research in Science Teaching, New Orleans.
Guala, F. (2003). Experimental localism and external validity. *Philosophy of Science, 70*, 1195-1205.
Gumperz, J. J. (1982). *Discourse strategies*. Cambridge: Cambridge University Press.
Gumperz, J. J., Cook-Gumperz, J. & Szymanski, M. H. (1999). *Collaborative practices in bilingual cooperative learning classrooms*. Santa Cruz, CA: Center for Research on Education, Diversity, and Excellence.
Guzzetti, B. J., Snyder, T. E., Glass, G. V. & Gamas, W. S. (1993). Promoting conceptual change in science: A comparative meta-analysis of instructional interventions from reading education and science education. *Reading Research Quarterly, 28*, 116-155.
Haas, C. & Flower, L. (1988). Rhetorical reading strategies and the recovery of meaning. *College Composition and Communication, 39*, 30-47.
Habermas, J. (1987). *The philosophical discourse of modernity*. (F. Lawrence, Trans.) Cambridge: MIT Press. (Original work published 1985).
Habermas, J. (1990). *Moral consciousness and communicative action*. (translated by C. Lenhardt & S. W. Nicholsen). Cambridge, MA: MIT press.
Halliday, M. A. K. & Martin, J. R. (1993). *Writing science: Literacy and discursive power*. Pittsburgh: University of Pittsburgh Press.
Hammer, D. (1994). Epistemological beliefs in introductory physics. *Cognition and Instruction, 12*, 151-183.
Hammer, D. (2004). The variability of student reasoning, lectures 1-3. In E. Redish & M. Vicentini (Eds.), *Proceedings of the Enrico Fermi Summer School, Course CLVI* (pp. 279-340). Italian Physical Society.
Hammer, D. & Elby, A. (2002). On the form of a personal epistemology. In B. K. Hofer & P. R. Pintrich (Eds.), *Personal epistemology: The psychology of beliefs about knowledge and knowing* (pp. 169-190). Mahwah, NJ: Lawrence Erlbaum.

Hammer, D., Elby, A., Scherr, R. E. & Redish, E. F. (2005). Resources, framing, and transfer. In J. Mestre (Ed.), *Transfer of learning from a modern multidisciplinary perspective* (pp. 89-119). Greenwich, CT: Information Age Publishing.

Hammer, D., Russ, R., Mikeska, J. & Scherr, R. (this volume). Identifying inquiry and conceptualizing students' abilities.

Hanson, N. R. (1958). *Patterns of discovery*. Cambridge: Cambridge University Press.

Harré, R. (1986). *Varieties of realism*. Oxford: Basil Blackwell.

Harris, P. L. (1994). Thinking by children and scientists: False analogies and neglected similarities. In L. A. Hirschfeld & S. A. Gelman (Eds.), *Mapping the mind: Domain specificity in cognition and culture* (pp. 294-315). Cambridge: Cambridge University Press.

Harte, J. (1988). *Consider a spherical cow*. Sausolito, CA: University Science Books.

Helm, H. & Novak, J. D. (Eds.). (1983). *Proceedings of the International Seminar on Misconceptions in Science and Mathematics*. Ithaca, NY: Cornell University.

Hempel, C. G. (1943). A purely syntactical definition of confirmation. *The Journal of Symbolic Logic, 8*, 122-43.

Hempel, C. G. (1965). *Aspects of scientific explanation and other essays in the philosophy of science*. New York: The Free Press; and London: Collier-Macmillan, Ltd.

Hempel, C. G. (1966). *Philosophy of natural science*. Englewood Cliffs, NJ: Prentice-Hall.

Hempel, C. G. (1970). On the 'standard conception' of scientific theories. In M. Radner & S. Winokur (Eds.), *Minnesota Studies in the Philosophy of Science IV* (pp. 142-163). Minneapolis: University of Minnesota Press.

Hempel, C. G. (1983a). Valuation and objectivity in science. In R. S. Cohen & L. Laudan (Eds.), *Physics, philosophy and psychoanalysis in honor of Adolf Grunbaum* (pp. 73-100). Dordrecht, Boston, Lancaster: D. Reidel Publishing Co.

Hempel, C. G. (1983b). Kuhn and Salmon on rationality and theory choice. *The Journal of Philosophy 80*, 570-572.

Hempel, C. G. (1988). Limits of a deductive construal of the function of scientific theories. In E. Ullman-Margalit (Ed.), *Science in Reflection* (pp. 1-15). The Israel Colloquium 3. Dordrecht: Kluwer Academic Publishers.

Herman, P., MacKenzie, S., Sherin, B. & Reiser, B. J. (2002). Assessing student learning in project-based science classrooms: Development and administration of written assessm. In P. Bell, R. Stevens & T. Satwicz (Eds.), *Keeping learning complex: The Proceedings of the Fifth International Conference of the Learning Sciences (ICLS)*. Mahwah, NJ: Erlbaum.

Herrenkohl, L. & Guerra, M. (1998). Participant structures, scientific discourse, and student engagement in fourth grade. *Cognition and Instruction, 16*(4), 431-473.

Herrenkohl, L. R., Palincsar, A. S., DeWater, L. S. & Kawasaki, K. (1999). Developing scientific communities in classrooms: A sociocognitive approach. *Journal of the Learning Sciences, 8*, 451-493.

Hesse, M. (1963). *Models and analogies in science*. London: Sheed and Ward

Hesse, M. (1966). *Models and analogies in science*. Notre Dame, IN: University of Notre Dame Press.

Hestenes, D. (1992). Modeling games in the Newtonian world. *American Journal of Physics, 60*, 732-748.

Hewson, P. W., Beeth, M. E. & Thorley, N. R. (1998). Teaching for conceptual change. In B. J. Fraser & K. G. Tobin (Eds.), *International handbook of science education* (pp. 199-218). Dordrecht, The Netherlands: Kluwer.

Hmelo-Silver, C. (this volume). Panel B.

Hodson, D. (1993), Re-thinking Old Ways: Towards a more critical approach to practical work in school science. *Studies in Science Education, 22*, 85-142.

Hofer, B. K. (2001). Personal epistemological research: Implications for learning and teaching. *Journal of Educational Psychology Review, 13*, 353-383.

Hofer, B. K. & Pintrich, P. R. (1997). The development of epistemological theories: Beliefs about knowledge and knowing and their relation to learning. *Review of Educational Research, 67*(1), 88-140.

Hofstein, A. & Lunetta, V. (1982). The role of the laboratory in science teaching: Neglected aspects of research. *Review of Educational Research, 52*, 201-217.

Hogan, K., Nastasi, B. & Pressley, M. (2000). Discourse patterns and collaborative scientific reasoning in peer and teacher-guided discussions. *Cognition and Instruction, 17*(4), 379-432.
Holland, D. & Quinn, N. (1987). Culture and cognition. In D. Holland & N. Quinn (Eds.), *Folk theories in language and thought*. Cambridge: Cambridge University Press.
Holmes, F. L. (1985). *Lavoisier and the chemistry of life: An exploration of scientific creativity*. Madison: University of Wisconsin Press.
Horwich, P., Ed. (1993). *World changes: Thomas Kuhn and the nature of science*. Cambridge, MA: MIT Press.
Hull, D. (1988). *Science as a process: An evolutionary account of the social and conceptual development of science*. Chicago: University of Chicago Press.
Hutchins, E. (1995) *Cognition in the wild*. Cambridge, MA: MIT Press.
Jackson, S., Krajcik, J. & Soloway, E. (2000). Model-it: A design retrospective. In M. Jacobson & R. Kozma (Eds.), *Advanced Designs For The Technologies Of Learning: Innovations in Science and Mathematics Education*. Hillsdale, NJ: Erlbaum.
Jasanoff, S., Markle, G. E., Petersen, J. C. & Pinch, T., Eds. (1995). *Handbook of science and technology studies*. Thousand Oaks: Sage Publications.
Johnson, M. (1987). *The body in the mind: The bodily basis of meaning, imagination, and reason*. Chicago: University of Chicago Press.
Kahneman, D., Slovic, P. & Tversky, A., Eds. (1982). *Judgments under uncertainty: Heuristics and biases*. Cambridge, UK: Cambridge University Press.
Kanis, I., Doran, R. & Jacobson, W. (1990). Assessing science process laboratory skills at the elementary and middle/Junior high levels. National Science Teachers' Association. Washington, DC.
Karmiloff-Smith, A. (1988). The child is a theoretician, not an inductivist. *Mind & Language, 3*, 183-195.
Karmiloff-Smith, A. (1992). *Beyond modularity*. Cambridge, MA: MIT Press.
Karmiloff-Smith, A. & Inhelder, B. (1975). "If you want to get ahead, get a theory". *Cognition, 3*, 195-212.
Karplus, R. & Thier, H. D. (1967). *A new look at elementary school science*. Chicago: Rand McNally.
Keesing, R. (1987). Models, folk and cultural. In D. Holland & N. Quinn (Eds.), *Folk theories in language and thought*. Cambridge: Cambridge University Press.
Keller, E. F. (1983). *A feeling for the organism: The life and work of Barbara McClintock*. New York: W. H. Freeman.
Kelly, G. J. (1997). Research traditions in comparative context: A philosophical challenge to radical constructivism. *Science Education, 81*, 355-375.
Kelly, G. J. (2000). The epistemological framing of a discipline: Writing science in university oceanography. *Journal of Research in Science Teaching, 37*(7), 691-718.
Kelly, G. J. (2006). Epistemology and educational research. In G. Camilli, P. Elmore & J. Green, (Eds.), *Handbook of complementary methods for educational research in education* (pp. 33-55). Mahwah, NJ: Lawrence Erlbaum Associates.
Kelly, G. J. & Bazerman, C. (2003). How students argue scientific claims: A rhetorical-semantic analysis. *Applied Linguistics, 24*(1), 28-55.
Kelly, G. J. & Brown, C. M. (2003). Communicative demands of learning science through technological design: Third grade students' construction of solar energy devices. *Linguistics & Education, 13*(4), 483-532.
Kelly, G. J., Chen, C. & Crawford, T. (1998). Methodological considerations for studying science-in-the-making in educational settings. *Research in Science Education, 28*(1), 23-49.
Kelly, G. J., Chen, C. & Prothero, W. (2000). The epistemological framing of a discipline: Writing science in university oceanography. *Journal of Research in Science Teaching, 37*, 691-718.
Kelly, G. J. & Crawford, T. (1997). An ethnographic investigation of the discourse processes of school science. *Science Education, 81*(5), 533-560
Kelly, G. J., Crawford, T. & Green, J. (2001). Common tasks and uncommon knowledge: Dissenting voices in the discursive construction of physics across small laboratory groups. *Linguistics & Education, 12*(2), 135-174.

BIBLIOGRAPHY

Kelly G. J., Drucker S. & Chen K. (1998). Students' reasoning about electricity: combining performance assessment with argumentation analysis. *International Journal of Science Education, 20*(7), 849-871.

Kelly, G. J. & Duschl, R. A. (2002). *Toward a research agenda for epistemological studies in science education*. Paper presented at the annual meeting of the National Association for Research in Science Teaching, New Orleans, LA.

Kelly, G. J. & Green, J. (1998). The social nature of knowing: Toward a sociocultural perspective on conceptual change and knowledge construction. In B. Guzzetti & C. Hynd (Eds.), *Perspectives on conceptual change: Multiple ways to understand knowing and learning in a complex world* (pp. 145-181). Mahwah, NJ: Lawrence Erlbaum Associates.

Kelly, G. J. & Takao, A. (2002). Epistemic levels in argument: An analysis of university oceanography students' use of evidence in writing. *Science Education, 86*, 314-342.

Kempton, W., Boster, J. S. & Hartley, J. A. (1995). *Environmental values in american culture*. Cambridge: MIT Press.

Khishfe, R. & Abd-El-Khalick, F. (2002). Influence of explicit and reflective versus implicit inquiry-oriented instruction on sixth graders' views of nature of science. *Journal of Research in Science Teaching, 39*(7), 551-578.

King, A. (1994). Inquiry as a tool in critical thinking. In D. F. Halpern (Ed.), *Changing college classrooms: New teaching and learning strategies for an increasingly complex world*. San Francisco: Jossey Bass.

Kitcher, P. (1993). *The advancement of science: Science without legend, objectivity without illusions*. New York: Oxford University Press.

Knorr-Cetina, K. (1999). *Epistemic cultures: How the sciences make knowledge*. Cambridge, MA: Harvard University Press.

Kolodner, J. L. (1993). *Case-based reasoning*. San Mateo, CA: Morgan Kaufmann.

Kornblith, H. (1994). A conservative approach to social epistemology. In F. Schmitt (Ed.), *Socializing epistemology: The social dimensions of knowledge* (pp. 93-110). Lanham MD: Rowman and Littlefield.

Kozma, R. (2003). The material features of multiple representations and their cognitive and social affordances for science understanding. *Learning and Instruction, 13*, 205-226.

Kozma, R., Chin, E., Russell, J. & Marx, N. (2000). The roles of representation and tools in the chemistry laboratory and their implications for chemistry learning. *The Journal of the Learning Sciences, 9*, 105-143.

Krajcik, J. (this volume). Commentary on Chinn's and Samarapungavan's paper.

Krajcik, J., Blumenfeld, B., Marx, R. & Soloway, E. (2000). Instructional, curricular, and technological supports for inquiry in science classrooms. In J. Minstell & E.Van Zee (Eds.), *Inquiry into inquiry: Science learning and teaching* (pp. 283-315). Washington, D.C.: American Association for the Advancement of Science Press.

Krajcik, J., Czerniak, C. & Berger, C. (1999). *Teaching children science: A project-based approach*. Boston: McGraw-Hill.

Kruglanski, A. W. & Webster, D. M. (1996). Motivated closing of the mind: "Seizing" and "freezing". *Psychological Review, 103*, 263-283.

Kuhn, D. (1989). Children and adults as intuitive scientists. *Psychological Review, 96*, 674-689.

Kuhn, D. (1993) Science as argument: Implications for teaching and learning scientific thinking. *Science Education, 77*(3), 319-337.

Kuhn, D. (1997). Constraints or guideposts? Developmental psychology and science education. *Review of Educational Research, 67*, 141-150.

Kuhn, D., Black, J., Keselman, A. & Kaplan, D. (2000). The development of cognitive skills to support inquiry learning. *Cognition and Instruction, 18*(4), 495-523.

Kuhn, T. S. (1957). *The Copernican revolution: Planetary astronomy in the development of Western thought* . Cambridge, MA: Harvard University Press.

Kuhn, T. S. (1962). *The structure of scientific revolutions*. Chicago: University of Chicago Press.

Kuhn, T. S. (1970). Reflections on my critics, originally in Lakatos and Musgrave, reprinted in Kuhn 2000.

Kuhn, T. S. (1977). *The essential tension: Selected essays in scientific tradition and change.* Chicago: University of Chicago Press.
Kuhn, T. S. (1993). Afterwords. In P. Horwich (Ed.), *World changes: Thomas Kuhn and the nature of science.* Cambridge: MIT Press.
Kuhn, T. S. (1996). *The structure of scientific revolutions* (4th ed.). Chicago: University of Chicago Press.
Kuhn, T. S. (2000). *The road since structure: Philosophical essays, 1970-1993 (with an autobiographical interview).* J. Conant & J. Haugeland (Eds.). Chicago: University of Chicago Press.
Kunda, Z. (1990). The case for motivated reasoning. *Psychological Bulletin, 108,* 480-498.
Kurz, E. M. &. Tweney, R. D. (1998). The practice of mathematics and science: From calculus to the clothesline problem. In M. Oakfield & N. Chater (Eds.), *Rational Models of Cognition.* Oxford: Oxford University Press.
Kurz-Milcke, E., Nersessian, N. J. & Newstetter, W. (2004). What has history to do with cognition? Interactive methods for studying research laboratories. *Journal of Cognition and Culture, 4,* 663-700.
Labov, J. (2004).
Lakatos, I. (1978). *The methodology of scientific research programmes.* J. Worrall & G. Currie (Eds.). Cambridge: Cambridge University Press.
Lakatos, I. & Musgrave, E. (Eds.) (1970). *Criticism and the growth of knowledge.* Cambridge: University Press.
Lakoff, G. (1987). *Women, fire, and dangerous things: What categories reveal about the mind.* Chicago: University of Chicago Press.
Lakoff, G. & Johnson, M. (1998). *Philosophy in the flesh.* New York: Basic Books.
Lansdown, B., Blackwood, P. & Brandwein, P. (1971). *Teaching elementary science: Through investigation and colloquium.* New York: Harcourt Brace Jovanovich, Inc.
Latour, B. (1986). Visualisation and cognition: Thinking with eyes and hands. *Knowledge and Society, 6,* 1-40.
Latour, B. (1987). *Science in action: How to follow scientists and engineers through society.* Cambridge, MA: Harvard University Press.
Latour, B. (1999). *Pandora's hope: Essays on the reality of science studies.* Cambridge, MA: Harvard University Press.
Latour, B. & Woolgar, S. (1986). *Laboratory life: The construction of scientific facts.* Princeton, NJ: Princeton University Press.
Laudan, L. (1977). *Progress and its problems: Towards a theory of scientific growth.* Berkeley, University of California Press.
Laudan, L. (1987). Progress or rationality? The prospects for normative naturalism. *American Philosophical Quarterly, 24,* 19-31.
Laudan, L. (1996). *Beyond positivism and relativism: Theory, method, and evidence.* Boulder, CO: Harper Collins.
Lave, J. (1988). *Cognition in practice: Mind, mathematics, and culture in everyday life.* New York: Cambridge University Press.
Lave, J. & Wenger, E. (1991). *Situated learning: Legitimate peripheral participation.* Cambridge: Cambridge University Press.
Lawson, A. E. (1995). *Science teaching and the development of thinking.* Belmont, CA: Wadsworth.
Layton, D., Jenkins, E., Macgill, S. & Davey, A. (1993). *Inarticulate Science? Perspectives on public understanding of science and some implications for science education.* Nafferton: Studies in Education, Ltd.
Le Grand, H. E. (1988). *Drifting continents and shifting theories,* New York: Cambridge University Press.
Lehrer, R. & Schauble, L. (2000). Inventing data structures for representational purposes: Elementary grade students' classification models. *Mathematical Thinking and Learning, 2,* 51-74.
Lehrer, R. & Schauble, L. (2002). Symbolic communication in mathematics and science: Co-constituting inscription and thought. In J. Byrnes & E. D. Amsel (Eds.), *Language, literacy, and*

BIBLIOGRAPHY

cognitive development: The development and consequences of symbolic communication (pp. 167-192). Mahwah, NJ: Lawrence Erlbaum Associates.

Lehrer, R. & Schauble, L. (2006). Scientific thinking and science literacy. In W. Damon, R. Lehrer, K.A. Renninger & I.E. Sigel (Eds.), *Handbook of child psychology, 6th edition, V4, Child psychology in practice. Hoboken, NJ:John Wiley.*

Lemke, J. (1983). *Classroom communication of science.* Final report to the US National Science Foundation. (ERIC Document Reproduction Service No. ED222346).

Lemke J. (1990). *Talking science: Language, learning, and values.* Norwood, NJ: Ablex.

Lemke, J. L. (2000). Across the scales of time: Artifacts, activities, and meanings in ecosocial systems. *Mind, Culture, and Activity, 7*(4), 273-290.

Leslie, A. (1984). Spatiotemporal continuity and the perception of causality in infants. *Perception, 13,* 287-305.

Leslie, A. M. (1987). Pretense and representation: The origins of "theory of mind." *Psychological Review, 94,* 412-426.

Linn, M. (2000). Designing the knowledge integration environment. *International Journal of Science Education, 22*(8), 781-796.

Lippmann, R. (2003). *Students' understanding of measurement and uncertainty in the physics laboratory: Social construction, underlying concepts, and quantitative analysis.* Unpublished Doctoral, University of Maryland.

Lising, L. & Elby, A. (2005). The impact of epistemology on learning: a case study from introductory physics. *American Journal of Physics, 73,* 372-382.

Loh, B., et al (2001). Developing reflective inquiry practices: A case study of software, the teacher, and students. In K. Crowley, C. Schunn & T. Okada (Eds.), *Design for science: Implications from everyday classroom, and professional settings* (pp.279-323). Mahwah, NJ: LEA, Inc.

Longino, H. E. (1990). *Science as social knowledge.* Princeton, NJ: Princeton University Press.

Longino, H. E. (1993). Subjects, power, and knowledge: Description and prescription in feminist philosophies of science. In L. Alcoff & E. Potter (Eds.), *Feminist epistemologies* (pp. 101-120). New York: Routledge.

Longino, H. E. (2002). *The fate of knowledge.* Princeton: Princeton University Press.

Longino, H. E. (this volume). Panel A.

Lottero-Perdue, P. S. (2005). *Critical analysis of science-related texts in a breastfeeding information, support, and advocacy community of practice.* Unpublished dissertation, University of Delaware.

Louca, L. (2004). *Case studies of fifth-grade student modeling in science through programming: Comparison of modeling practices and conversations.* Unpublished Doctoral, University of Maryland, College Park.

Louca, L., Elby, A., Hammer, D. & Kagey, T. (2004). Epistemological resources: Applying a new epistemological framework to science instruction. *Educational Psychologist, 39*(1), 57-68.

Louca, L., Hammer, D. & Bell, M. (2002). Developmental versus context-dependant accounts of abilities for scientific inquiry: A case study of 5-6th grade student inquiry from a discussion about a dropped pendulum. In P. Bell, R. Stevens & T. Satwicz (Eds.), *Keeping learning complex: The proceedings of the Fifth International Conference of the Learning Sciences* (pp. 261-267). Mahwah, NJ: Erlbaum.

Louisell, W. H. (1973). *Quantum statistical properties of radiation.* New York: Wiley.

Lucas, D., Broderick, N., Lehrer, R. & Bohanan, R. (under review). Making the grounds of science inquiry and evidence visible in the classroom.

Lunetta, V. & Tamir, P. (1979). Matching lab activities with teaching goals. *The Science Teacher, 46,* 22-24.

Lunetta, V., Hofstein, A. & Giddings, G. (1981). Evaluating science laboratory skills. *The Science Teacher, 48,* 22-25.

Lynch, M. (1993). *Scientific practice as ordinary action: Ethnomethodology and the social studies of science.* Cambridge: Cambridge University Press.

Lynch, M. & Macbeth, D. (1998). Demonstrating physics lessons. In J. Greeno & S. Goldman (Eds.), *Thinking practices in mathematics and science learning* (pp. 269-297). Mahwah, NJ: Lawrence Erlbaum.

BIBLIOGRAPHY

Lynch, M. & Woolgar, S., Eds. (1990). *Representation in scientific practice.* Cambridge, MA: MIT Press.

MacLachlan, G. L. & Reid, I. (1994). *Framing and interpretation.* Melbourne: Melbourne University Press.

Mach, E.

McCormick, B. H., DeFanti, T. A. & Brown, M. D. (1987, November). Special Issue on Visualization in scientific computing. *Computer Graphics, 21.*

Machamer, P., Darden, L. & Craver, C. F. (2000). Thinking about mechanisms. *Philosophy of Science, 67,* 1-25.

Magnani, L., Nersessian, N. & Thagard, P., Eds. (1999). *Model-based reasoning in scientific discovery.* New York: Kluwer Academic/Plenum Publishers.

Martin, J. R. (1993). Literacy in science: Learning to handle text as technology. In M. A. Halliday & J. R. Martin (Eds.), *Writing science: Literacy and discursive power* (pp. 166-202). Pittsburgh, PA: University of Pittsburgh Press.

Masterman, M. (1974). The Nature of a Paradigm. In I. Lakatos & A. Musgrave (Eds.), *Criticism and the growth of knowledge* (59-89). Cambridge: University Press.

Masuda, T. & Nisbett, R. E. (2001). Attending holistically versus analytically: Comparing the context sensitivity of Japanese and Americans. *Journal of Personality and Social Psychology, 81,* 922-934.

Matthews, M. (1994). *Science teaching: The role of history and philosophy of science.* London:Routledge.

Maxwell, J. C. (1861-2). On physical lines of force. In W. D. Niven (Ed.). *Scientific papers.* Cambridge: Cambridge University.

May, D. B., Hammer, D. & Roy, P. (under review). Children's analogical reasoning in a 3rd-grade science discussion.

Mayo, D. G. (1996). *Error and the growth of experimental knowledge.* Chicago: University of Chicago Press.

Meichtry, Y. J. (1993). The impact of science curricula on student views about the nature of science. *Journal of Research in Science Teaching, 30*(5), 429-443.

Meldrum, A., Boatner, L. A. & Ewing, R. C. (2002). Nanocrystalline zirconia can be amorphized by ion radiation. *Physical Review Letters, 88,* 025503-1 – 025503-4.

Menard, H. W. (1986). *The ocean of truth: A personal history of global tectonics.* Princeton, NJ: Princeton University Press.

Merton, R. K. (1973). The normative structure of science. In R. K. Merton (Ed.), *The sociology of science* (pp. 267-278). Chicago: University of Chicago Press (originally published 1942).

Metz, K. E. (1995). Reassessment of developmental constraints on children's science instruction. *Review of Educational Research, 65,* 93-127.

Metz, K. E. (1997). On the complex relation between cognitive developmental research and children's science curricula. *Review of Educational Research, 67,* 151-163.

Metz, K. E. (2000). Young children's inquiry in biology: Building the knowledge bases to empower independent inquiry. In J. Minstrell & E. H. van Zee (Eds.), *Inquiring into inquiry teaching and learning in science* (pp. 371-404). Washington, DC: American Association for the Advancement of Science.

Metz, K. E. (2004). Children's understanding of scientific inquiry: Their conceptualization of uncertainty in investigations of their own design. *Cognition and Instruction, 22*(2), 219-290.

Metz, K. E. (to appear). The knowledge building enterprises in science and elementary school science classrooms: Analysis of problematic differences and strategic leverage points. In L. B. Flick & N. G. Lederman (Eds.), *Scientific inquiry and the nature of science: Implications for teaching, learning, and teacher education.* Dordrecht, the Netherlands: Kluwer.

Michotte, A. (1963). *The perception of causality.* New York: Basic Books.

Mikeska, J. (2006). Falling Objects. In D. a. v. Z. Hammer & E. H. (Ed.), *Seeing the science in children's thinking: Case studies of student inquiry in physical science* (71-95). Portsmouth, NH: Heinemann.

Miller, J. D. (1998). The measurement of civic scientific literacy. *Public Understanding of Science, 7,* 203-223.

Millar, R. & Driver, R. (1987). Beyond processes. *Studies in Science Education, 14,* 33-62.

BIBLIOGRAPHY

Millar, R., Leach, J. & Osborne, J., Eds. (2000). *Improving science education: The contribution of research.* Philadelphia: Open University Press.

Millar, R. & Osborne, J., Eds. (1998). *Beyond 2000: Science education for the future.* London: King's College.

Millikan, R. A. (1965). The electron and the light-quant from the experimental point of view. *Nobel Lectures – Physics 1922-41.* Amsterdam.

Minsky, M. L. (1986). *Society of mind.* New York: Simon and Schuster.

Minstrell, J. & Van Zee, E., Eds. (2000). *Teaching in the inquiry-based science classroom.* Washington, DC: American Association for the Advancement of Science.

Mitchell, S. (1996). *Improving the quality of argument in higher education. Interim report.* London: Middlesex University, School of Education.

Morgan, M. S. & Morrison, M., Eds. (1999). *Models as mediators.* Cambridge: Cambridge University Press.

Morgan, W. J. (1968). Rises, trenches, great faults, and crustal blocks. *Journal of Geophysical Research, 73* (March 15), 1959-1982.

Myers, G. (1991). Lexical cohesion and specialized knowledge in science and popular science texts. *Discourse Processes, 14,* 1-26.

Myers, G. (1992). Textbooks and the sociology of scientific knowledge. *English for Specific Purposes, 11,* 3-17.

Myers, G. (1997). Words and pictures in a biology textbook. In T. Miller (Ed.), *Functional approaches to written text: Classroom applications* (pp. 93-104). Paris: USIA.

National Public Radio (Broadcast in March, 2002). *Teaching curiosity and the scientific method.*

National Research Council. (1996). *National science education standards.* Washington, DC: National Academy of Sciences.

National Research Council (1996). *National standards in science education.* Washington, DC: National Academy Press. Retrieved from http://www.nap.edu.

National Research Council. (2000). *Inquiry and the national standards in science education.* Washington, DC: National Academy Press. Retrieved from http://www.nap.edu.

National Research Council. (2007). Taking science to school: Learning and teaching science in grades K-8. Washington, DC: National Academy Press. Retrieved from http://www.nap.edu.

Nelson, L. H. (1993). Epistemological communities. In L. Alcoff & E. Potter (Eds.) *Feminist epistemologies* (pp. 121-159). New York: Routledge.

Nelson, L. H. & Nelson, J., Eds. (2003). *Feminist interpretations of W.V. Quine.* Penn State University Press.

Nersessian, N. J. (1984a). Aether/or: The creation of scientific concepts. *Studies in the History & Philosophy of Science, 15,* 175-212.

Nersessian, N. J. (1984). *Faraday to Einstein: Constructing meaning in scientific theories.* Dordrecht: Martinus Nijhoff/Kluwer Academic Publishers.

Nersessian, N. J. (1992a). Constructing and instructing: The role of 'abstraction techniques' in developing and teaching scientific theories. In R. Duschl & R. Hamilton (Eds.), *Philosophy of Science, Cognitive Science & Educational Theory and Practice.* Albany, NY: SUNY Press.

Nersessian, N. J. (1992b). How do scientists think? Capturing the dynamics of conceptual change in science. In R. Giere (Ed.), *Minnesota Studies in the Philosophy of Science.* Minneapolis: University of Minnesota Press.

Nersessian, N. J. (1992c). In the theoretician's laboratory: Thought experimenting as mental modeling. In D. Hull, M. Forbes & K. Okruhlik (Eds.), *PSA 1992.* East Lansing, MI: PSA.

Nersessian, N. J. (1995). Should physicists preach what they practice? Constructive modeling in doing and learning physics. *Science & Education, 4,* 203-226.

Nersessian, N. J. (1999). Model-based Reasoning in Conceptual Change. In L. Magnani, N. J. Nersessian & P. Thagard (Eds.), *Model-based reasoning in scientific discovery.* New York: Kluwer Academic/Plenum Publishers.

Nersessian, N. J. (2002a). The cognitive basis of model-based reasoning in science. In P. Carruthers, S. Stich, & M. Siegal (Eds.), *The cognitive basis of science.* Cambridge: Cambridge University Press.

Nersessian, N. J. (2002b). Kuhn, conceptual change, and cognitive science. In T. Nickles (Ed.), *Thomas Kuhn.* Cambridge: Cambridge University Press.

Nersessian, N. J. (2002c). Maxwell and the "method of physical analogy": Model-based reasoning, generic abstraction, and conceptual change. In D. Malament (Ed.), *Reading natural philosophy: Essays in the history and philosophy of science and mathematics*. Lasalle, IL: Open Court.
Nersessian, N. J. (this volume). Inquiry workshop topic: How science works?
Nersessian, N. J. (2005). Interpreting scientific and engineering practices: Integrating the cognitive, social, and cultural dimensions. In M. Gorman, R. D. Tweney, D. Gooding & A. Kincannon (Eds.), *Scientific and technological thinking*. Hillsdale, NJ: Lawrence Erlbaum.
Newton, D. P. & Newton, L. D. (1999). Knowing what counts as understanding in different disciplines: some 10-year-old children's conceptions. *Educational Studies, 25*(1), 35-54.
Newton, D. P. & Newton, L. D. (2000). Do teachers support causal understanding through their discourse when teaching primary science? *British Educational Research Journal, 26*, 599-613.
Newton, L. D., Newton, D. P., Blake, A. & Brown, K. (2002). Do primary school science books for children show a concern for explanatory understanding? *Research in Science & Technological Education, 20*, 227-240.
Newton, P., Driver, R. & Osborne, J. (1999). The place of argumentation in the pedagogy of school science. *International Journal of Science Education, 21*(5), 553-576.
Nisbett, R., Peng, K., Choi, I. & Norenzayan, A. (2001). Culture and systems of thought: Holistic v. analytic cognition. *Psychological Review, 108*(2), 291-310.
Nisbett, R. & Ross, L. (1980). *Human inference: Strategies and shortcomings of social judgment*. Englewood Cliffs, NJ: Prentice-Hall.
Norris, S. P. (1985). The philosophical basis of observation in science and science education. *Journal of Research in Science Teaching, 22*, 817-833.
Norris, S. P. (1992). Practical reasoning in the production of scientific knowledge. In R. A. Duschl & R. J. Hamilton (Eds.), *Philosophy of science, cognitive psychology, and educational theory and practice* (pp. 195-225). Albany: State University of New York Press.
Norris, S. P. (1995). Learning to live with scientific expertise: Toward a theory of intellectual communalism for guiding science teaching. *Science Education, 79*, 201-217.
Norris, S. P. & Phillips, L. M. (1987). Explanations of reading comprehension: Schema theory and critical thinking theory. *Teachers College Record, 89*, 281-306.
Norris, S. P. & Phillips, L. M. (1994a). The relevance of a reader's knowledge within a perspective view of reading. *Journal of Reading Behavior, 26*, 391-412.
Norris, S. P. & Phillips, L. M. (1994b). Interpreting pragmatic meaning when reading popular reports of science. *Journal of Research in Science Teaching, 31*, 947-967.
Norris, S. P. & Phillips, L. M. (2003). How literacy in its fundamental sense is central to scientific literacy. *Science Education, 87*, 224-240.
Norris, S. P. & Phillips, L. M. (this volume). Reading as inquiry.
Norris, S. P., Phillips, L. M. & Korpan, C. A. (2003). University students' interpretation of media reports of science and its relationship to background knowledge, interest, and reading difficulty. *Public Understanding of Science, 12*, 123-145.
O'Neill, D. K. & Polman, J. L. (2004). Why educate "little scientists?" Examining the potential of practice-based scientific literacy. *Journal of Research in Science Teaching, 41*(3), 234-266.
Ogborn, J., Kress, G., Martins, I. & McGillicuddy, K. (1996). *Explaining Science in the Classroom*. Buckingham: Open University Press.
Ohlsson, S. (1995). Learning to do and learning to understand? A lesson and a challenge for cognitive modelling. In P. Reimann & H. Spads (Eds.), *Learning in Humans and Machines* (pp. 37-62). Oxford: Elsevier.
Olson, D. R. (1994). *The world on paper*. Cambridge: Cambridge University Press.
Olson, D. R. (1996). Literate mentalities: Literacy, consciousness of language, and modes of thought. In D. R. Olson & N. Torrance (Eds.), *Modes of thought* (pp. 141-151). Cambridge: Cambridge University Press.
Osborne, J. F. (2001). Promoting argument in the science classroom: A rhetorical perspective. *Canadian Journal of Science, Mathematics, and Technology Education, 1*(3), 271-290.
Osborne, J. F., Erduran, S., Simon, S. & Monk, M. (2001). Enhancing the quality of argument in school science. *School Science Review, 82*(301), 63-70.

BIBLIOGRAPHY

Osborne, R. & Freyberg, P. (1985). *Learning in science: The implications of children's science.* Auckland, New Zealand: Heinemann.

Palincsar, A. S. & Magnusson, S. J. (2001). The interplay of first-hand and text-based investigations to model and support the development of scientific knowledge and reasoning. In S. Carver & D. Klahr (Eds.), *Cognition and instruction: Twenty-five years of progress* (pp. 151-194). Mahway, NJ: Lawrence Erlbaum Associates.

Palmquist, B. C. & Finley, F. N. (1997). Preservice teachers' vies of the nature of science during a postbaccalaureate science teaching program. *Journal of Research in Science Teaching, 34*, 595-615.

Pea, R. D. (1993a). Distributed multimedia learning environments: The Collaborative Visualization Project. *Communications of the ACM, 36,* 60-63.

Pea, R. D. (1993b). Learning scientific concepts through material and social activities: Conversational analysis meets conceptual change. *Educational Psychologist, 28,* 265-277.

Pea, R. D. (1993c). Practices of distributed intelligence and designs for education. In G. Salomon (Ed.), *Distributed cognitions: Psychological and educational considerations* (pp. 47-87). Cambridge: Cambridge University Press.

Peacock, A. & Weedon, H. (2002). Children working with text in science: disparities with 'literacy hour' practice. *Research in Science & Technological Education, 20,* 185-197.

Pellegrino, J., Chudowsky, J. & Glaser, R., Eds. (2001). *Knowing what student know: The science and design of educational assessment.* Washington, DC: National Academy Press. Retrieved from http://www.nap.edu.

Penner, D. E. (2000). Explaining systems: Investigating middle school students' understanding of emergent phenomena. *Journal of Research in Science Teaching, 37,* 784-806.

Penney, K., Norris, S. P., Phillips, L. M. & Clark, G. (2003). The anatomy of junior high school science textbooks: An analysis of textual characteristics and a comparison to media reports of science. *Canadian Journal of Science, Mathematics and Technology Education, 3,* 415-436.

Perkins, D. N. & Grotzer, T. A. (2000). *Models and moves: Focusing on dimensions of causal complexity to achieve deeper scientific understanding.* Paper presented at the annual conference of the American Educational Research Association, New Orleans, LA.

Perry, W. G. (1970). *Forms of intellectual and ethical development in the college years.* New York: Holt, Rinehart & Winston.

Pfundt, H. & Duit, R. (1991). *Bibliography: Students' alternative frameworks and science.* Kiel, Germany: IPN.

Phelan, J. (in preparation). Chaos in the Corridor. In D. a. v. Z. Hammer, E. H. (Ed.), *Video Case Studies of Children's Inquiry in Physical Science.*

Phillips, L. M. (1988). Young readers' inference strategies in reading comprehension. *Cognition and Instruction, 5,* 193-222.

Phillips, L. M. (1992). The generalizability of self-regulatory thinking strategies. In S. P. Norris (Ed.), *The generalizability of critical thinking* (pp. 138-156). New York: Teachers College Press.

Phillips, L. M. & Norris, S. P. (1999). Interpreting popular reports of science: What happens when the reader's world meets the world on paper? *International Journal of Science Education, 21,* 317-327.

Phillips, L. M., Smith, M. L. & Norris, S. P. (2005). Commercial reading programs: What's replacing narrative? In B. Maloch, J. V. Hoffman, D. L. Schallert, C. M. Fairbanks, J. Worthy (Eds.), *54th Yearbook of the National Reading Conference* (pp. 286-300). Oak Creek, WI: National Reading Conference.

Pickering, A., Ed. (1992). *Science as practice and culture.* Chicago: University of Chicago Press.

Pintrich, P. R., Marx, R. W. & Boyle, R. A. (1993). Beyond cold conceptual change: The role of motivational beliefs and classroom contextual factors in the process of conceptual change. *Review of Educational Research, 63,* 167-199.

Pitts, V. M. & Edelson, D. C. (2004). Role, goal, and activity: A framework for characterizing participation and engagementin project-based learning environments. In Y. B. Kafai, W. A. Sandoval, N. Enyedy, A. S. Nixon & F. Herrera (Eds.), *Proceedings of the Sixth International Conference of the Learning Sciences, Santa Monica, CA, June 23-26, 2004* (pp. 420-426). Mahwah, NJ: Erlbaum.

Polman, J. L. (2000). *Designing project-based science: Connecting learners through guided inquiry.* New York: Teachers College Press.

Polya, G. (1954). *Induction and analogy in mathematics*, Vol. 1. Princeton, NJ: Princeton University Press.
Poper, K. R. (1959). Back to the pre-socratics. *Proceedings of the Aristotelian Society, 59*, 1-24.
Posner, G. J., Strike, K. A., Hewson, P. W. & Gertzog, W. A. (1982). Accommodation of a scientific conception: Toward a theory of conceptual change. *Science Education, 66*, 211-227.
Pressley, M. & Wharton-McDonald, R. (1997). Skilled comprehension and its development through instruction. *School Psychology Review, 26*, 448-467.
Putnam, R. & Borko, H. (2000). What do new views of knowledge and thinking have to say about research on teacher learning? *Educational Researcher, 29*(1), 4-15.
Quine, W. V. (1969). *Ontological relativity and other essays*. New York: Columbia University Press.
Rahm, J., Miller, H. C., Hartley, L. & Moore, J. C. (2003). The value of and emergent notion of authenticity: examples from two student/teacher-science partnerships programs. *Journal of Research in Science Teaching, 40*, 737-756.
Redish, E. F. (2004). A theoretical framework for physics education research: Modeling student thinking. In E. Redish & M. Vicentini (Eds.), *Proceedings of the Enrico Fermi Summer School, Course CLVI*. Italian Physical Society.
Reichenbach, H. (1938). *Experience and prediction: An analysis of the foundations and the structure of knowledge*. Chicago: University of Chicago Press.
Reif, F. & Larkin, J. H. (1991). Cognition in scientific and everyday domains: Comparison and learning implications. *Journal of Research in Science Teaching, 28*, 733-760.
Reiner, M., Slotta, J. D., Chi, M. T. H. & Resnick, L. B. (2000). Naive physics reasoning: A commitment to substance-based conceptions. *Cognition and Instruction, 18*, 1-34.
Reinvention Center at Stonybrook (2001). *Reinventing undergraduate education: Three years after the Boyer Report*. Available at www.sunysb.edu/reinventioncenter/boyerfollowup.pdf.
Reiser, B. J. (2004). Scaffolding complex learning: The mechanisms of structuring and problematizing student work. *Journal of the Learning Sciences, 13*(3), 273-304.
Reiser, B. J., Tabak, I., Sandoval, W. A., Smith, B. K., Steinmuller, F. & Leone, A. J. (2001). BGuILE: Strategic and conceptual scaffolds for scientific inquiry in biology. In S. M. Carver & D. Klahr (Eds.), *Cognition and instruction: Twenty-five years of progress* (pp. 263-305). Mahwah, NJ: Erlbaum.
Renner, J. W. & Stafford, D. G. (1972). *Teaching science in the secondary school*. New York: Harper & Row.
Resnick, M. (1996). Beyond the centralized mindset. *The Journal of the Learning Sciences, 5*, 1-22.
Reveles, J. M., Cordova, R. & Kelly, G. J. (2004). Science literacy and academic identity formulation. *Journal for Research in Science Teaching, 41*, 1111-1144.
Rheinberger, H. J. (1997). *Toward a history of epistemic things: Synthesizing proteins in the test tube*. Stanford, CA: Stanford University Press.
Rivard, L. P. & Straw, S. B. (2000). The effect of talk and writing on learning science: An exploratory study. *Science Education, 84*, 566-593.
Rogers, E. M. (1995). *Diffusion of innovations* (4th ed.). New York: The Free Press.
Rogoff, B. (1990). *Apprenticeship in thinking: Cognitive development in social context*. Oxford: Oxford University Press.
Roseberry, A., Warren, B. & Conant, F. (1992). Appropriating scientific discourse: Findings from language minority classrooms. *The Journal of the Learning Sciences, 2*, 61-94.
Rosenberg, S. A., Hammer, D. & Phelan, J. (2006). Multiple epistemological coherences in an eighth-grade discussion of the rock cycle. *Journal of the Learning Sciences, 15*(2), 261-292.
Roth, W. M. (1995). *Authentic school science*. Boston: Kluwer Academic.
Roth, W. M. (1999). *Scientific research expertise from middle school to professional practice*. Paper presented at the Annual Meeting of the American Educational Research Association, Montreal, Quebec.
Roth, W. M. & Lee, S. (2002). Scientific literacy as collective praxis. *Public Understanding of Science, 11*, 1-24.
Roth, W. M. & Lee, S. (2004). Science education as/for participation in the community. *Science Education, 88*, 263-291.

BIBLIOGRAPHY

Roth, W. M. & McGinn, M. K. (1998). Inscriptions: Toward a theory of representing as social practice. *Review of Educational Research, 68*, 35-59.
Roth, W. M., McGinn, M. K., Woszczyna, C. & Boutonné, S. (1999). Differential participation during science conversations: The interaction of focal artifacts, social configuration, and physical arrangements. *The Journal of the Learning Sciences, 8*, 293-347.
Rowell, P. M. & Ebbers, M. (2004). School science constrained: print experiences in two elementary classrooms. *Teaching and Teacher Education, 20*, 217-230.
Rudolph, J. (2002). *Scientists in the classroom: The cold war reconstruction of American science education.* New York: Palgrave Macmillian.
Rudolph, J. (2003). Portraying epistemology: School science in historical context. *Science Education, 87*, 64-79.
Rudolph, J. (this volume). Commentary on "Inquiry, activity, and epistemic practice," by Gregory J. Kelly.
Rudwick, M. J. S. (1976). The emergence of a visual language for geological science. *History of Science, 14*, 149-195.
Rumelhart, D. E. (1980). Schemata: The building blocks of cognition. In R. J. Spiro, B. Bruce & W. F. Brewer (Eds.), *Theoretical issues in reading and comprehension.* Hillsdale, NJ: Erlbaum.
Russ, R. S. (2006). A framework for recognizing mechanistic reasoning in student scientific inquiry. Unpublished doctoral dissertation, University of Maryland, College Park.
Russ, R., Mikeska, J., Scherr, R. E. & Hammer, D. (2005). *Recognizing and distinguishing scientific aspects of first graders' reasoning about motion.* Paper presented at the Annual Meeting of the American Educational Research Association, Montreal.
Russell, J. (1992). The theory theory: So good they named it twice? *Cognitive Development, 7*, 485-519.
Russell, T. (1983). Analyzing arguments in science classroom discourse: Can teachers' questions distort scientific authority? *Journal of Research in Science Teaching, 20*, 27-45.
Rutherford, F. J. & Ahlgren, A. (1990). *Science for all Americans.* New York: Oxford University Press.
Salmon, W. (1990). Rationality and objectivity in science, or Tom Kuhn meets Tom Bayes. *Scientific Theories, V. 14 of Minnesota Studies in Philosophy of Science* (175-204). University of Minnesota Press.
Samarapungavan, A. (1992). Children's judgments in theory choice tasks: Scientific rationality in childhood. *Cognition, 45*, 1-32.
Samarapungavan, A., Vosniadou, S. & Brewer, W. F. (1996). Mental models of the earth, sun, and moon: Indian children's cosmologies. *Cognitive Development, 11*, 491-521.
Sandoval, W. A. (this volume). Exploring children's understanding of the purpose and value of inquiry.
Sandoval, W. A. (2005). Understanding students' practical epistemologies and their influence on learning through inquiry. *Science Education, 89*, 634-656.
Sandoval, W. A. & Morrison, K. (2003). High school students' ideas about theories and theory change after a biological inquiry unit. *Journal of Research in Science Teaching, 40*(4), 369-392.
Sandoval, W. A. & Reiser, B. J. (2004). Explanation-driven inquiry: integrating conceptual and epistemic scaffolds for scientific inquiry. *Science Education, 88*, 345-372.
Santa Barbara Classroom Discourse Group (1992). Do you see what I see? The referential and intertextual nature of classroom life. *Journal of Classroom Interaction, 27*, 29-36.
Sawyer, R.K., Ed., (2006). *The Cambridge handbook of the learning sciences.* New York: Cambridge University Press.
Schank, R. C. (1982). *Dynamic memory.* Cambridge: Cambridge University Press.
Schank, R. C., Fano, A., Bell, B. & Jona, M. (1993/1994). The design of goal-based scenarios. *The Journal of the Learning Sciences, 3*(4), 305-346.
Schauble, L. (1990). Belief revision in children: The role of prior knowledge and strategies for generating evidence. *Journal of Experimental Child Psychology, 49*, 31-57.
Schauble, L. (this volume). Three questions about development.
Schauble, L., Glaser, R., Duschl, R., Schultz, S. & John, J. (1995). Students' understanding of the objectives and procedures of experimentation in the science classroom. *The Journal of the Learning Sciences, 4*(2) 131-166.
Schraw, G. (2001). Current themes and future directions in epistemological research: A commentary. *Educational Psychology Review, 13*, 451-464.

Schwab, J. (1962). The teaching of science as inquiry. In J. Schwab & P. Brandwein (Eds.), *The teaching of science* (pp1-104) Cambridge, MA: Harvard University Press.

Schwab, J. J. (1978). Education and the structure of the disciplines. In J. J. Schwab (Ed.), *Science, curriculum, and liberal education* (pp. 229, 272, excerpts). Chicago, IL.

Schwab, J. J. & Brandwein, P. F. (1966). *The teaching of science*. Cambridge, MA: Harvard University Press.

Schwartz, R. S., Lederman, N. G. & Crawford, B. A. (2004). Developing views of nature of science in an authentic context: An explicit approach to bridging the gap between nature of science and scientific inquiry. *Science Education, 88*, 610-645.

Schwitzgebel, E. (1999). Children's theories and the drive to explain. *Science & Education, 8*, 457-488.

Shapere, D. (1964). The structure of scientific revolutions. *Philosophical Review, LXXIII*, 383-394.

Shapere, D. (1982). The concept of observation in science and philosophy. *Philosophy of Science, 49*, 485-525.

Shapin, S. & Schaffer, S. (1985). *Leviathan and the air-pump*. Princeton: Princeton University Press

Shavelson, R., Baxter, G. & Pine, J. (1992). Performance assessments: Political rhetoric and measurement reality. *Educational Researcher, 21*, 22-27.

Shelley, C. (1996). Visual abductive reasoning in archeology. *Philosophy of Science, 63*, 278-301.

Shelley, C. (1999). Multiple analogies in evolutionary biology. *Stud. Hist. Phil. Biol. & Biomed. Sci., 30*(2),143-180.

Shore, B. (1997). *Culture in mind: Cognition, culture and the problem of meaning*. New York: Oxford University Press.

Shultz, T. R. (1982). Rules of causal attribution. *Monographs of the Society for Research in Child Development*, Serial 194, Vol. 47, No. 1.

Shymansky, J. A., Yore, L. D. & Good, R. (1991). Elementary school teachers' beliefs about the perceptions of elementary school science, science reading, science textbooks, and supportive instructional factors. *Journal of Research in Science Teaching, 28*, 437-454.

Siegel, H. (1995). Why should educators care about argumentation. *Informal Logic, 17*(2), 159-176.

Siegel, H. (this volume). Panel A.

Siegler, R. S. (1996). *Emerging minds: The process of change in children's thinking*. New York: Oxford University Press.

Simon, H. A. (1980). Problem solving and educaton. In D. T. Tuma & R. Reif (Eds.), *Problem solving and education: Issues in teaching and research* (pp. 81-96). Hilldale, NJ: Erlbaum.

Slotta, J. D., Chi, M. T. H. & Joram, E. (1995). Assessing students' misclassifications of physics concepts: An ontological basis for conceptual change. *Cognition and Instruction, 13*, 373-400.

Smith, C., Snir, J. & Grosslight, L. (1992). Using conceptual models to facilitate conceptual change: The case of weight-density differentiation. *Cognition and Instruction, 9*, 221-283.

Smith, C. & Unger, C. (1997). What's in dots-per-box? Conceptual bootstrapping with stripped-down visual analogs. *Journal of the Learning Sciences, 6*, 143-181.

Smith, C., Wiser, M., Anderson, C.W., Krajcik, J. & Coppola, B. (2005). Implications of research on children's learning for assessment: Matter and atomic molecular theory classroom-based assessment system for science: A model. Paper commissioned by the Committee on Test Design for K-12 Science Achievement, Center for Education, National Research Council, http://www7.nationacademies.org/bota/Test_Design_K-12_Science.html

Smith, C. L., Maclin, D., Houghton, C. & Hennessey, M.G. (2000). Sixth-grade students' epistemologies of science: The impact of school science experiences on epistemological development. *Cognition and Instruction, 18*(3), 349-422.

Smith, F. (1978). *Understanding reading*. New York: Holt, Rinehart & Winston.

Smith, J. P., III, diSessa, A. A. & Roschelle, J. (1993/1994). Misconceptions reconceived: A constructivist analysis of knowledge in transition. *The Journal of the Learning Sciences, 3*, 115-163.

Smolin, L. (2006). *The trouble with physics: The rise of string theory, the fall of a science, and what comes next*. Boston: Houghton Mifflin Co.

Sodian, B., Zaitchik, D. & Carey, S. (1991). Young childrens differentation of hypothetical beliefs from evidence. *Child Development, 62*(4), 753-766.

Solomon, J. (1987). Social influences on the construction of pupils' understanding of science. *Studies in Science Education, 14*, 63-82.

BIBLIOGRAPHY

Solomon, M. (1992). Scientific rationality and human reasoning. *Philosophy of Science*, 59(3), 439-455.
Solomon, M. (2001). *Social empiricism*. Cambridge, MA: MIT Press.
Solomon, M. (this volume). Social epistemology of science.
Soloway, E., Guzdial, M. & Hay, K. E. (1994). Learner-centered design: The challenge for HCI in the 21st century. *Interactions, 1*(2), 36-47.
Sosa, E. (1991). *Knowledge in perspective*. Cambridge: Cambridge University Press.
Souza Lima, E. (1995). Culture revisited: Vygotsky's ideas in Brazil. *Anthropology and Education Quarterly, 26*(4), 443-457.
Spelke, E. S., Breinlinger, K., Macomber, J. & Jacobson, K. (1992). Origins of knowledge. *Psychological Review, 99*, 605-632.
Spiro, R. J. & Myers, A. (1984). Individual differences on underlying cognitive processes in reading. In P. D. Pearson, R. Barr, M. L. Kamil & P. Mosenthal (Eds.), *Handbook of reading research* (pp. 471-501). New York: Longman.
Star, S. L. & Griesemer, J. G. (1989). Institutional ecology, 'translations' and boundary objects: Amateurs and professionals in Berkeley's Museum of Vertebrate Zoology, 1907-39. *Social Studies of Science, 19*, 387-420.
Stevens, R., Wineburg, S., Herrenkohl, L. R. & Bell, P. (2005). The comparative understanding of school subjects: Past, present, and future research agenda. *Review of Educational Research, 75*(2), 125-157.
Stewart, J., Hafner, R., Johnson, S. & Finkel, E. (1992). Science as model-building: Computers and high-school genetics. *Educational Psychologist, 27*, 317-336.
Stich, S. & Nichols, S. (1998). Theory theory to the max. *Mind & Language, 13*, 421-449.
Stich, S. (this volume) Panel A.
Strike, K. A. (1995). Epistemology and education. In T. Husén & T. N. Postlethwaite (Eds.), *International encyclopedia of education*, Vol. 4, 2nd ed. (pp. 1996-2001). Tarrytown, NY: Elsevier Science.
Strike, K. A. & Soltis, J. F. (1992). *The ethics of teaching* (2nd ed.). New York: Teachers College Press.
Suchman, L. A. (1987). *Plans and Situated Actions: the Problem of Human-Machine Communication*. Cambridge and New York: Cambridge University Press.
Suppe, F. (1989). *The semantic conception of theories and scientific realism*. Chicago: University of Illinois Press.
Suppe, F. (1998). The structure of a scientific paper. *Philosophy of Science*, 65(3), 381-405.
Suppe, F., Ed. (1977). *The structure of scientific theories*. Champagne-Urbana, IL: University of Illinois Press.
Suppes, P. (1969). Models of data. In *Studies in the methodology and foundations of science*: Selected papers from 1951 to 1969 (pp. 24-35). Dordrecht: D. Reidel.
Sweller, J. (1994). Cognitive load theory, learning difficulty, and instructional design. *Learning and Instruction, 4*, 295-312.
Takao, A. Y. & Kelly, G. J. (2003). Assessment of evidence in university students' scientific writing. *Science & Education. 12*, 341-363.
Takao, A. Y., Prothero, W. & Kelly, G. J. (2002). Applying argumentation analysis to assess the quality of university oceanography students' scientific writing. *Journal of Geoscience Education, 50*(1), 40-48.
Thagard, P. (1993). Societies of minds: Science as distributed computing. *Studies in the History and Philosophy of Science, 24*, 49-67.
Thagard, P. (1994). Mind, society, and the growth of knowledge. *Philosophy of Science, 61*(4), 629-645.
Thagard, P. (2000). *How scientists explain disease*. Princeton: Princeton University Press.
Thomson, W., (Lord Kelvin). (1884). *Lectures on molecular dynamics*. Baltimore, MD: Johns Hopkins University.
Tishman, S. & Perkins, D. (1997). The language of thinking. *Phi Delta Kappan, 78*, 368-374.
Tomasello, M. (1999). *The cultural origins of human cognition*. Cambridge, MA: Harvard University Press.
Toulmin, S. (1958). *The uses of argument*. New York: Cambridge University Press.
Toulmin, S. (1961). *Foresight and understanding*. New York: Harper and Row.

Toulmin, S. (1972). *Human understanding, Vol. 1: The collective use and evolution of concepts*. Princeton: Princeton University Press.
Toulmin, S. (1979). The inwardness of mental life. *Critical Inquiry, 6*(1), 1-16.
Traweek, S. (1988). *Beamtimes and lifetimes: The world of high energy physicists*. Cambridge, MA: Harvard University Press.
Trumbull, D. & Kerr, P. (1993). University researchers' inchoate critiques of science teaching: Implications for the content of preservice science teacher education. *Science Education, 77*(3), 301-317.
Trumpler, M. (1997). Converging images: Techniques of intervention and forms of representation of sodium-channel proteins in nerve cell membranes. *Journal of the History of Biology, 20*, 55-89.
Tuminaro, J. (2004). *A cognitive framework for analyzing and describing introductory students' use and understanding of mathematics in physics*. Unpublished Doctoral, University of Maryland, College Park.
Tweney, R. D. (1985). Faraday's discovery of induction: A cognitive approach. In D. Gooding & F. A. J. L. James (Eds.), *Faraday Rediscovered*. New York: Stockton Press.
Tweney, R. D. (2002). Epistemic artifacts: Michael Faraday's search for the optical effects of gold. In L. Magnani & N. J. Nersessian (Eds.), *Model-Based Reasoning: Science, Technology, Values*. New York: Kluwer Academic/Plenum Publishers.
Tyson, L. M., Venville, G. J., Harrison, A. G. & Treagust, D. F. (1997). A multidimensional framework for interpreting conceptual change events in the classroom. *Science Education, 81*(4), 387-404.
Tytler, R. & Peterson, S. (2004). From "try it and see" to strategic exploration: Characterizing young children's scientific reasoning. *Journal of Research in Science Teaching, 41*(1), 94-118.
van Eemeren, F. H., Grootendorst, R., Henkemans, F. S., Blair, J. A., Johnson, R. H., Krabbe, E. C. W., Plantin, C., Walton, D. N., Willard, C. A., Woods, J. & Zarefsky, D. (1996). *Fundamentals of argumentation theory: A handbook of historical backgrounds and contemporary developments*. Mahwah, NJ: Lawrence Erlbaum Associates.
van Waterschoot van der Gracht, W. A. J. (Ed.) (1928). *Theories of continental drift: A symposium*, Tulsa: American Association of Petroleum Engineers.
Van Zee, E. H., Iwasyk, M., Kurose, A., Simpson, D. & Wild, J. (2001). Student and teacher questioning during conversations about science. *Journal of Research in Science Teaching, 38*, 159-190.
von Glasersfeld, E. (1991). Cognition, construction of knowledge, and teaching. In M. Matthews (Ed.), *History, philosophy, and science teaching: Selected readings* (pp.117-132). Toronto: OISE Press.
Vosniadou, S. & Brewer, W. F. (1992). Mental models of the earth: A study of conceptual change in childhood. *Cognitive Psychology, 24*, 535-585.
Vosniadou, S. & Brewer, W. F. (1994). Mental models of the day/night cycle. *Cognitive Science, 18*, 123-183.
Vosniadou, S., Ioannides, C., Dimitracopoulou, A. & Papademetriou, E. (2001). Designing learning environments to promote conceptual change in science. *Learning and Instruction, 11*, 381-419.
Walton, D. (1996). *Argumentation schemes for presumptive reasoning*. Mahwah, NJ: Erlbaum Press.
Watson, B. & Konicek, R. (1990). Teaching for conceptual change: Confronting children's experience. *Phi Delta Kappan, 71*, 680-685.
Wellington, J. & Osborne, J. (2001). *Language and literacy in science education*. Buckingham, UK: Open University Press.
Wellman, H. M. (1990). *The child's theory of mind*. Cambridge: MIT Press.
Wellman, H. M. & Gelman, S. A. (1992). Cognitive Development: foundational theories of core domains. *Annual Review of Psychology, 43*, 337-375
Wenger, E. (1998). *Communities of practice*: Cambridge: Cambridge University Press.
Wenk, L. & Smith, C. (2004). *The impact of first-year college science courses on epistemological thinking: A comparative study*. Paper presented at Annual Meeting of the National Association of Research in Science Teaching, Vancouver, BC, Canada.
Wertsch, J. (1991). *Voices of the Mind*. New York: Harvester.
White, B. Y. & Frederiksen, J. R. (1998). Inquiry, modeling, and metacognition: Making science accessible to all students. *Cognition and Instruction, 16*, 3-118.
Whitehead, A. H. (1929). *The aims of education*. Cambridge: Cambridge University Press.

BIBLIOGRAPHY

Wilensky, U. (1999). NetLogo. http://ccl.northwestern.edu/netlogo. Center for Connected Learning and Computer-Based Modeling. Northwestern University, Evanston, IL.

Wilensky, U. & Resnick, M. (1999). Thinking in levels: A dynamic systems approach to making sense of the world. *Journal of Science Education and Technology, 8*, 3-18.

Windschitl, M. (2001). Independent inquiry projects for pre-service science teachers: Their capacity to reflect on the experience and to integrate inquiry into their own teaching. *Paper presented at annual Conference of the National Association of Research in Science Teaching*. St. Louis.

Windschitl, M. (2003). Inquiry projects in science teacher education: What can investigative experiences reveal about teacher thinking and eventual classroom practice? *Science Education, 87*(1), 112-143.

Windschitl, M. (2004). Caught in the cycle of reproducing folk theories of "Inquiry": How pre-service teachers continue the discourse and practices of an atheoretical scientific method. *Journal of Research in Science Teaching, 45*(5), 481-512.

Windschitl, M. (this volume). Our challenge in disrupting popular folk theories for "Doing Science".

Windschitl, M. & Thompson, J. (2006). Transcending simple forms of school science investigation: The impact of preservice instruction on teachers' understandings of model-based inquiry. *American Educational Research Journal*, 43(4), 783-835.

Wiser, M. (1995). Use of history of science to understand and remedy students' misconceptions about time and temperature. In D. Perkins (Ed.), *Software Goes to School*. New York: Oxford University Press.

Wiser, M. & Amin, T. (2001). "Is heat hot?" Inducing conceptual change by integrating everyday and scientific perspectives on thermal phenomena. *Learning and Instruction, 11*, 331-355.

Wittgenstein, L. (1958). *Philosophical investigations* (3rd ed.). (G. E. M. Anscombe, Trans.). New York: Macmillan Publishing.

Woit, Peter. (2006). *Not even wrong: The failure of string theory and the continuing challenge to unify the laws of physics*. London : Jonathan Cape,

Woods, D. D. (1997). Towards a theoretical base for representation design in the computer medium: Ecological perception and aiding human cognition. In J. Flach, P. Hancock, J. Caird & K. Vincente (Eds.), *The Ecology of Human-Machine Systems*. Hillsdale, NJ: Lawrence Erlbaum.

Wortham S. (2003). Curriculum as a resource for the development of social identity. *Sociology of Education, 76*, 229-247.

Yore, L. D., Bisanz, G. L. & Hand, B. M. (2003). Examining the literacy component of science literacy: 25 years of language arts and science research. *International Journal of Science Education, 25*, 689-725.

Yore, L. D., Craig, M. T. & Maguire, T. O. (1998). Index of science reading awareness: An interactive-constructive model, text verification, and grades 4-8 results. *Journal of Research in Science Teaching, 35*, 27-51.

Zaritsky, R., Kelly, A. E., Flowers, W., Rogers, E. & O'Neill, P. (2003). Clinical design sciences: A view from sister design efforts. *Educational Researcher, 32*(1), 32-34.

Zhang, J. (1997). The nature of external representations in problem solving. *Cognitive Science, 21*(2), 179-217.

Zhang, J. & Norman, D. A. (1995). A representational analysis of numeration systems. *Cognition, 57*, 271-295.

Ziman, J. (1984). *An introduction to science studies*. Cambridge: Cambridge University Press.

Zimmerman, C. (2000). The development of scientific reasoning skills. *Developmental Review, 20*, 99-149.

Zuckerman, H. (1988). The sociology of science. In N. Smelser (Ed.), *Handbook of sociology* (pp. 511-574). Newbury Park, CA: Sage.

AUTHOR BIOS

Fouad Abd-El-Khalick is Associate Professor of Science Education at the University of Illinois, Urbana-Champaign. His interests include investigating science teachers' pedagogical content knowledge, and global and specific subject matter structures, and the use of concept maps as learning and assessment tools. Current research focuses on the teaching and learning about nature of science (NOS) in grades K-12, and in pre-service and in-service science teacher education settings, including characterizing learners' views of NOS, developing and assessing the effectiveness of instructional approaches targeting learners' NOS views, using history and philosophy of science to promote learners' understandings of NOS, and investigating factors that facilitate or impede the translation of science teachers' NOS views into instructional practice. (Ph.D., Science Education, Oregon State University)

Philip Bell is an Associate Professor of Cognitive Studies in Education with appointments in Educational Psychology and Curriculum & Instruction at the University of Washington. He is a learning scientist researching science education, learning technologies, argumentation, and everyday science and technology. In one sense, he is a computer scientist sitting in a college of education. He draws upon his training as an engineer to construct new technological tools and environments that support learning and interaction. In another sense, he is a social scientist studying qualities of human learning, thinking, and collaboration. In most of his research these two strands are interwoven into a form of mixed-method, design-based research that explores new approaches for orchestrating learning and interaction in educational settings. He is co-editor and contributor to *Internet Environments for Science Education*. A reasonable subtitle for the book would be "how information technologies can support the learning of science." (Ph.D. University of California, Berkeley)

Janice Bordeaux is Associate Dean for Program Development, George R. Brown School of Engineering, Rice University and a member of the Adjunct Faculty in Psychology. She has a long-standing interest in the role of schooling in cognitive and psychosocial development. Early in her career, she worked as a pediatric psychologist on interdisciplinary research teams within the medical community that focused on learning disabilities caused by disease and treatment. Currently, Janice helps faculty meet the educational standards framed by the national Engineering Accreditation Criteria 2000, which mandates formal and continuous program assessment and evaluation. She also contributes research design and assessment expertise to several projects, mostly on engineering curriculum and educational technology. These include a workgroup coordinating Science and Engineering laboratory instruction across the curriculum, a virtual mechanical design lab sequence, problem-based learning in introductory bioengineering, a cognitive apprenticeship model in advanced bioinformatics, curricula and technology for logic-based and challenge computer programming, and an open-source, free educational technology that promotes authoring collaborations within educational communities. (Ph.D. Developmental Psychology, Duke University)

William Brewer is a Professor in the Department of Psychology, at the University of Illinois at Urbana-Champaign, a research professor in the Institute of Communications Research, and a part-time faculty member in the Beckman Institute Cognitive Science. He carries out research in four broad areas: (a) knowledge representation (schemas, mental models and naive theories); (b) human memory (autobiographical memory, memory and knowledge, memory confidence); (c) psychology of science (psychological responses to anomalous data, distortions in scientific texts, the nature of scientific models, the role of explanations in scientific theories, the theory-laden nature of observation); and (d) knowledge acquisition (large scale conceptual change). (Ph.D. in Experimental Psychology and Neuropsychology University of Iowa)

Nancy Brickhouse is Professor of Science Education and Director of the School of Education at the University of Delaware. She is the immediate past editor of international research journal *Science Education*. Her research interests focus on girls' engagement in science in and out of school, girls' construction of identities in science and understanding the multiple and dynamic nature of identities over time and in a variety of environments, teaching about the nature of the sciences, multiculturalism/feminism and the sciences. (Ph.D., Science Education from Purdue University)

Clark Chinn is Associate Professor of Educational Psychology in the Graduate School of Education at Rutgers University. He is director of the Ph.D. Program in Education. His research addresses theoretical and applied issues in learning and reasoning in science education and in reading. He is interested in how students come to understand and believe theories and in the development of reasoning. His (2002) *Science Education* article with B. Malhotra "Epistemologically authentic reasoning in schools: A theoretical framework for evaluating inquiry tasks." is widely cited by scholars. Currently with NSF funding he is employing microgenetic research methods to investigate step-by-step changes in knowledge as middle school children construct scientific models and theories over an extended period of time. He is also interested in collaborative learning and problem solving in classroom contexts and has been a recipient of an NSF Early Career Research Award as well as the 2001 Richard E. Snow Award for Early Career Contributions,

Division 15 (Educational Psychology), American Psychological Association (Ph. D. Educational Psychology University of Illinois at Urbana Champaign)

Richard Duschl is Professor of Science Education in the Graduate School of Education, Rutgers University and a member of the Rutgers Center for Cognitive Science. Prior to joining the Rutgers faculty, Richard held academic positions, at King's College London, Vanderbilt University, University of Pittsburgh, Hunter College-CUNY, and University of Houston and was a high school earth science teacher in Charles Co. Maryland public schools. One focus of research examines how the history and philosophy of science can be applied to science education. The research agenda conducted in collaboration with Richard Grandy is to better understand the social and cognitive dynamics for making science classrooms inquiry and epistemic communities. A second focus of research is the design of instructional sequences that promote assessment for learning. He also has expertise in informal science education and in earth science education. Richard has served as the editor of *Science Education* and chaired the National Research Council Study Committee that produced the 2007 report *Taking Science to School: Learning and Teaching Science in Grades K-8.* (Ph.D., Science Education, University of Maryland - College Park)

Daniel C. Edelson is Director of Education for the National Geographical Society. Formerly he was Associate Professor at Northwestern University with joint appointments in the School of Education & Social Policy and the Department of Computer Science. Dr. Edelson conducts research on the design of software, curriculum materials, and teacher professional development to support inquiry learning. Since 1992, he has directed a series of projects exploring the use of technology as a catalyst for pedagogical reform in science education. As part of that research, he has led the development of a number software environments including: My World, a geographic information system for inquiry-based learning; the Progress Portfolio, an inquiry-support tool; and CreANIMate, a case-based teaching system for elementary school biology. He has also developed numerous inquiry-based science curriculum units for Earth and environmental science courses for students from middle school through college. Since 1995, he has been the PI or co-PI on more than a dozen NSF-funded educational research and development grants, and he is on the leadership team of the NSF-funded Center for Curriculum Materials in Science. (Ph.D. in Computer Science (Artificial Intelligence) from Northwestern University)

Eugenia Etkina is Associate Professor of Science Education in the Graduate School of Education at Rutgers University. She has extensive teaching experience in physics and astronomy instruction at middle school, high school and university levels. She teaches science methods courses for prospective elementary teachers, physics teaching methods courses for high school teachers and conducts professional development programs for in-service science teachers. She co-directs

Astrophysics Summer Institute that started in 1997. With Al Van Heuvelen she has developed the ISLE Project as an approach to teaching physics where students construct their understanding using processes similar to those used by scientists in real world research. This research has received support from NSF grants. She also researches the use of students' reflective Weekly Reports as a mechanism for advancing learning. (Ph.D. in physics education from Moscow State Pedagogical University)

Drew Gitomer is Distinguished Presidential Appointee in the Policy Evaluation and Research Center at Educational Testing Service, Princeton, NJ. Formerly he was Senior Vice-President for Research and Development at Educational Testing Service. Gitomer's research interests include developing assessment models that have direct implications for instruction. His other current research interests include policy and evaluation issues related to teacher education, licensure, induction, and professional development. Gitomer was the project co-director of Arts PROPEL, a portfolio assessment effort involving middle and high school teachers and students in music, visual arts, and writing. He co-directed several NSF sponsored projects focusing on alternative conceptions of assessment in science education. Gitomer also led an effort to develop an interactive video-based intelligent tutoring system to help users develop skill in technical troubleshooting. (Ph.D. in Cognitive Psychology at the University of Pittsburgh)

Richard Grandy is Professor of Philosophy and Cognitive Sciences at Rice University. He taught at Princeton (where he was a colleague of Thomas Kuhn's) and North Carolina before moving to Rice in 1980. Richard's central interest is in the special human abilities to create and acquire knowledge of logic, mathematics and science. These interests in philosophy of logic, mathematics and science have led him to related questions in education, epistemology, metaphysics, and language as well. He is the editor of *Theories and Observations*, *Readings in the Philosophy of Science* (with B. Brody) and numerous articles in philosophy of science, philosophy of logic, cognitive science and related areas. He was the first director (and one of the founders) of the Cognitive Sciences Program at Rice. He and Richard Duschl have been collaborating on various projects for two decades. (Ph.D. from the History and Philosophy of Science Program at Princeton)

David Hammer is Professor of Physics and Curriculum & Instruction at the University of Maryland in College Park. He studied physics in college, took a break to teach high school mathematics and physics, then went on to graduate school, first in physics then to complete a doctorate in Science and Mathematics Education with Prof. Andrea diSessa at UC-Berkeley. His first faculty position was at Tufts University, and in 1998 he moved to the University of Maryland. His research focuses on students' intuitive "epistemologies" (knowledge about knowing and learning) and on how instructors interpret and respond to student thinking. He is currently pursuing these interests with students and colleagues in both departments at levels from elementary school through college physics majors. (Ph.D. Science and Mathematics Education, University of California-Berkeley)

Cindy Hmelo-Silver is Associate Professor of Educational Psychology in the Graduate School of Education at Rutgers University. She directs the GSE Educational Psychology Ph.D. program. She is interested in how people construct knowledge and develop reasoning strategies through problem-solving situations as well as the early acquisition of expertise, and the role of technology in the construction of knowledge and reasoning strategies. She studies the role of technology to support knowledge construction and collaborative learning and problem-solving. Her research investigates scaffolded support for problem-based learning and complex systems understanding. Cindy has been the recipient of an NSF Early Career Research Award and received research funding from the National Academy of Education Spencer Postdoctoral Fellowship program. (Ph.D. Cognitive Psychology from Vanderbilt University)

Gregory J. Kelly is a Professor of Science Education at Penn State University. Previously, he worked as a faculty member at the University of California Santa Barbara. Greg studied physics at the State University of New York at Albany and completed his. He is a former physics and mathematics teacher and served as a Peace Corps Volunteer in Togo for four years. His research examines issues of knowledge and discourse in science education settings. Recent studies have examined communicative demands of technological design for third grade students, the role of dissent in student small group discourse, and uses of argumentation analysis for assessing students' use of evidence. This work has been supported by grants from Spencer Foundation, National Science Foundation, and a postdoctoral fellowship National Academy of Education. Presently, Greg teaches courses concerning the uses of history, philosophy, sociology of science in science teaching, qualitative research methods, and teaching and learning science in secondary schools. He serves editor of the international research journal *Science Education*. (PhD in science education at Cornell University)

Joseph Krajcik, is Associate Dean for Research and Professor of Science Education in the School of Education at the University of Michigan. He focuses his research on designing science classrooms so that learners engage in finding solutions to meaningful, real world questions through inquiry, collaboration and the use of learning technologies. Along with his colleagues from Northwestern University, AAAS and Michigan State, Joe through funding from the NSF, is designing and testing the next generation of middle school curriculum materials to engage students in developing deep understandings of science content and practices. Joe also co-directors the Center for Curriculum Materials in Science, a NSF funded Center for Teaching and Learning, in which he and his graduates students are exploring various ways to scaffold students in doing complex tasks, such as writing evidence-based scientific explanations. Professor Krajcik has authored and co-authored over 100 manuscripts. He is an active member of the National Association of Research in Science Teaching and in 1998 was elected President of NARST. In 2001, the American Association for the Advancement of Science recognized Professor Krajcik for his accomplishments in science education by inducting him as a fellow of AAAS. He began his science education career teaching high school chemistry for eight years in Milwaukee, Wisconsin. (Ph.D. in Science Education from the University of Iowa)

Helen Longino is Professor of Philosophy and of Women's Studies at Stanford University. She held academic positions at Mills College, Rice University and the University of Minnesota. She is the author of *Science as Social Knowledge* (Princeton) and of *The Fate of Knowledge* (Princeton) and of articles in philosophy of science, philosophy of biology, and feminist philosophy. Her scholarly writings argue for the significance of values and social interactions to scientific inquiry. She is also a coeditor of *Scientific Pluralism*, a recent volume in the *Minnesota Studies in Philosophy of Science* series. (Ph.D. Philosophy, Johns Hopkins University)

Nancy J. Nersessian is Regents Professor and Professor of Cognitive Science at Georgia Institute of Technology where she is appointed jointly in the School of Public Policy and the School of Interactive Computing. Her research focuses on creativity, innovation and conceptual change in science. She tries to understand the cognitive and cultural mechanisms that lead up to scientific innovation, both theoretical and experimental. A major theme of this research is conceptual innovation and change in physics and in physics education, specifically the role of analogical and visual modeling and thought experimenting (simulative modeling). She is author of numerous publications on these and other topics, including the books *Faraday to Einstein: Constructing Meaning in Scientific Theories* (Kluwer, 1984, 1990); *Model-Based Reasoning in Scientific Discovery* (ed. with L. Magnani and P. Thagard; Plenum 1999), *Model-Based Reasoning: Science, Technology, and Values (edited with L. Magnani; Kluwer 2002)* and is currently at work on a book, *Creating Concepts: A cognitive-historical approach to conceptual change*, to be published by MIT Press. Her current research includes investigating reasoning and representational procedures in interdisciplinary research laboratories. This research examines the nature and role of physical and computational models researchers construct to simulate biological phenomena in problem solving and in learning. In her spare time she sings opera and other classical music. (Ph.D. in Philosophy, Case Western Reserve University)

Stephen Norris is Professor and Canada Research Chair in Scientific Literacy and the Public Understanding of Science in the Department of Educational Policy Studies at the University of Alberta. His research interests lie mainly in the philosophy of educational research, the philosophy of reading, science education policy, and reading in science education. Major problems addressed by his current research include understanding and enhancing the role of scientific knowledgeability in citizens lives; specifying more clearly and completely the relationship between the fundamental sense of literacy, being able to read and write, and scientific literacy; and finding and employing more engaging forms of communication, such as the narrative, for increasing interest in and understanding of science. (Ph.D. in Philosophy of Education. University of Illinois at Urbana-Champaign)

John L. Rudolph is Associate Professor in the School of Education and a faculty affiliate in the Program of Science and Technology Studies at the University of Wisconsin. His work focuses on the history of science education in the United States. In addition to this, he has written on issues related to the nature of science in the present-day school curriculum and on how the history and philosophy of science has been used as a theoretical framework in science education research. He is currently at work on a book-length historical study that examines the varied way scientific epistemology has been portrayed in classrooms over the past 125 years. He is currently engaged in a project that looks at how the high school laboratory and instructional apparatus developed in the 1960s embodied particular ideas of scientific epistemology. He is the author of *Scientists in the Classroom: The Cold War Reconstruction of Science Education (2002)* and currently serves as co-editor of the "Science Studies and Science Education" section of *Science Education.* This book is part of a larger historical study that will trace changing portrayals of scientific process in school classrooms from the late 1800s to the mid 1990s. (Ph.D. in the history of science and curriculum and instruction, University of Wisconsin-Madison).

Ala Samarapungava is Associate Professor in Educational Psychology in the Department of Education at Purdue. Her research focuses on reasoning and learning in science and mathematics from childhood through adulthood. There are three lines to her research: (1) the content and structure of children's concepts in scientific domains; (2) the epistemic dimensions of scientific reasoning; (3) how specific learning environments influence student learning. She is interested in developmental, cultural, and epistemic aspects of knowledge acquisition in the sciences and mathematics. Ala publishes extensively in both cognitive science and science education research journals. She was a 1993 recipient of the prestigious National Academy of Education/Spencer Postdoctoral Fellowship award. (Ph. D. in cognitive psychology, University of Illinois at Urbana-Champaign)

William Sandoval is Associate Professor in the Graduate School of Education at UCLA. His research interests include (1) Children's ideas about how science is done and what scientific knowledge is, and how these ideas influence their efforts to learn science. (2) Teachers' ideas about how science is done and what scientific knowledge is, and how these ideas influence their science teaching. (3) The role technology can play in mediating science learning and teaching in classrooms, especially in supporting meaningful scientific inquiry. (4) Understanding how innovative designs for education can help us develop better theories of learning and better educational practice. Bill was a member of the National Academies study committee report *American High School Laboratory: Role and Vision*. He has been a leading voice on learners' epistemological development, a topic on which he has published widely. (Ph.D., Learning Sciences, Northwestern University)

Leona Schauble is Professor of Education in the Department of Teaching and Learning, Vanderbilt University. She is a cognitive developmental psychologist with research interests in the relations between everyday reasoning and more formal, culturally-supported, and schooled forms of thinking, such as scientific and mathematical reasoning. In 1991, she received a National Academy of Education Spencer Fellowship to investigate developmental changes in how children and adults understand the goals and strategies of scientific experimentation. This work generated findings concerning how people learn to design informative experiments and "read" patterns of evidence, including covariation, lack of covariation, and correlations between variables and outcomes. A second important theme in her research is the design and study of instruction. At the University of Wisconsin and subsequently at Vanderbilt University, she has continued studies of learning in both informal and formal educational settings. Her current research focus, in collaboration with Professor Richard Lehrer, is on the origins and development of model-based reasoning in school mathematics and science. In this project, they work collaboratively with teachers on an extended basis to generate reform in teaching and learning of mathematics and science, at levels from elementary through middle school. As participating teachers collectively develop an educational agenda that emphasizes representational competence and modeling, researchers conduct studies that track the long-tem development of forms of epistemology that would otherwise be very difficult or impossible to study. (Ph.D. Developmental and Educational Psychology, Columbia University)

Harvey Siegel is Professor of Philosophy and Chair of the Department of Philosophy at the University of Miami. His areas of specialization include epistemology, philosophy of science, and philosophy of education, and he has long been interested in philosophical questions concerning science education. He is the author of many papers in these areas, and of *Relativism Refuted: A Critique of Contemporary Epistemological Relativism* (Kluwer, 1987), *Educating Reason: Rationality, Critical Thinking, and Education* (Routledge, 1988), and *Rationality Redeemed?: Further Dialogues on an Educational Ideal* (Routledge, 1997). He edited *Reason and Education: Essays in Honor of Israel Scheffler* (Kluwer, 1997). His papers on science education have appeared in *Science Education, Science & Education, Interchange, Journal of College Science Teaching, Phi Delta Kappa, Journal of Research in Science Teaching, Studies in History and Philosophy of Science, Educational Studies, Synthese*, and in other journals and edited volumes. (Ph.D. Philosophy Harvard University)

AUTHOR BIOS

Miriam Solomon is Professor of Philosophy at Temple University and Director of Graduate Studies. Her research interests are in philosophy of science, history of science, epistemology, biomedical ethics and gender and science. She is the author of *Social Empiricism* (2001) as well as numerous scholarly articles in philosophy of science and sociology of science. Professor Solomon is past president of the Philosophy of Science Association, serves on the editorial board of Philosophy of Science and was program chair for the Philosophy of Science Association's 2004 meetings. (Ph.D., Harvard University)

Stephen Stich is Board of Governors Professor of Philosophy and Cognitive Science at Rutgers University with joint appointments in the Department of Philosophy and Rutgers Center for Cognitive Science. He is the recipient of the Jean-Nicod Prize for 2007. He is also Honorary Professor of Philosophy at the University of Sheffield. At Rutgers he directs the Research Group on Evolution and Higher Cognition. At Sheffield he was actively involved in the Innateness and the Structure of the Mind Project and is currently on the Organizing Committee for the Culture and the Mind Project. He has done work in the philosophy of mind, the foundations of cognitive science, naturalized epistemology, theory of mind and moral psychology. His current research interests include moral psychology, theory of mind, evolutionary psychology and gene-culture co-evolution. Recent publications include *Mindreading* (2003), *The Cognitive Basis of Science* (2002) and co-editor of *The Blackwell Guide to Philosophy of Mind* (2003). (Ph.D. Philosophy, Princeton University)

Mark Windschitl is Associate Professor in Science Education at the University of Washington and a former middle school teacher. He does much of his investigative work with new secondary science teachers. His research group Teachers' Learning Trajectories Project examines how the epistemologies and methodologies of science are understood by science educators and how they are enacted in the discourse and activity of the K-12 science classroom. His most extended line of research involves how pre-service teachers conceptualize and enact canonical scientific practices (posing questions, developing testable models, designing investigations, constructing arguments that relate evidence back to models). Four of these studies have used, as their central feature, authentic empirical investigations that various groups of pre-service teachers have conducted. Mark has drawn upon multiple theory bases to understand better what he calls a "folk theory" of science inquiry that is reinforced and reproduced through particular linguistic and cultural practices within the science education community. His recent writings include a paper commissioned by the National Academies for the report *American High School Laboratory: Role and Vision,* his 2006 *Phi Delta Kappa* article "Why we can't talk to one another about science education reform", and his 2006 *American Educational Research Journal* article with J. Thompson. (Ph.D. Curriculum and Instruction, Iowa State University)

INDEX

AAAS Benchmarks 31, 80, 101
abstraction 18, 67-69, 78, 179, 182, 212-213, 253
academic forums 222, 299
activity 24, 61-62, 99, 103-117, 118, 118, 139, 153-155, 158, 160-162, 210, 264, 269-270, 294, 300
    activity theory 99, 105-108, 113
    collective activity 105, 106, 264
agency 27, 105, 149, 192, 288
analytic philosophy 8, 33, 86, 127
analytical fluency 180
argumentation 16, 22, 24, 28, 31-36, 51, 54, 107-108, 112, 118-119, 121, 129-130, 132, 153-154, 184, 200, 222-225, 226, 228, 230, 236-239, 244, 249, 266, 281-282, 289-290, 292, 296-297
    construction of argument 28, 33-34, 36, 107, 113, 222, 249, 265, 282, 289, 290-291, 297
    critical arguments 34-35
    regular arguments 34
artificial intelligence 25
assessment 1-4, 7, 23, 29-32, 90, 108, 241, 270, 275-276, 300
    alternative assessments 29
    authentic tasks 30
    dynamic 1, 30
    models 1, 108
    performance-based 29-30
    procedures 29
assumptions 12, 16, 33, 46, 50, 80-85, 103, 191, 203, 217, 276, 292
astronomy 6, 15, 45, 48
"authentic science" 80-83, 96, 284, 298, 300
"authentic science inquiry" 80-81, 83, 129, 284-285, 295, 298
authoritative documents 279, 283, 293, 299, 301-302
authority 88, 96, 104, 200, 218, 226
    authority of science 14, 32
    intellectual 92, 93, 104, 110
    teachers' 34, 104
automating operations with minimal pedagogical value 169
axiomatic system 8
background knowledge 21, 171, 197, 247, 249, 250, 254, 256, 295, 299

Bayesianism 91, 234
belief change 45, 88, 103, 191, 192, 194-203, 216, 225-226
belief-understanding distinction 191, 195, 198, 203-204, 223, 225
bias 86, 87, 88, 91, 92, 93, 124, 205, 240
"biasing factors" 88, 94
"big ideas" 19, 226, 230, 232
Biology Guided Inquiry Learning Environment (BGuILE) 231
British Empiricists 41
case-based approaches 98, 179
causality 2, 17, 35, 52, 67, 142, 151, 154, 186, 192, 194, 205-207, 215
    causal judgments of children 52-53, 55, 142, 148, 156, 192, 206
    causal mechanism(s) 67, 146, 148, 206, 215
    origins of causal knowledge 52
Center for Learning Technologies in Urban Schools (LeTUS) 2, 172, 176-178
"certainty bias" 240, 265
ChemStudy 4, 21
child as little scientist analogy 38-39, 44, 46-47
child as scientific theory generator 45-46, 48
children's
    abilities to engage in inquiry 5, 28, 101, 103, 105, 121, 138-156, 157-158, 184
    development of resources 78, 153-156, 159-161
    expressed beliefs about what scientists do 158
    reasoning 17, 23, 139, 149, 152, 159
    scientific thinking 32, 35, 159
    theories of astronomy 45, 48
claims of science 10, 28, 31, 288-291
classroom
    classroom community 96, 106, 118-120, 161-162, 178, 285
    classroom inquiry 32, 80, 82-83, 96, 118-120, 157, 162, 171, 194, 201, 215-218, 264-266, 277, 278, 285, 292, 294, 300-302
    elementary school 55, 239, 243
    pre-college 4, 80, 82-84, 93, 98, 301-302

360

norms 55, 93, 98, 118-120, 132, 283, 301
cognitive apprenticeships 27, 62
cognitive-cultural systems 70-71, 74, 78
cognitive development psychology 17, 27, 38-39, 43, 45, 50-52, 60, 126, 184
    knowledge-based approaches 38-39
cognitive load 211, 222, 224-225
cognitive processes 3, 8, 11, 21, 28, 71, 84, 111, 200, 288, 290
Cognitive Reconstruction of Knowledge Model 191
cognitive sciences 1, 26, 39, 59-64, 70, 86, 172, 186
cognitive structure 151, 154, 159
    unitary ontologies 152, 159
    manifold ontologies 152
collaboration 82, 84, 91, 101, 136-137, 224, 273
college science 80, 276, 281, 297
community of learners 70, 102, 118, 272
compartmentalism 29
comprehension strategies instruction 243
computer-based modeling programs 9, 20, 64, 68, 218, 232, 273
computer science 1, 179, 222
concept acquisition 13, 32, 48, 61, 125-126, 229, 289
conceptual change 5, 7, 11-16, 17, 26, 58-61, 63-69, 81, 83-84, 104, 111, 191-225, 226-232, 278, 281, 293
    goals of a theory of 59, 194-195, 203
    problem of 58, 83, 191-225, 226
    teaching 16, 26, 193, 201, 203, 213-222, 226, 228, 278
    theory 12, 69, 81, 111, 191-194, 203
conceptual structures 3, 26, 44, 60, 101
conference consensus points 305-307
    scientific method 307-310
    school science 312-321
    teacher professional development 319-321
conference dissensus points 307, 310-311
Connexions Project (CNX) 184
consensus 1, 34, 80, 83, 88-92, 98, 112, 120, 136, 138-140, 142, 145, 155, 157, 26, 268, 289-291
consequential validity problem 31
constructivism 23-25, 61, 100, 104, 111-112, 159, 255-256
    cognitive constructivism 23-25, 61
    metaphysical constructivism 23-24
"context of discovery" 10-11, 88
"context of justification" 10-11, 88

contextualization 27, 106, 174, 176, 229, 251, 256
continental drift theory 87-88, 217
continuum hypothesis 61
core concepts 52, 163, 270
cosmology 48
    early childhood cosmological models 48
criteria for theory choice 14, 36, 40, 99, 198
critical contextual empiricism (CCE) 92-93, 95, 98
criticism 12, 91-93, 98, 103-104, 110
crucial experiment 89
cultural artifacts 60, 70-72, 105, 107-108, 279, 303
cultural institutions of science 38-39, 42, 47-49, 82, 293, 299
    historical development of 39
cultural transmission 57, 84, 123-126
curiosity 47, 175, 185
curriculum 1-2, 4-5, 7, 23, 29-30, 33, 63-64, 78, 131, 138-139, 164-165, 167, 170-172, 176-178, 180, 182-184, 201, 226, 228, 239, 248, 266, 269, 275, 292-293, 300
    design of 1, 2, 4, 5, 7, 63, 131, 164-165, 167, 170-172, 177, 180, 182-183, 201, 266
    inquiry-based curriculum 2, 5, 23, 138, 165, 170-172, 180, 182, 201, 228, 248, 275, 292-293
    introductory 64, 78, 183
data 2, 8-11, 16-17, 22-24, 30-33, 41-42, 49, 52-53, 55, 60, 88-89, 107, 109, 119, 138-140, 164, 166-171, 176, 198-200, 216-218, 232-234, 249-250, 264, 276, 280-281, 289, 295-299
    responses to anomalous data 10-11, 41, 191, 199
data analysis techniques 2, 5, 55, 60, 166, 169, 171, 234, 281
data collection 2, 6, 22, 32, 40-41, 47, 55, 60, 166, 176, 233-234, 236, 268, 273, 276, 280
data modeling 2, 8-9, 17, 22-23, 32, 68, 164, 166, 169, 212, 217-218
data revision 11, 23, 41, 83
decision making 15, 24, 86-87, 105, 127
decision vectors 87-91, 94-95
democracy 92-93, 97-98, 103
design
    framework 3, 7, 164-166, 172-175, 177, 179, 182, 186
    of learning environments 1, 11, 23, 26-28, 31-32, 36, 74, 99, 166, 171, 180, 182, 278
    software 107-108, 164-168, 171

361

to support teacher learning 165, 178
developmental change 55
developmental psychology 17, 38, 43, 45, 50-52, 126, 184
dialogic communication 22, 35, 120
dialogic justification 11, 102
dialogic practices, learner & teacher 311
dialogic process of inquiry 10, 16, 32, 101-102, 139, 302-315
disciplinary matrix 7, 12-13, 16
discourse ethics 100
discovery 9-11, 20, 33, 62, 84, 88, 100, 103, 112, 134, 229, 232, 241, 244
dissent 88-90, 92, 103, 110
division of labor 87, 97, 108, 116
"doing science" 6, 33, 36, 39, 123, 228, 231, 244, 256, 270, 279, 283, 290-294, 298-299, 301-302
earth science 107, 132
educational design research 166, 180, 182
educational psychology 28
educational standards 28, 31, 64, 185, 227, 248
electromagnetism 64, 66-67
embedding the tacit knowledge of experts 167
empirical hypothesis testing 280, 294
enculturation, process of 95, 119-121
engineering research 59, 69, 74, 78, 165-166, 172
"enlightenment project" 278, 292
'enquiry into enquiry', see also inquiry 4, 32, 284
epistemic
    activities 32-33, 150, 153-155, 159-160
    agent 100, 105, 112-113
    beliefs
        communities 101-102, 118
        cultures 58
        fairness 91
        forms 154, 159-160
        framework/ context 3, 101
        goals 32-33, 93, 98, 157-158
        practices 99-100, 104, 106, 108, 110, 112, 117-118, 121-122, 288, 291, 304-305
    students' "practical epistemic beliefs"
epistemological
    "resources" 95, 153-154, 159, 160
    "practical" beliefs 158
    understanding 197, 202-203, 265
epistemology 54, 56, 86-113, 117-118, 126-127, 132, 138, 152, 157-162, 197-198, 218, 261, 265, 288-289
    Cartesian epistemology 104
    of science 86-100
    "personal" epistemology 99, 112, 157, 162, 265
    social epistemology 86-98, 101-102, 106, 109, 112-113, 118, 132, 288
ethnography 60, 63, 75, 79, 100, 105, 107, 110, 263-264
evidence 3-4, 6, 10-11, 14-15, 22-23, 26, 28, 31-34, 36, 41-42, 55, 95, 99, 101-104, 107-108, 110, 121, 131, 138, 141-142, 151, 155, 196-198, 201, 215-218, 220, 222-223, 227-228, 230, 248-249, 256, 265, 286-292, 296
    acquisition of 3, 141, 286, 296
    evaluation of 3, 32, 103, 107, 110, 155, 197, 201, 248, 265, 290
    ex post facto 10
evidence-explanation (E-E) continuum 3, 31, 101, 313
exemplar 12-13, 61-62, 222
expanded scientific method (Sme) 280-281, 309-310
    dissensus view 310-311
    nature of science 7 tenets 317-318
    school science 317-318
experiment 2, 5-6, 8-9, 16, 21, 23-24, 30, 40, 54, 59, 65, 71, 74-76, 89, 131, 135, 142-148, 150-151, 199, 212, 216, 220, 263-265, 272, 276, 281, 298, 301
    design of 5, 16, 21, 30, 131, 135, 151, 276, 281
    modeling 2, 59, 272
experimental repeatability 12
explanation 2-6, 14, 23, 28, 31-34, 36, 44, 48, 51, 55, 101, 103-104, 134, 142, 148, 150, 153-154, 156, 160, 175, 184, 186, 195, 205-206, 214-215, 219-220, 224-226, 234, 236, 248, 254, 272-273, 275, 295-296
    causal 2, 142, 156, 186, 205-206, 272
    mechanistic 142, 160
    statistical/ probabilistic 2
expressed degree of certainty 240
facts 15, 25, 36, 238-239, 242, 247-248, 300
falsificationism 234
feedback 2, 4, 29-30, 101, 223
feminism 86, 92-94, 98, 100
feminist epistemology 92, 100
focusing of attention 222
folk
    psychology 46
    theories 279, 283, 292-295, 297, 299, 301-302
    theory of doing science 279, 292-295, 301-302

formal science 286
full-inquiry lesson units 3, 31
Full Option Science Study (FOSS) kits 275
general relativity 16
Geographical Information Systems (GIS) 2, 22, 176
geoscience 164, 179
Gestalt psychology 13
global climate change 9, 176
goals
    domain specific 182
    domain general 182
    goal-based scenarios 172, 174, 176
    of science 103-104, 118-119, 244, 295-296, 298, 300
    of science education 1, 3-4, 6, 11, 23, 27-28, 48, 63-64, 83, 98, 101, 121-122, 140, 164, 174, 178, 198, 202-204, 222-224, 231, 233, 238, 289
"god's eye view of reading" 254-25
group participation 27, 161-162
groupwork 103, 222, 224
hands-on activities 1-2, 5, 30, 139, 216, 301
heuristics 15, 19, 86-87, 91, 107-108, 115, 165
    negative 15
    positive 15
    representativeness heuristic 86-87, 91
    salience heuristic 86-87
history of science 1, 13, 15, 19, 41, 57-59, 62, 79, 91, 93, 216
human genome project 9
hypothesis 8, 16, 30, 60, 63, 65, 110, 151, 153, 155, 280-281, 294-299
hypothetico-deductive view of science/ Hempel-Oppenheimer's Deductive-Nomological Explanation Model 5, 7-8, 33, 42, 62, 263, 284
immersion units 2-4, 23, 31-32, 36, 135, 228, 231, 268-270, 272, 315-319
incommensurability 13-14, 60, 192
independent thinking 5, 189, 294
individual cognition 157-158
individual knower 99
induction, problem of 131
infants 46-47, 50-51, 81, 130
    behavior 47, 50, 81
    beliefs 46, 51, 130
inference to the best explanation 234
inquiry
    discipline-based 292
    evaluating inquiry 29, 201, 275
    identifying inquiry 23, 30, 62, 80, 97, 101-102, 138-158, 278, 284-287, 294

    modes of inquiry 4, 22, 63, 71, 263-266, 278, 286, 292
    open-ended 166-167, 180
    process of inquiry 22, 28, 32, 62, 139, 158, 164, 243, 248
    purpose of inquiry 1-6, 29-31, 99, 104, 118, 122, 138, 157-163, 295-300
    re-framing inquiry 4, 6, 21, 28, 157
    scientific practices of inquiry 304-305
    students' beliefs of inquiry 81, 97
    students' perceptions of inquiry 81, 105, 120-121, 158, 161-162, 184
    views of 22, 57, 95, 101, 104, 112, 129, 134, 139, 288, 298, 300-301
"inscription" devices 22, 272-273, 282
inscriptional work 101-102, 107-109, 263-267, 284
instruction 2-4, 7, 20, 26-27, 31, 33, 55-56, 74, 84, 100-101, 108, 118-119, 121-122, 138, 155, 157, 174-175, 184, 187, 193, 201, 211, 213-215, 219, 222, 225-226, 233, 242-243, 266-267, 272, 284-285, 291, 300
    thematic sequences 3, 315-319
instructional sequences 312-319
    learning progressions 315
instrumentalism 112
instruments 8, 11-12, 16, 21-22, 51, 75, 250-251, 277
integrated science learning 3, 165, 272
interaction pattern 192
internet 265
interpretive tasks 244, 246
intersubjectivity 99-102, 108, 111
"intrinsic motivation" assumption 80-81
intuition 27, 126-127, 129, 131-133, 254
    epistemic intuitions 126-127
    "Gettier intuitions" 127
    "intuition mongering" 126, 129, 131-133
judgment 243, 254, 256
justification 8-11, 55, 88, 99, 102, 112, 130, 132, 134, 169, 227, 233-234, 236, 240, 281, 288
K-12 science education 1, 64, 138, 190. 205, 221, 301
knowledge
    children's resources for understanding knowledge 153-162, 278
    construction of 28, 57, 84, 102, 118, 172-175, 182, 186, 191, 203, 273
    content knowledge 25, 118, 171, 177, 183, 277
    declarative 26, 28, 173

INDEX

domain-specific 57, 183-184
as "fabricated stuff" 154, 159
organization 173, 184
prior 21, 27, 30, 182, 186, 188, 193, 214-215, 222, 226-227, 229
procedural 36, 186, 290
as "propagated stuff" 154, 159
propositional 290
refinement 172-173, 191
schematic 26
specificity of 183-185
strategic 197
as "temporary truths" 5
"knowledge-about-science" assumption 81-82
knowledge system learning 197
laboratories 2, 4, 23, 28-33, 35-36, 59-61, 69-72, 74-75, 79-82, 84, 119-120, 132-133, 182, 248, 271, 276, 301
as evolving cognitive-cultural systems 70-71, 74
bioengineering 59, 79, 84
confirmation lab 28, 301
high school 23, 28-32, 36
interdisciplinary research laboratories 59-60, 74
university 59, 84, 119-120
language 9, 11, 16, 21-22, 27-28, 31, 102, 125, 193, 203, 233-242, 248-249, 251, 258-259, 291
of instruction in science classrooms 237, 242
observation language 9, 11, 16, 21-22
problem of 21
of science 2, 9, 21, 28, 31, 102, 193, 233-237, 242
of science teaching 2, 193, 233, 237-242
of science textbooks 237, 239
of scientific journal articles 238
theoretical language 9, 21
laws 11, 15, 17, 21, 43, 63, 290
learning
agentive learning environments 72, 74
as a passive individualistic process 1, 242
as an active individual and social process 1, 23, 27, 52, 80-81, 122, 215, 218, 222, 227, 290
as conceptual change 16-17, 26, 83-84, 104, 191-198, 202-203, 213-215, 224-226
content/process learning 3, 100-101, 112
discovery learning 100, 112, 244
laboratory learning 32, 59, 71-74
Learning By Design 2

learning environments 1, 3, 11, 23, 26-29, 31-32, 36, 72, 74, 78, 81, 83, 99, 120, 132, 167, 182, 185, 188-190, 227, 230-231, 265, 278
learning sciences 1, 3-4, 23, 26-29, 99-100, 102, 167, 179, 263
learning theory 26, 81, 108, 289
progressions 316-319
strategies 78, 162, 202, 222, 272
Learning-for Use Design framework 172-175, 178, 182-183, 185, 190
Learner-Centered Psychological Principles 167, 171, 186
Learning through Collaborative Visualization (CoVis) 164, 180
lectures 28, 214, 301
LETUS 2, 172, 176-178
levels of causal mechanism 206
intentional causes 206
intrinsic causes 206
simple surface level causes 206
linguistics 1, 79, 154, 266
"little scientists" (see also "infants") 44, 46-47, 51, 81, 130
logic 8-11, 13, 33, 62, 92
deductive 10
inductive 9-10, 13
symbolic 8
logical positivism 7-11, 16, 34, 41-43, 100, 250
Looking at the Environment high school environmental science curriculum 174, 176
management
of information 1-2
of materials 1-2, 167
manifold resources 139, 157-162, 278
mappings 17, 208-209, 211, 221-222
mathematics 8, 19-20, 24, 46, 76, 217, 227
algebra 19, 168
calculus 19
education 19-20, 244
measurement techniques 6, 9, 23, 54, 251
mechanics 18-19, 64, 69, 151, 191
mechanism 43-45, 48, 52, 61, 67, 74, 86, 128, 146-152, 155-157, 192, 204, 206, 227, 280-281, 296-297, 300
mechanistic
accounts of natural phenomena 150, 157-158, 160, 162-163
arguments 152
explanation 142, 160
inquiry 150, 160, 162
reasoning 142, 147, 150, 155
sense-making 152, 155

364

understanding 142, 151
media messages 279, 283, 299, 301, 303
meta-knowledge about theories 46-47
metalanguage 236
metalinguistic terms 266
metaphysical
    assumptions 12, 203
    anti-realism 25
    realism 24-25
"mob psychology" 14
models
    as iconic 18
    as mathematical descriptions 18
    as physical analogues 18
    analogical models 60, 78
    belief in 213-219, 225, 228
    causal 55, 165, 194, 204-206, 225
    computer models 9, 20, 57, 63-64, 68, 218, 232, 273
    complexity of 20, 206, 211-214, 219, 221, 230, 232
    contrastive 212
    "copy" theories of models 53
    family of 18, 201, 212-213
    functional 77, 84, 205
    goodness of fit 68, 78
    interconnection of 202, 212-213, 222, 225
    mathematical models 18, 20, 55, 57, 63, 66-67, 124
    mental models 18, 20, 64, 75-76, 202, 208, 292
    model-based discourse 13, 17, 22-23, 31, 43, 62-70, 77-78, 282, 296, 298
    model construction 22, 55, 63, 66-67, 134
    model-data relations 2, 8-9, 17, 22-23, 55, 199, 203, 212, 216-218, 229
    "the modeling game" 53
    physical models 20, 57, 60, 63, 65-66, 76
    purpose-based 205
    regularity-based 205
    relations with the world 61, 203, 207-208, 211
    sampling 205
    situation-action 205
    structure-function 205
    of student epistemologies 152
    taxonomy of 21, 219
    teleological 205
    testing 6, 95, 124, 214-215, 218-220, 228, 232, 261, 281, 296-297
    topographical or spatial 205

understanding of 63, 74, 78, 212-218, 221, 223-225, 228-230
use of 63, 191, 213-218, 221, 225, 228, 230, 232, 292, 299
value of 229
visual models 20, 60
modified Dragnet theory of teaching 25
motivation 22, 24-26, 80-81, 96, 171-174, 176, 178, 182, 197-198, 209, 270, 272
motivating inquiry 6, 80-81, 134, 155, 171, 174-175, 185
multiple-dimensions theory 191, 194, 211
multivariate linear relations 19
National Committee on Science Education Standards and Assessment 64
National Science Education Standards (NSES) 1, 5, 28, 31, 64, 129, 227, 248
naturalism 93, 132
nature of science 13, 21, 30, 32, 38, 56, 98, 110, 129, 131, 158, 244, 269, 288-291
    creativity 290
    empirical basis 290
    seven (7) tenets 8-11, 13-14, 21, 308, 314-315, Appendix Chapter 21
    sociocultural embeddedness 290
    subjectivity 290
    tentativeness 10, 290
nature of science 7 tenets 8-11, 13-14, 21, 308, 314-315, Appendix Chapter 21
nonexperimental testing 220
non-linear equations 19
norms 38, 48, 50-51, 55-56, 92-93, 95, 97-98, 100, 102-103, 105, 109-112, 118-120, 123-128, 132, 136
    cultural norms 92, 102, 119-120, 126
    moral norms 125, 128
    norms of science 48, 95, 119, 127-128
normal science 7, 11, 34, 61
normative judgments 87-89, 95
normative status 35-36
NSF funded science curriculum 1, 4, 7, 21
objectivity 13-14, 16, 33, 91-92, 112
observational-theoretical (OT) distinction 21
oceanography 88-89, 106-108, 110, 113, 115-117, 122, 166, 289-291
ontology 75, 112, 139, 151-152, 154, 156, 159, 192-193, 205, 207-208
oral tradition 260
organized skepticim 38, 46
p-prims 152, 159, 194
paradigms 7, 11-12, 61, 100, 179, 279, 302
pedagogical content knowledge (PCK) 177, 277
peer review 16, 105, 108, 110, 260, 266, 288-289

365

INDEX

"perceptual symbols" 43
personal epistemology 112, 157, 162, 265
personal relevance of science 263, 266, 285-286
phenomena 4, 7, 13, 22, 34, 40-41, 43-45, 48, 53-55, 64, 68, 74, 76-77, 101, 150, 155, 158, 167, 175, 202, 205, 207-212, 216-222, 224-230, 239, 281, 295-297, 303
philosophy of mind 1
philosophy of science 1, 8, 14, 17, 23-26, 40-42, 62, 86, 99, 100, 112, 131-132
Piagetian stage theories 38
plate tectonics, theory of 15, 43, 88-89, 107, 110, 217, 234, 282
pluralism 92, 100
policy 4, 93, 122, 135-137, 174-176, 243, 264, 268-269, 275-278
post-behaviorist cognitive revolution 13
post-positivism 41, 58
post-World War II America 4
power relations 14-15, 119-120
pre-Socratic science 40, 47, 49, 134
prediction 14-16, 18, 61, 63, 65, 141-147, 152, 157, 198, 219, 221, 275
pretend play 53
prior conceptions 193-196, 202, 204, 211, 214, 225
  entrenchment of 193, 196-197, 201
prior knowledge 21, 27, 30, 182, 186, 188, 193, 214-215, 222, 226-227, 229
  importations from 27, 186, 214-215, 222
probability 2, 192, 207-208, 242
problem-based learning (PBL) 3, 6, 183, 268, 272-274
problem set 61
problem solving 5, 13, 15, 33, 59, 61, 63-65, 70-72, 77-78, 82, 134, 136
  one parameter problems 18
  two parameter problems 18
problematizing 53, 229
productive science 147
professional development 83, 164-165, 177-179, 183, 188, 268-270
professional science 95-98, 138, 151, 158, 285-286
project-based science 3, 174, 183, 272, 274
public standards 103-104, 110
"puzzle solving" 7
questions 5, 30, 32, 35, 55, 83, 106, 132, 134-135, 139, 154, 162, 164, 171, 179-180, 224, 229, 241-249, 253, 269, 271-272, 280-281, 284-285, 292, 294-298, 300

generating 5, 55, 134-135, 164, 171, 224, 248, 269, 280-281, 284-285, 294-296, 298, 300
value of asking 134, 139, 162, 284-285
rational degree of confirmation 10
rationality 10, 13-16, 32-35, 86-87, 91, 100, 130, 132, 146
  as the ultimate aim of education 32-35, 130, 132
reading 39, 83, 200-201, 226-227, 233-267, 284, 287
  default theories of 263
  in the science classroom 83, 233-263
  model of 250, 256, 263
  'reading as inquiry' 83, 233-264, 266-267, 284
  simple view of reading 83, 233, 242-247, 253, 256-257, 262, 265
  strategies 247-248, 261
reasoning
  analysis of students' reasoning 32, 140-141, 149-150, 152, 155, 157, 276
  anthropomorphic 155
  children's development of reasoning 4, 23, 29, 31, 101, 152
  dialogical 102, 112
  features of children's reasoning 27, 139, 150-152, 155, 157-158
  mode of 152, 154-155
  model-based 23, 57-79, 84, 96, 130, 134, 280, 296-298, 303
  phenomenon-based 303
  relation-based 280, 296, 303
  scientific 2, 10, 23, 25, 29, 62, 68, 90, 101, 234, 238, 240, 286
  subject-centered reason 100
reality 14
reducing cognitive load 222, 224-225
Reform Teaching Observation Protocol 277
relevant discourse community 11, 97, 100, 102-105, 108, 111, 118
representation 4, 10, 12, 17-18, 20, 23-26, 42-44, 53, 55-56, 59-61, 63-71, 77, 101, 169-170, 175, 202, 208, 213, 221, 224, 253, 264, 273, 294, 296
  model-like 18, 43-44, 55, 63-66, 77, 202, 208, 213, 221, 224
  representing functions 12, 55-56, 66-68, 221, 294
  representing tools 51, 53, 55, 169, 273
representation theorem 17-18
"research traditions" 15-16
resource allocation 167, 174, 264, 268-269

366

revolutionary science 7
reward structure 120
rhetoric of conclusions approach 4, 28, 247, 265, 284
rolling cylinders 17-19
Scenario-Based Inquiry Learning (SBIL) 174, 176-177
"school science" 1-5, 7, 28, 30-31, 36, 62, 82, 91, 96-97, 99, 131, 158, 164, 233, 237-240, 242, 249, 280, 287-288, 292-294, 301-302, 312-321
    nature of science 7 tenets 317-318
    teacher professional development 319-321
science
    aims of science 31, 248
    as an experiment-driven enterprise 1-2, 6, 8, 24
    as experimentation 1
    as explanation/ model building and revision 1, 284
    as inquiry 80, 138, 247-248
    as a model-driven enterprise 8
    as a social construction 39, 47
    as social knowledge 101, 136
    as a theory-driven enterprise 2, 8, 22, 83
    professional 38, 40, 95-98, 138, 151, 158, 285-286
"science for scientists" 4-5, 7
science studies 1, 3-4, 7, 13, 16, 18, 57, 62, 102, 110, 125, 284
scientific
    articles 54, 106, 109, 117, 200, 238, 266, 281
    community 7-8, 11, 13, 16, 21, 92, 96-97, 103, 105, 110, 136, 162, 237, 251, 262, 286
    expertise 72, 82, 92, 102, 179, 226
    investigation 28, 30, 112, 158, 166-167, 248, 281, 294, 300
    knowledge 3, 6-7, 10-11, 13, 15, 21, 28, 32, 36, 40-41, 49, 57-58, 63, 80-81, 92, 102-103, 112, 117, 124, 136-137, 158, 163, 182, 194, 201, 226, 247, 250-251, 258, 260, 265, 284, 286, 289
    literacy 80, 97, 102, 239, 243-244, 249
    method 5, 8, 16, 25, 48, 62, 179, 268-269, 279-281, 292-293, 295, 298-299, 301-302, 307-311
    method expanded 8, 280-281, 309-310
    processes 5, 8, 10, 13, 16, 30, 218, 304-305
    realism 25, 112
    revolutions 7, 14, 57, 60-61, 192

scientific inquiry 1-3, 5-6, 8, 22-23, 25, 28, 58, 62, 80-81, 83, 95-98, 101-103, 121, 134, 138-140, 145-146, 148, 151, 155-157, 160, 162, 166, 176, 184-185, 200, 218, 227-228, 233, 244, 249, 260-261, 263-264, 267, 285-286, 289, 294-295, 298-299, 304-305, 308-310
    "working definition" 149-151, 157
    "philosophical definition" 157
SCOPE 2
semantic realism 21
sense perception 2, 6, 21-22, 83, 290
seven tenets 8-11, 13-14, 21, 309-310, 317-318, 322-325
single lesson approach 3
skepticism 38, 46, 100
social
    empiricism 27, 33, 38-39, 58, 81, 86-93, 95-98, 100-101, 104, 106, 109-113, 132, 288-289
    problems 6, 135
    process 1, 3, 11, 16, 27-28, 39, 47, 91, 93, 101, 105, 111, 118-119, 186, 266, 288
    psychology 1
    values 30, 92, 102
sociocultural practice 99, 104-108, 116, 118, 120, 160-161
sociology of science 1, 14, 33, 86, 91
software 108, 120, 164-168, 170-171
speech acts 234-237
Sperberian approach 125-126
standardized tests 245
"state transition model" 17
strategic selection of functionality 168
students
    high school 28, 31-32, 91, 163-165, 171, 174, 180, 182, 217, 231, 240-241, 247, 300, 303
    university 19, 59, 107, 119, 154, 183, 240-241, 246-247, 285
student centered learning 3
systematic observation 47, 55, 263-264
teacher 2, 4, 18-19, 23, 25-28, 34, 36, 56, 82-84, 101, 104, 111, 119-120, 140, 161, 164-165, 177-184, 194-195, 201, 213-215, 223, 227-228, 239, 242-243, 257, 261-262, 272-283, 290, 292-302, 319-321
    as facilitator 273
    knowledge 18-19, 27-28, 56, 104, 177, 183, 213, 278-279
    learning 165, 177-178, 181, 295
    pre-service teachers 140, 273-274, 278, 280, 290, 294, 296-299, 301-302
    preparation 82, 165, 177, 180, 182, 275-277

professional development models 83, 164-165, 177-179, 183, 319-321
secondary science teachers 290, 292, 294
training 4, 184, 262, 275
technology 1-2, 4, 6, 22, 28, 38, 57, 60, 70-72, 74, 104, 107-109, 111, 117, 123, 136-137, 164-167, 170-172, 177, 179, 184, 186, 229, 243, 268, 273
school technology infrastructures 170
testability 275, 298
text 4, 20, 31, 67, 102, 109, 214-216, 233-234, 237-239, 241-247, 249-251, 254-266, 272, 274-275, 286
nature of scientific text 233-234, 237, 257
refutation text 214
role of 4, 242, 258
students' interpretation of 233, 249, 251, 254-257, 264
textbook 4, 10, 32, 39, 61-62, 120, 177, 200, 214, 216-217, 226-227, 237-239, 242, 244, 248, 261, 265, 268-269, 275, 283, 293-294, 300-301
theory
as causally possible collections 17
as formal objects 42
as mental representation 43-44
as models 42-43, 52
change 10-11, 23, 26, 45, 49, 191, 198-199
development 5, 10, 13, 15, 52
evaluation 9-11, 16, 24, 47, 138
ladenness of data 41-42
of mind 38, 45, 54
of speech 257
plausibility of 56, 218
revision 11, 16, 47, 83, 138, 219
theory/observation language (O/T) distinction 9
theory-theory 17, 43-45, 52
theories of conceptual change 5, 111, 191-194, 203
belief-focused theories 191
theories of knowledge refinement 191, 193-194, 212

theories of restructuring 191-192
thermodynamics 6, 76, 192
thought experiment 59-60, 63
time scales 105-106, 108-109, 113, 117, 121
mesolevel time scales 105-106, 109, 113, 117
ontogenetic time scales 105-106, 109, 113, 117
sociohistorical time scales 105-106, 109, 113, 117, 121
uncertainties 151, 207, 254, 276
underdetermination of theories 216
underlying theoretical processes 35, 206, 227, 280-281, 297-298, 300
understanding 1, 3, 5, 9, 17, 20-21, 28-33, 39-40, 44-45, 50, 52, 59, 62, 64, 70, 82, 98, 105, 112, 118, 121, 129, 131, 153, 157, 161, 164-165, 172-177, 191-192, 194-215, 218-219, 221-232, 239, 246-247, 249, 253, 256, 261, 275, 288, 291
universal gravitation, theory of 16
"universal novice" status of children 40, 54
universality 48
unobservable mechanism 21, 297
uptake 103, 247
values 12-13, 30, 32, 55-56, 92-93, 97-100, 104, 218, 261, 287
vascular construct model system 74, 76
venues 103, 110
Vienna Circle 8, 37
vocabulary 146, 150, 203, 226, 237, 239, 243
WISE (Web-based Inquiry for Science Education) 2
working memory 211-212, 224-225
"workshop science" 286
world-view 34, 36, 111
WorldWatcher 166, 168-172, 176
writing 33, 40, 99, 106-109, 117, 183, 233, 239-240, 242, 248-249, 257-261, 263, 289, 308-309
scientific 40, 99, 106-109, 117, 239-240, 257
student 108, 239, 260-261, 289, 307-308

Printed in the United States
201934BV00006B/34-36/A